Assembly of Enveloped RNA Viruses

Monique Dubois-Dalcq
Kathryn V. Holmes
Bernard Rentier

Editorial Assistance: David W. Kingsbury

Springer-Verlag Wien New York

Dr. *Monique Dubois-Dalcq*

Head of Section on Neural and Molecular Ultrastructure, Laboratory of Molecular Genetics,
National Institute of Neurological and Communicative Disorders and Stroke,
National Institutes of Health, Bethesda, Maryland, U.S.A.

Professor Dr. *Kathryn V. Holmes*

Department of Pathology, Uniformed Services, University of the Health Sciences,
Bethesda, Maryland, U.S.A.

Dr. *Bernard Rentier*

"Premier Assistant" and "Maître de Conférences", Laboratory of General and Medical Microbiology,
University of Liege, Belgium

Professor Dr. *David W. Kingsbury*

St. Jude Children's Research Hospital, Memphis, Tenn., U.S.A.

With 94 partly coloured Figures

Library of Congress Cataloging in Publication Data. Dubois-Dalcq, Monique. Assembly of enveloped RNA viruses.
Includes bibliographical references and index. 1. Viruses, RNA-Reproduction. I. Holmes, Kathryn V. II. Rentier,
Bernard. III. Kingsbury, David W. IV. Title. V. Title: Assembly of enveloped R.N.A. viruses. QR395.D83. 1984.
576'.64. 84-14111.

ISBN-13:978-3-7091-8758-6   e-ISBN-13:978-3-7091-8756-2
DOI: 10.1007/978-3-7091-8756-2

Foreword

This book is a collection of critical reviews about a diverse group of virus families with two features in common: the stable repository of genetic information in each virus is RNA, and each virus modifies and appropriates a particular patch of the eukaryotic cell membrane system to complete its structure. The reviews take the reader from the level of virus genome structure and expression through the quaternary interactions between virus-specified elements and cellular components that cooperate to produce virus particles. There are spectacular illustrations in this volume, but it is much more than a picture gallery. Reading widely in this book can be an effective antidote to overspecialization: in these pages, you are likely to learn much about viruses and about cells that you didn't know before; you'll discover illuminating parallels between diverse virus families; you'll come away with a sharpened awareness of important things that are still to be learned.

Memphis, Tenn., Summer 1984   *David W. Kingsbury*

Preface

This book was written at the suggestion of Dr. David W. Kingsbury made at a workshop on viruses organized by the Multiple Sclerosis Society in Aspen, Colorado, U.S.A., three years ago. Originally, we had thought to focus on the morphological aspects of viral assembly. Later, during our discussions on the process of budding of enveloped RNA viruses, it became evident that we should include biochemical data in our review and correlate them with the structural aspects of virus maturation. To highlight the way in which various viruses use the cellular machinary for maturation, we have composed a series of schemes. We also decided to add to the well known budding RNA viruses a description of the rotaviruses, since budding appears essential to their maturation. Dr. Kingsbury accepted the difficult task of critically reading and editing each one of our ten chapters. His broad views and in-depth knowledge of the virological literature were most valuable. We are extremely grateful to Dr. Kingsbury for his dedication and promptness in editing.

Summer, 1984   *The Authors*

Acknowledgements

Many scientists have contributed in various manners to this book. We are very thankful to all of them. Their specific contributions are listed below. We thank:

F. V. Alonzo, University of Alabama, Birmingham, for Fig. 4-5 a.

J. C. Armstrong, European Molecular Biology Organization, Heidelberg, Federal Republic of Germany, for data on coronavirus RNA in advance of publication.

H. Arnheiter, Laboratory of Molecular Genetics, National Institute of Neurological and Communicative Disorders and Stroke, National Institutes of Health, Bethesda, Maryland, for critical review of Chapter I, for advice on rhabdoviruses and orthomyxoviruses and for Fig. 2-2 and 4-2 b.

T. Bächi, Institute for Immunology and Virology, Zürich, Switzerland, for Fig. 3-5 and 4-2 a, c and d.

J. N. Behnke, Uniformed Services University of the Health Sciences, Bethesda, Maryland, for helpful discussions on coronaviruses and Fig. 7-6 a.

W. J. Bellini, Laboratory of Molecular Genetics, National Institute of Neurological and Communicative Disorders and Stroke, National Institutes of Health, Bethesda, Maryland, for advice and for sharing data on paramyxoviruses.

J. Boyle, Uniformed Services University of the Health Sciences, Bethesda, Maryland, for advice on coronaviruses and rotaviruses.

W. Brandt, Walter Reed Army Institute of Research, Washington, D. C., for advice on Flaviviruses and Figs. 8-11 and 8-12.

D. T. Brown, University of Texas, Austin, for sharing data on Alphaviruses in advance of publication.

J. C. Brown, University of Virginia, Charlottesville, for advice on rhabdoviruses and Fig. 2-8 c.

M. J. Buchmeier, Scripps Clinic and Research Foundation, La Jolla, California, for advice on arenaviruses and for Figs. 6-2 b and 6-3 a and b.

C. M. Calberg-Bacq, University of Liege, Belgium, for advice on retroviruses and for Figs. 3-10, 9-1 e, 9-7, and 9-11 b.

E. O. Caul, Public Health Laboratory, Bristol, England, for Figs. 7-1 b and 7-4.

D. Cavanagh, Houghton Poultry Research Station, England, for sharing data on coronavirus RNA in advance of publication.

R. W. Compans, University of Alabama, Birmingham, for Fig. 4-5 a.

S. Dales, University of Western Ontario, London, Canada, for Fig. 7-5.

E. de Harven, University of Toronto, Canada, for Fig. 9-1 a.

E. W. Doller, Uniformed Services University of the Health Sciences, Bethesda, Maryland, for helpful discussions on coronaviruses and Fig. 7-3.

M. K. Estes, Baylor University, Houston, Texas, for advice on rotaviruses and information in advance of publication.

M. F. Frana, Uniformed Services University of the Health Sciences, Bethesda, Maryland, for advice on coronaviruses.

H. Frank, Max-Planck Institute for Virus Research, Tübingen, W. Germany, for sharing data on orthomyxoviruses and for Figs. 4-5 b and c, 4-6, 9-8, 9-10 b, and 9-11 a, c, d and f.

P. M. Grimley, Uniformed Services University of the Health Sciences, Bethesda, Maryland, for advice on togaviruses and Figs. 8-3, 8-4, 8-5, 8-13, 8-14, 8-15, and 8-16.

A. K. Harrison, Centers for Disease Control, Atlanta, Georgia, for Figs. 2-5, 2-6, and 2-9 b.

R. N. Hogan, Laboratory of Molecular Genetics, National Institute of Neurological and Communicative Disorders and Stroke, National Institutes of Health, Bethesda, Maryland, for Figs. 3-2 a and b and 3-6 a.

C. R. Howard, London School of Hygiene and Tropical Medicine, for advice on arenaviruses and for Figs. 6-3 c and 6-4.

A. Kapikian, National Institute of Allergy and Infectious Diseases, National Institutes of Health, Bethesda, Maryland, for advice on rotaviruses and for Fig. 10-1.

D. Kolakofsky, University of Geneva, Switzerland, for sharing data on paramyxoviruses and for Fig. 2-4.

S. Kozma, University of Liege, Belgium, for Fig. 9-5.

C.-J. Lai, Laboratory of Infectious Diseases, National Institute of Allergy and Infectious Diseases, National Institutes of Health, Bethesda, Maryland, for advice and sharing data on orthomyxoviruses.

R. A. Lamb, Northwestern University, Evanston, Illinois, for sharing preprints on orthomyxoviruses.

R. A. Lazzarini, Laboratory of Molecular Genetics, National Institute of Neurological and Communicative Disorders and Stroke, National Institutes of Health, Bethesda, Maryland, for advice on rhabdoviruses.

R. B. Luftig, University of South Carolina, Columbia, for Fig. 9-4.

L. Markoff, Laboratory of Infectious Diseases, National Institute of Allergy and Infectious Diseases, National Institutes of Health, Bethesda, Maryland, for sharing data on orthomyxoviruses.

B. Murphy, Laboratory of Infectious Diseases, National Institute of Allergy and Infectious Diseases, National Institutes of Health, Bethesda, Maryland, for advice on orthomyxoviruses.

F. A. Murphy, Colorado State University, Fort Collins, for Figs. 2-5, 2-9 b, 6-2 a and c, and 6-3 d.

W. Odenwald, Laboratory of Molecular Genetics, National Institute of Neurological and Communicative Disorders and Stroke, National Institutes of Health, Bethesda, Maryland, for advice on rhabdoviruses and Figs. 2-8 a and b.

L. Oshiro, California State Department of Health, Berkeley, for Figs. 7-1 a, 7-7 a, and 7-8 a.

B. Petrie, Baylor University, Houston, Texas, for advice on rotaviruses and Figs. 10-2, 10-3, and 10-4.

A. Pinter, Memorial Sloan-Kettering Cancer Center, New York, for Fig. 9-3.

C. S. Raine, Albert Einstein College of Medicine, Yeshiva University, New York, for Figs. 3-3 a, 3-4, and 3-9.

C. D. Richardson, Laboratory of Molecular Genetics, National Institute of Neurological and Communicative Disorders and Stroke, National Institutes of Health, Bethesda, Maryland, for advice on paramyxoviruses and retroviruses.

F. Rickaert, Laboratory of Molecular Genetics, National Institute of Neurological and Communicative Disorders and Stroke, National Institutes of Health, Bethesda, Maryland, for Figs. 3-2 c and d.

M. Reginster, University of Liege, Belgium, for advice on orthomyxoviruses.

S. G. Robbins, Uniformed Services University of the Health Sciences, Bethesda, Maryland for advice on coronaviruses.

E. Rodriguez-Boulan, State University of New York, Brooklyn, for Figs. 1-3 b and 1-4.

M. Schubert, Laboratory of Molecular Genetics, National Institute of Neurological and Communicative Disorders and Stroke, National Institutes of Health, Bethesda, Maryland, for advice on rhabdoviruses.

J. L. Sever, Infectious Diseases Branch, National Institute of Neurological and Communicative Disorders and Stroke, National Institutes of Health, Bethesda, Maryland, for his sustained support to the electron microscopic studies of viral assembly by M.D.D. and B.R.

J. K. Smith, U.S. Army Medical Research Institute of Infectious Diseases, Fort Dietrick, Maryland for advice on bunyaviruses and Fig. 5-2.

E. Strauss, California Institute of Technology, Pasadena, for helpful advice on togaviruses and for sharing data in advance of publication.

J. Strauss, California Institute of Technology, Pasadena, for helpful advice on togaviruses and for sharing data in advance of publication.

V. Stollar, Rutgers Medical School, Piscataway, New Jersey, for sharing data on Alphaviruses in advance of publication.

L. S. Sturman, New York State Department of Health, Albany, for advice on coronaviruses.

M. M. Sveda, Laboratory of Molecular Genetics, National Institute of Neurological and Communicative Disorders and Stroke, National Institutes of Health, Bethesda, Maryland, for advice and sharing data on orthomyxoviruses.

J. Turner, New York State Department of Health, Albany, for assistance with high voltage electron microscopy and with Fig. 7-7 e.

W. W. Newcomb, University of Virginia, Charlottesville, for advice on rhabdoviruses and Fig. 2-8 c.

G. Warren, European Molecular Biology Organization, Heidelberg, Federal Republic of Germany, for Figs. 8-2, 8-6 and 8-7.

J. S. Wolinsky, University of Texas, Houston, for Fig. 3-8.

R. G. Wyatt, National Institute of Allergy and Infectious Diseases, National Institutes of Health, Bethesda, Maryland, for advice on rotaviruses.

P. R. Young, London School of Hygiene and Tropical Medicine, for advice on arenaviruses and Figs. 6-3 c and 6-4.

We gratefully acknowledge the excellent collaboration of Judy Hertler. She organized the bibliography and correspondence, spent many hours working on the word processor and actively participated in the final editing of the book. Her contribution to the index was invaluable. The diagrams designed by B. Rentier were skillfully executed by Trudy Nickelson. Ray Rusten printed the micrographs from our laboratory.

This work was supported in part by grant # 17899 from the National Institutes of Health and by grant # 2 H-13 from the United States Agency for International Development and by grants from the "Fonds National de la Recherche Scientifique" of Belgium.

The opinions expressed in this report are the private views of the authors and should not be construed as official or as necessarily reflecting the views of the Uniformed Services University or the Department of Defense.

Contents

1

An Overview of the Assembly
of Enveloped RNA Viruses

I. Introduction

All enveloped RNA viruses have in common an RNA genome surrounded by an envelope made of virus-modified host membrane. These viruses are grouped into separate families on the basis of differences in their mechanisms of penetration into cells, their strategies of replication and transcription, their assembly processes and the structure of their virions. This book will review the assembly of the following eight families of enveloped RNA viruses: *Rhabdoviridae, Paramyxoviridae, Orthomyxoviridae, Bunyaviridae, Arenaviridae, Coronaviridae, Togaviridae (Alpha-, Flavi-, Rubi, Pestiviruses)*, and *Retroviridae* (Table 1-1). The ultrastructural aspects of most of these viruses have been described in detail (Dalton and Haguenau, 1973). The molecular aspects of replication and maturation of these viruses have been extensively studied in many cases but much less in others (pestiviruses, rubiviruses).

We have also included a discussion of *Rotaviridae* in this book. Although these are generally not considered to be budding enveloped RNA viruses, enveloped forms of rotaviruses bud into the rough ER during replication. The formation of these enveloped forms probably involves the same mechanism of assembly as the enveloped RNA viruses which are discussed in this book.

Viruses are obligate intracellular parasites that depend heavily on cellular metabolism for their reproduction and survival. The genome of an enveloped RNA virus is always encapsidated by proteins in a complex called the nucleocapsid (NC). These proteins play essential roles in virus replication and maintain the integrity and symmetry of their respective NCs. NCs need viral envelopes to exit the cells that produce them. These envelopes also mediate interaction with the surfaces of host cells so that the NC can infect these cells and replicate again. Envelopes are lipid bilayers into which one to three virus glycoproteins are inserted, giving each virus unique antigenic determinants (Rott and Klenk, 1977). The glycoproteins form surface projections (peplomers) with the shape of spikes or knobs.

The lipid composition of a virus envelope usually reflects that of the host cell (Rott and Klenk, 1977). Cellular plasma membrane lipids are arranged asymmetrically across the membrane (Karnofsky *et al.*, 1982). The same asymmetry of lipids, such as phospholipids and cholesterol, is also seen in virus envelopes (reviewed by Lenard, 1978; Simons and Garoff, 1980).

The cellular compartments and organelles required for entry, replication and assembly of enveloped RNA viruses are schematized in Fig. 1-1. Clathrin coated pits

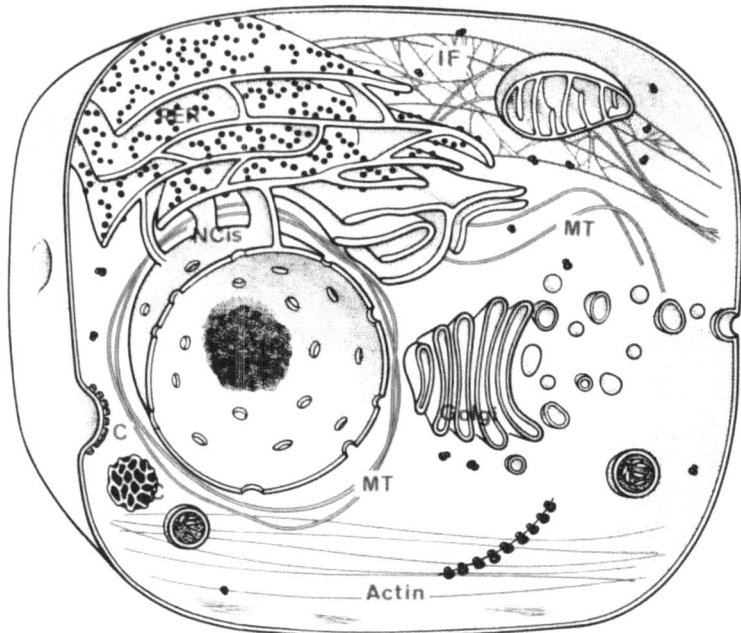

Fig. 1-1 Representation of cellular compartments and organelles used by enveloped RNA viruses for entry, replication and assembly. Each compartment is depicted in a different area of the cell for clarity. *C* clathrin basket around coated pits and vesicles. *RER* rough endoplasmic reticulum communicates with nuclear cisterns *(NCis)*. From the Golgi apparatus, transport vesicles travel to the cell surface and fuse with the plasma membrane. Cytoskeletal elements are represented in color. Their role in viral assembly is not clearly understood. *MT* microtubules, 24 nm in diameter, originate in an organizing center near the Golgi apparatus (not shown). Two MTs are shown around the nucleus. *IF* intermediate filaments are 10 nm in diameter. Five different IF subclasses (keratin, desmin, vimentin, neurofilaments, gliofilaments) are found in different cell types. Actin filaments are 5–7 nm in diameter and form stress fibers or bundles under the membrane

are the sites where most of these viruses enter. Clathrin is a highly conserved protein which forms a characteristic polyhedral sheet or basket on the surface of coated pits or vesicles (Harrison and Kirckhausen, 1983). The rough endoplasmic reticulum (ER) and ribosomes are involved in synthesis of viral proteins, and the Golgi apparatus in post-translational modifications and transport of the viral glycoproteins. From the Golgi apparatus, glycoproteins are shuttled to the cell surface in transport vesicles. The nucleus is necessary for replication of some segmented viruses. Cytoskeleton elements (microtubules, intermediate filaments, actin filaments) are involved in numerous cellular functions (reviewed by Amos, 1979; Lazarides, 1980) and may also play a role in viral assembly and release (see below).

II. Virus Entry

Some enveloped RNA viruses, such as *Paramyxoviridae* enter a cell by fusion of their envelopes with the cell surface, while many others, such as *Orthomyxo-, Rhabdo-,* and *Alphaviridae* enter via "coated pits" (reviewed by Howe *et al.,* 1980). A coated pit is a specific membrane region which is covered by clathrin on the cytoplasmic side and is a site where hormones and other ligands enter cells. A smooth vesicle containing the virus will form from this coated pit (Pastan and Willingham, 1981; Dickson *et al.,* 1981). This smooth vesicle becomes acidified (pH 5–6) (Tycko and Maxfield, 1982) and has been called a receptosome, endosome or prelysosomal endocytic vacuole (Willingham and Pastan, 1980; Helenius and Marsh, 1982; Simons *et al.,* 1982; Marsh *et al.,* 1983). Endosome is the preferred designation. At this point, the viral envelope fuses with the vesicle membrane, allowing the NC to penetrate into the cytoplasm. The importance of acid pH for virus fusion is illustrated by the antiviral effects of drugs that raise vesicular pH (Helenius *et al.,* 1982; Marsh *et al.,* 1982) and by experiments showing that viruses which do not normally fuse with the plasma membrane may do so when the extracellular pH is lowered (Matlin *et al.,* 1982).

III. Virus Genomes

The sizes of the genomes of enveloped RNA viruses vary from 5000 bases (retroviruses), to up to 18,000 bases (corona-, paramyxoviruses) (Varmus, 1982; Strauss and Strauss, 1983) (Table 1-1). The genome may be of negative or positive polarity. Positive-sense genomes can act as messenger RNA (mRNA) and are therefore usually infectious, even when devoid of viral proteins. Negative-sense genomes are never infectious by themselves; they need virus-specified RNA polymerase proteins to transcribe their positive-sense mRNAs (Baltimore, 1971). These mRNAs are usually capped at their 5' ends and have poly(A) tails at their 3' ends. Among the negative strand virus families are *Rhabdo-, Paramyxo-, Orthomyxo-, Bunya-,* and *Arenaviridae.* The enveloped positive strand group contains the *Corona-* and *Togaviridae (Alpha-, Flavi-, Pesti-, Rubivirus). Retroviridae* constitute a separate entity because of the uniqueness of their replicative cycle, involving reverse transcription of the viral genome into a complementary DNA (cDNA) and integration of the double-stranded DNA into the host cell genome. In addition, some genomes are segmented and others are not: *Rhabdo-, Paramyxo-, Corona-, Toga-,* and *Retroviridae* have nonsegmented genomes, while *Orthomyxo-, Bunya-,* and *Arenaviridae* contain eight, three, and two RNA segments, respectively.

Rhabdo- and *Paramyxoviridae* are very similar in their genome organization and their replication strategy and it was, therefore, suggested that *Rhabdoviridae* emerged from *Paramyxoviridae* (Strauss and Strauss, 1983). *Orthomyxo-, Arena-,* and *Bunyaviridae* also share some characteristics, since they have segmented genomes enabling each of them to undergo high-frequency genetic reassortment. *Arenaviridae* possibly require for their replication an intranuclear step of yet unknown

Table1-1. *Properties of enveloped RNA viruses*

Virus families (-viridae)	Genome			Nucleocapsid		Virus Budding Site[b]	Virion		
	Size (×10³ bases)	Polarity	Seg-ments	Site of Assembly[a]	Inclusions		NC symmetry	Shape[c]	Size
Rhabdo-	11	−	1	C	C	cell sur-face (ER)	helical	generally bacilliform	70–170 nm
Paramyxo-	15 to 18	−	1	C	C & N	surface	helical	P	200–500 nm
Orthomyxo-	13.5	−	8	N?	C & N?	surface	helical	P	80–100 nm
Bunya-	14 to 17.5	−	3	C	?	Golgi	helical	S	80–110 nm
Arena-	9.6 to 12.3	−	2	C (?)	?	surface	helical	S	100–300 nm
Corona-	18	+	1	C	C	ER, Golgi (nuclear envelope)	helical	S	60–160 nm
Alpha-	12	+	1	C	C & rarely N	surface (ER)[v] ER[i]	icosahedral	S	60– 65 nm
Flavi-	12	+	1	C	C & rarely N	ER, Golgi	icosahedral	S	35– 45 nm
Retro-	5 to 9 (diploid)	+ (and DNA provirus)	1	C	C (A particles)	surface (ER)	complex	S	~100 nm
Rota-	36	ds[d]	11	C	C	ER	icosahedral	S	~ 25 nm

[a] C = cytoplasm, N = nucleus.
[b] ER = endoplasmic reticulum, v = in vertebrate cells, i = in invertebrate cells.
[c] S = spherical, P = pleomorphic.
[d] ds = double stranded.

nature. Two unique events of *Orthomyxoviridae* transcription occur in the nucleus: the transfer of a 5′ terminal cap structure from freshly synthesized cellular mRNAs to viral mRNAs, and the splicing of some of the viral mRNAs. Retrovirus transcription from the integrated DNA necessarily occurs in the nucleus, and splicing may generate one of the mRNAs (*gag-pol*, see Chapter 9).

IV. Synthesis of Virus Components

Viruses of most families presented here have three kinds of virus proteins: those associated with the genome, others associated with the envelope, and nonstructural proteins, found only in the cell. The designations of these proteins are given for each virus family in Table 1-2. All viruses have at least one main structural NC protein, often phosphorylated, and all negative strand viruses have an RNA polymerase associated with the genome. Some viruses have yet another, more highly phosphorylated NC-associated protein which appears to play a role in RNA synthesis. Virus envelopes may contain, in addition to the lipid bilayer, one to three glycoproteins, and in some cases a nonglycosylated protein such as the matrix or membrane protein of Orthomyxo-, Paramyxo-, and Rhabdoviridae.

Sites of Virus Protein Synthesis

Virus glycoproteins are synthesized at the rough ER and are directly inserted into its membrane, while other virus proteins are synthesized on free ribosomes in the cytoplasm. Some proteins migrate to other cellular compartments, such as the nucleus (influenza virus). Eventually, they bind to either genomic RNA or to the inside of the membrane at sites where viruses bud. The intracellular transport of proteins synthesized on free ribosomes is still poorly understood. Is it directed by cellular cytoskeletal elements or do the proteins passively diffuse to their destinations the genome or the membrane?

Synthesis of Virus Glycoproteins

Virus glycoproteins are synthesized and processed on intracellular membranes and transported to the cell surface like cellular membrane glycoproteins. In fact, virus glycoproteins have served as models for the study of synthesis, transport and processing of cellular glycoproteins. Here, we will briefly describe these processes, which have been extensively reviewed (Lodish and Rothman, 1980; Sabatini, 1982). Translation of the mRNAs for glycoproteins begins on free ribosomes. Usually, an amino-terminal insertion signal (signal peptide) emerges from the ribosome and then binds the mRNA-ribosome complex to a signal receptor in the ER membrane (Fig. 1-2 a) (Blobel, 1980; Sabatini, 1982). The signal peptide directs the insertion of the nascent polypeptide chain through the rough ER membrane. Usually, signal peptides are then cleaved off, which creates a new free amino-terminus in the ER lumen. A hydrophobic domain spanning the lipid bilayer anchors most viral proteins in the membrane and is called the halt transfer signal. Only a short hydro-

Table 1-2. *Proteins of enveloped RNA viruses*[a]

Virus families (-viridae)	Genome Associated			Envelope		Nonstructural (Intracellular)	
	Main structural	Polymerase & other enzymes	Others	Nonglycosylated	Glycosylated (spikes)	Precursors of structural proteins	Unrelated to viral structural proteins
Rhabdo-	N	L	NS	M	G		C (or V)
Paramyxo-	NP	L	P	M	HN or HA, F1, F2	Fo	NS1, NS2, M2 (M3?)
Orthomyxo-	N(p)[b]	PA, PB1 PB2	—	M1	HA1, HA2 NA	HA	Several
Bunya-	N	L?	—	—	G1, G2		?
Arena-	NP	L	P	—	GP1 (and 2)	GPC	NS 200 NS 35 NS 14
Corona-	N(p)	—	—	—	E1, E2	—	6K, NS 70, NS 86, NS 72 (NS 60)
Alpha-	C	—	—	—	E1, E2, E3	p62	9?
Flavi-	V2	—	—	V1	V3	?	
Retro-							
Avian C: ASV	p27	pol, p15	pp19[c], p10, p12	pp19	gp85, 37[d]	env	
Murine C: MuLV	p30	pol	pp12, p10	p15, p15E	gp70	gag	
Murine B: MMTV	p28	pol	pp21, p14	p10	gp52, 36	gag-pol	NS 53, NS 34, NS 35
Rota-	VP6	—	VP2, VP1	VP3	VP7[d]	NS 28[d]	

[a] Proteins are designated according to current usage (see individual chapters). Retroviridae prototypes are abbreviated as in Chapter 9.
[b] (p) = phosphorylated.
[c] pp = phosphoprotein.
[d] gp = glycoprotein.

philic domain at the carboxyl-terminus remains on the cytoplasmic side of the membrane (Fig. 1-2 a). Deletion of the membrane attachment regions of the influenza virus hemagglutinin (HA) (Gething and Sambrook, 1981; Sveda et al., 1982) or the G glycoprotein of the rhabdovirus vesicular stomatitis virus (VSV) (Rose et al., 1982) by genetic manipulation causes secretion of these normally integral membrane glycoproteins (Fig. 1-2 b).

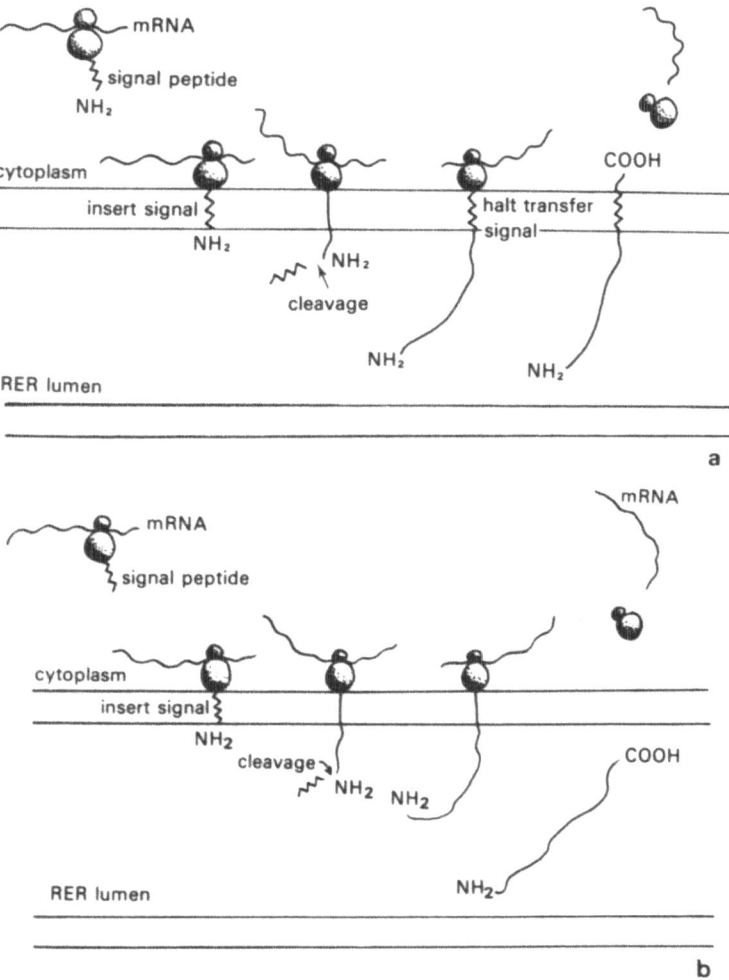

Fig. 1-2. Synthesis and insertion of virus glycoproteins into the rough endoplasmic reticulum (RER). (a) A signal peptide is synthesized first and binds to a membrane receptor protein on the RER membrane (symbolized by two lines). The inserted polypeptide chain grows through the ER membrane. The signal peptide is then cleaved off by a cellular signal peptidase in most cases. A stretch of amino acids serves as a halt transfer signal and prevents the passage of the carboxyl-terminal portion of the polypeptide chain through the membrane (modified from Sabatini et al., 1982). (b) Effect of a deletion of a halt transfer signal by genetic manipulation. Insertion and cleavage of the signal peptide are normal. However, the absence of the halt transfer signal results in the release of the polypeptide chain into the ER lumen, from which it will be secreted into the surrounding medium

Table 1-3. *Cellular functions involved in modification and transport of viral components*[a]

Virus families (-viridae)	Glycosylation N-linked	O-linked	Use of transport vesicles[b]	Proteolytic cleavage Target	Location	Phosphorylation	Acylation
Rhabdo-	+		G	—	—	NS, M	G (Indiana VSV)
Paramyxo-	+		HN or HA, F	Fo	plasma membrane in RER (NDV)[c] in virion (Sendai)	P	F
Orthomyxo-	+		HA, NA	HA	plasma membrane	N, M_1, NS_1	HA
Bunya-	+		virions	110 K	intracellular		
Arena-	+		GPC?	GPC	intracellular	P	
Corona-	E_2	E_1	E_2, virions	E_2	plasma membrane?	N	E_2
Alpha-	+		p 62, E_1^v, virions[i]	p62	plasma membrane		E_1, E_2
Flavi-	+		virions	?	?		V_3
Retro-	+		*env*, sometimes virions virions	*gag-pol* *gag* *env*	virion at budding site intracellular (in ER)	pp19 (ASV); pp12 (MuLV); pp21 (MMTV)	
Rota-	+		virions	VP3	outer capsid	?	?

[a] Virus proteins are designated as in Table 1-2.
[b] v=vertebrate cells, i=invertebrate cells.
[c] NDV = Newcastle Disease Virus.

Glycosylation

As these proteins are being inserted into the rough ER, mannose-rich oligosaccharides are transferred to asparagine residues on the nascent polypeptide chain (Li *et al.*, 1978; Kornfeldt and Tabas, 1978; Tabas and Kornfeldt, 1978; Lennarz, 1980). These N-linked core oligosaccharides can be cleaved by endoglycosidase-H. Short chains of oligosaccharides can be added in the Golgi apparatus to serine or threonine residues via O-glycosidic linkages. Only E_1 of coronaviruses (Chapter 7) has been shown to be an O-linked glycosylated polypeptide (Table 1-3). Viral proteins, like other integral membrane proteins, are thought to enter the Golgi complex on the side proximal to ER cisternae lacking ribosomes, also called the "*cis* face" of the Golgi. Probably small transport vesicles bud from these transitional elements and fuse with dilated ends of multiple Golgi cisternae (Farquhar and Palade, 1981). A direct connection between rough ER and Golgi has been described in some cells (Lindsey and Ellisman, in press). Glycoproteins apparently move from the proximal Golgi, where mannose residues are trimmed, to the distal Golgi, also called "*trans* face" of the Golgi, where glycosyl transferases convert simple oligosaccharides into complex ones by addition of N-acetyl-glucosamine, galactose, sialic acid and fucose (Rothman, 1981; Green *et al.*, 1981 a; Griffiths *et al.*, 1982, 1983). These complex oligosaccharides are not cleavable by endoglycosidase H. Glycosylation and sugar content of viral glycoproteins may vary. For instance, the HA glycoprotein of influenza A and B viruses has no sialic acids because of the activity of their other viral glycoprotein, the neuraminidase (reviewed by Compans and

Table 1-4. *Some reagents used to study viral assembly*

Reagent	Acting upon	Mechanism
Tunicamycin[1]	Glycosylation of proteins	Blocks addition of mannose-rich oligosaccharides
D-Glucosamine[2]	Glycosylation of proteins	Depletes the UTP pool essential for activation of sugars which are the donors in the glycosylation process
2-deoxyglucose	Glycosylation of proteins	Mannose analog
Concanavalin A[3]	Crosslinking of cell surface glycoproteins	Binds to αD-glucose and αD-mannose
Monensin[4]	Endocytosis Membrane recycling Transport of glycoproteins Protein secretion, etc.	Carboxylic ionophore which intercalates into membranes and abolishes proton gradient. Similar effect obtained by calcium depletion
Cytochalasin B[5]	Microfilaments, shape Glycosylation	Inhibits polymerization of actin monomer; alters hexose transport
Cytochalasin D[5]	Microfilaments, shape	Inhibits polymerization of actin monomer

[1] Takatsuki *et al.*, 1975; Stallcup and Fields, 1981.
[2] Keppler *et al.*, 1970.
[3] Noonan and Burger, 1974.
[4] Marsh *et al.*, 1982; Tartakoff, 1983.
[5] Tannenbaum, 1978; Lin *et al.*, 1980; Stallcup *et al.*, 1983.

Klenk, 1979). Glycosylation may change protein conformation, help plasma membrane insertion of glycoproteins after transport and/or mask protease digestion sites.

Alterations in glycosylation have various effects on viruses of different groups or subgroups. To study these effects, inhibitors of glycosylation, such as tunicamycin, D-glucosamine (Keppler *et al.*, 1970) and 2-deoxyglucose have been used. The mechanisms of action of these drugs are summarized in Table 1-4 (Takatsuki *et al.*, 1975; Stallcup and Fields, 1981). Studies of the effects of tunicamycin on assembly of rhabdo- and alphaviruses have shown that glycosylation is often essential for the normal assembly of these viruses (Leavitt *et al.*, 1977 a). In fact, the lack of carbohydrates may impair the ability of a glycoprotein to move through the intracellular membranous system (Leavitt *et al.*, 1977 b). Glycosylation probably influences the folding and conformation of glycoproteins (Gibson *et al.*, 1979). Surprisingly, the rare VSV particles formed in the presence of tunicamycin have normal infectivity in spite of the absence of G protein glycosylation (Gibson *et al.*, 1978).

Transport of Virus Glycoproteins

This transport probably occurs via vesicles pinched off from Golgi cisternae which move towards the cell surface by an unclear mechanism, perhaps involving the cytoskeleton and clathrin (Figs. 1-3) (Rodriguez-Boulan *et al.*, in press). Disruption of the microtubular system by drugs such as nocodazole, which disperse the Golgi elements, does not appear to interfere with the efficient vesicular transport of the G protein of VSV to the plasma membrane (Rogalsky *et al.*, 1982). The vesicles fuse with the plasma membrane, exposing the glycosylated part of the polypeptide to the extracellular space (Lodish and Rothman, 1979). On the virion surface, viral glycoproteins are grouped into spikes made of 3 to 6 glycoprotein molecules. It is not clear whether these multimers are first formed in the Golgi, in transport vesicles or at the cell surface at the sites of viral budding. In infected cells, these spikes are not observed outside the viral buds.

Cleavage of Precursor Proteins

In some viruses, large intracellular precursors of viral structural proteins must be cleaved by host cell proteolytic enzymes before they become biologically active (Table 1-3). As mentioned above, a signal peptidase removes the signal sequence from many viral glycoproteins (not specifically mentioned in Table 1-3). One group of retroviruses (avian C) has a virus-encoded protease. However, a cellular protease is required to cleave and activate it (see Chapter 9). The properties of the viral precursor proteins, as well as the sites where cleavage occurs, vary from one family to another (Table 1-3). Cleavage may occur in the ER, at the plasma membrane, at budding sites or in virions. In retroviruses, each of the three precursor proteins is probably cleaved at a different site: *env* in the ER, *gag* at the budding site and *gag-pol* in the released virions (see Chapter 9). Thus, in this case, cleavage appears to be tightly coordinated with viral assembly events, each protein being released and becoming functional when it is needed.

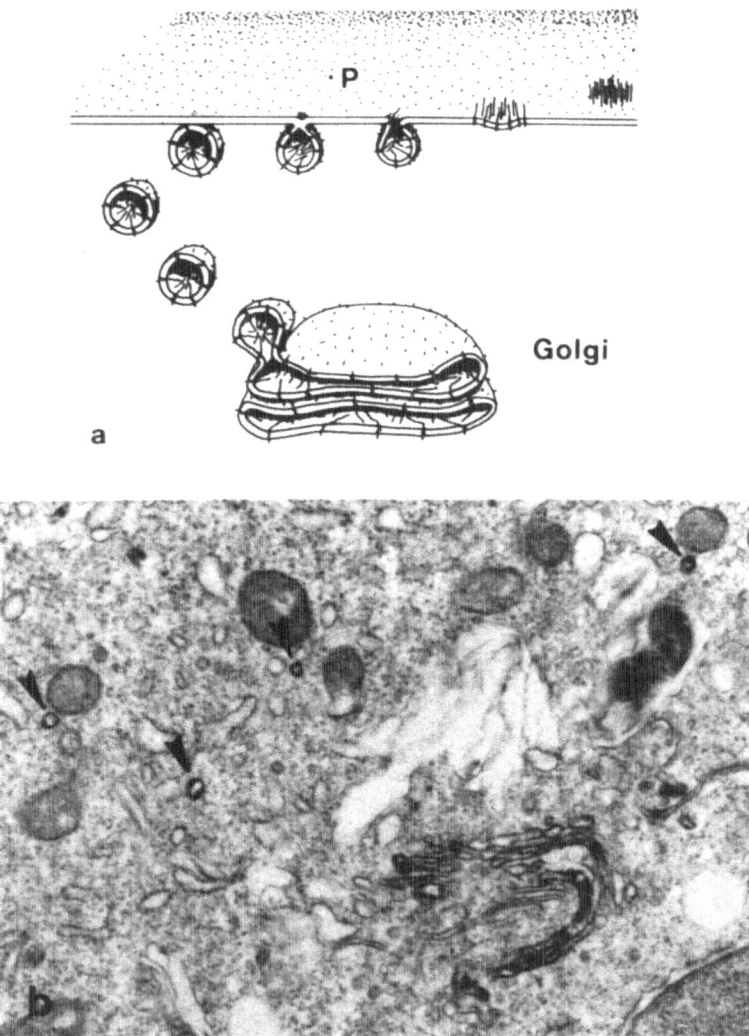

Fig. 1-3. *(a)* Scheme of the transport of transmembrane glycoproteins from the Golgi to the cell surface. Vesicles containing the transmembrane glycoproteins are pinched off from the proximal cisternae of the Golgi apparatus where glycosylation is completed. The vesicles then move to the surface where they fuse with the plasmalemma *(P)*, exposing the glycosylated portion of each glycoprotein to the extracellular space while the rest of the molecule remains attached to the membrane. *(b)* Transmission electron microscopic demonstration of a virus glycoprotein in the Golgi apparatus of Madin Darby canine kidney cells. The hemagglutinin (HA) glycoprotein of a temperature-sensitive mutant of the influenza virus strain A/WSN, with the mutation in the hemagglutinin gene, has been labeled by the immunoperoxidase technique 40 min after shifting the cells from the non-permissive to the permissive temperature. HA is detected in the Golgi membranes and in four vesicles larger than coated vesicles (arrowheads). Magnification: × 19,000. (Courtesy of Dr. E. Rodriguez-Boulan)

Other Post-Translational Modifications

Another frequently observed modification of virus proteins is phosphorylation. Almost all phosphorylated proteins are genome-binding proteins. Two other modifications of virus proteins may also occur: attachment of fatty acids and, occasionally, sulfation. The glycoproteins HA and NA of influenza viruses, as well as the glycoprotein G of VSV, are sulfated (Compans and Klenk, 1979; Hsu and Kingsbury, 1982 a). At least one species of spike glycoprotein of *Orthomyxo-, Rhabdo-, Paramyxo-, Alpha-,* and *Coronaviridae* is acylated (Schmidt and Schlesinger, 1980; Schmidt, 1982 a and b). The fatty acids are linked to the hydrophobic tail of the glycoprotein associated with the lipid bilayer, suggesting a possible anchoring function. However, it does not appear that this is a prerequisite for anchorage of all viral glycoproteins (see Chapter 2).

To study the role of several cellular functions involved in the synthesis of biologically active virus components and their assembly, several drugs which interfere with these functions have been applied to infected cells. In addition to the drugs mentioned above, monensin has been especially useful in the study of transport of glycoproteins (Tartakoff, 1983), and concanavalin A (con A) in the study of late events of viral assembly. The effects of these and the other drugs mentioned are summarized in Table 1-4.

V. Assembly of Virus Components

Virus assembly consists of a series of events governed by interactions between virus proteins with each other and with genomic RNA, as well as by interaction of the NC with the virus-modified membrane at the budding site. The specificity and nature of some interactions occurring during virus assembly have been studied in mixed infections and by genetic manipulations. Reassortants of viruses with segmented genomes, studies with temperature-sensitive mutants, and phenotypically mixed pseudotypes (reviewed in Zavada, 1982) have been especially useful in this regard. NC assembly generally occurs within the cytoplasm, but in some cases, such as in the *Orthomyxoviridae,* the nucleus may be the site of assembly of NCs.

NC assembly can be reconstituted *in vitro* with some viruses, although with a lower efficiency than in cells (Patton *et al.,* 1983). In contrast, the formation of the envelope around the nucleocapsid needs the intact cell and its membranes and has not yet been achieved in the test tube.

Nucleocapsid (NC) Assembly (Encapsidation)

NC assembly can be independent of virus budding or can require the interaction with the membrane. For instance, in the case of retroviruses type C, NC assembly appears concomitant with viral budding, while *Paramyxo-, Orthomyxo-,* and *Rhabdoviridae* can assemble their NCs independently from budding. NC assembly can only occur if there is an encapsidation signal on the virus genome where the capsid protein(s) will first bind. In some cases, such a signal is present at or near the 5' terminus of RNA: for instance, at the leader sequence in VSV (Blumberg *et al.*, 1983), at a 13 nucleotide sequence in each of the 8 fragments of influenza viruses, and at the 5' end of the alphavirus genome, which contains a sequence not found in the subgenomic mRNAs or the negative strand template (Strauss and Strauss, 1983). In VSV, all the information required for NC assembly is contained in the RNA and the NC major structural protein (Blumberg *et al.*, 1983). The assembly is cooperative and progresses from the 5' end towards the 3' end.

Structure of NCs

Once assembled, the NC acquires its final structure: helical, icosahedral or complex (Table 1-1). This may occur at different intracellular sites, depending on the virus. For instance, *Paramyxoviridae* NCs appear to acquire a stable, helical organization (with fixed spacing and width of the helix) inside the cytoplasm before they interact with the membrane. If the M protein (or another viral protein) attaches to the NC, it may play an important role in stabilizing the NC structure (see Chapter 3). Later, this organized helical NC will be progressively folded inside the budding virion. This attests to a large degree of flexibility of the helical NC. In other cases, such as in the *Rhabdoviridae,* loose, thin strands of NC can form a helix of variable turn-to-turn spacing and width in the cytoplasm. In contrast, inside the viral bud, NC is organized in a tight coil of fixed spacing and width. It is at this stage that the M protein becomes associated with the NC close to the envelope. This association may determine the final configuration of NC (see Chapter 2). In a third category of viruses, such as retroviruses type C and *Togaviridae,* the icosahedral structure of the NC probably forms only at the membrane and no free NCs are seen in the cytoplasm. In *Retroviridae,* the major core protein forms a shell around a NC strand, probably organized in a loose helical structure (see Chapter 9).

Sites of Budding

How NC and envelope proteins find their way to a specific membrane site is still a mystery. As shown in Table 1-1, viruses of different families mature at distinct membrane sites, such as the rough and smooth ER, the nuclear envelope, the Golgi apparatus, or the plasma membrane. Some viruses can assemble either at the surface or within intracellular vesicles, depending on the host cell. In particular, transport of glycoproteins to specific membrane destinations has been shown to affect the location of virus budding. Although the signals by which cellular glycoproteins are assorted into the nuclear envelope, rough ER, Golgi membranes, and apical or baso-lateral plasma membranes of polarized cells (see below) have not yet been identified, it is clear that individual cellular glycoproteins may be directed specifically to each of these sites. Virus glycoproteins may make use of these intracellular transport pathways by mimicking the signals of host cellular glycoproteins destined to similar locations. Viruses which bud from the Golgi membranes, such as corona- and bunyaviruses, may have an essential glycoprotein (such as the E_1

glycoprotein of coronaviruses) which cannot be transported past the Golgi (see Chapters 5 and 7).

Different virions bud at different domains of the plasma membrane of polarized epithelial cells (reviewed by Rodriguez-Boulan, 1983). Such cells have their apical pole facing the external medium and their basal pole facing the internal milieu of the body or the substrate. There is an asymmetrical distribution of enzymes, proteins and transporting system between these two poles. In polarized cells, such as Madin Darby canine kidney cells (MDCK), some viruses, such as *Rhabdo-* and *Retroviridae* will bud only at the baso-lateral pole (Roth *et al.*, 1983 a), whereas others, such as *Orthomyxo-*, *Paramyxo-* and *Alphaviridae*, will bud only from the apical pole (Fig. 1-4) (Rodriguez-Boulan and Sabatini, 1978). The virus budding site on the plasma membrane is determined by the membrane localization of the virus glycoproteins (reviewed in Rodriguez-Boulan, 1983). Thus, the distribution mechanisms of epithelial cells must recognize specific structural features in viral glycoproteins as signals for delivery to the apical or baso-lateral domain. When polarized cells are infected with a recombinant SV40 vector carrying cloned cDNA of the HA glycoprotein gene of influenza viruses, the HA is found preferentially at the apical surface (Roth *et al.*, 1983 b). Thus, the structural features of the HA glycoprotein itself are sufficient for the normal sorting of virus glycoproteins and their incorporation into a specific surface site; neither other virus proteins nor the viral genome are required. When epithelial cells are treated with the ionophore monensin, HA transport to the apical pole is only delayed, whereas VSV G transport to the baso-lateral side is arrested in the Golgi, so that the virus now buds into intracellular vacuoles (Alonso and Compans, 1981).

Early Interactions Between NC and Envelope Proteins

What are the first interactions between NC and envelope proteins? This question has been extensively reviewed by Simons and Garoff (1980). In *Alpha-* or *Retroviridae*, cleavage of a protein precursor at the plasma membrane is a prerequisite for virus budding. Before or during virus budding, host cell membrane proteins are excluded from the envelope of the budding virion. In addition, in *Rhabdo-* and *Retroviridae*, there is a selective packaging of genomic RNA and, therefore, there must be a mechanism to sort out NCs containing the genomic RNA strand from NCs containing full-length RNA of the wrong polarity. This sorting out implies that the NCs must contain a signal for envelopment, either in the NC protein or in the RNA molecule. A specific nucleotide sequence involved in RNA packaging in a retrovirus C (murine leukemia virus) has been identified recently (Mann *et al.*, 1983) (see Chapter 9).

During virus assembly, various types of interactions occur between the different viral components at the membrane. Transmembrane interactions occur between the viral components on opposite sides of the membrane and lateral aggregation of viral components within or under the membrane also takes place. Two types of transmembrane interactions have been seen. The first consists of the binding of NC protein to the cytoplasmic domain of a transmembrane glyco-

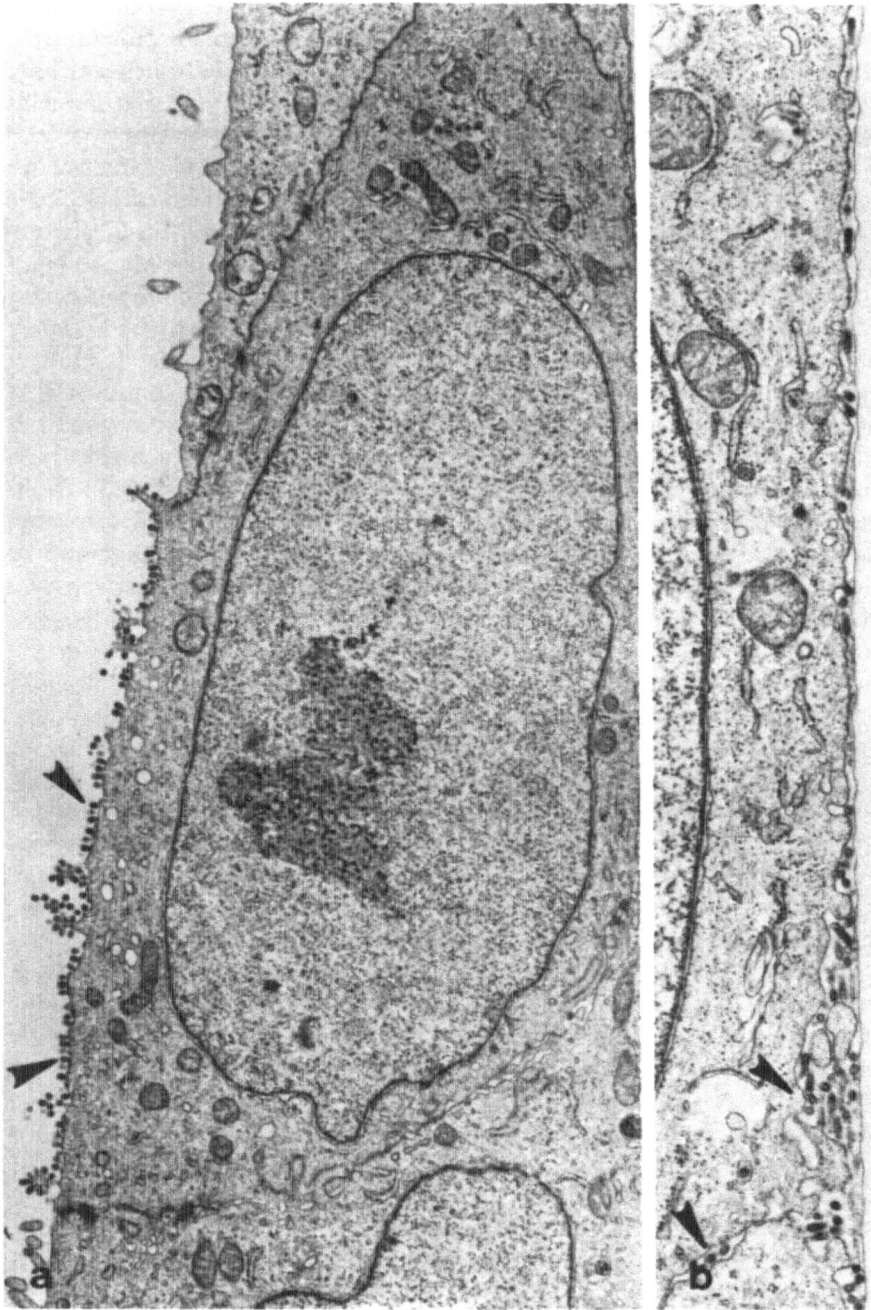

Fig. 1-4. Budding of two different enveloped RNA viruses at distinct poles of Madin Darby canine kidney cells. *(a)* Influenza virions bud from the apical pole (arrowheads). *(b)* Vesicular stomatitis virus buds mostly from the basal pole and the lateral side of the cell (arrowheads). Magnification *(a)* 10,600, *(b)* 18,000 [Courtesy of Dr. E. Rodriguez-Boulan from Rodriguez-Boulan and Sabatini, 1978 *(a)* and of Dr. E. Rodriguez-Boulan *(b)*]

protein in a specific one-to-one ratio, as shown for togaviruses. As this specific interaction occurs, there is a rigorous exclusion of host membrane proteins from this area of the lipid bilayer. Similarly, in *Bunya-* and *Coronaviridae,* which both have a helical NC, interactions between NC and transmembrane virus glycoproteins probably occur at budding sites in the Golgi and smooth ER vesicles.

The second type of transmembrane interaction occurs when NCs interact in a specific way with an intermediate matrix or assembly protein which may also bind to the cytoplasmic side of the membrane. Examples are the lipid-binding hydrophobic protein of retroviruses (a *gag* product, see Chapter 9) and the M proteins of the *Rhabdo-, Orthomyxo-,* and *Paramyxoviridae.* However, M protein properties may vary from one virus to another (Table 1-5). The influenza virus M protein has a very large hydrophobic domain (Gregoriades, 1980) and may traverse the lipid bilayer (Reginster *et al.,* 1976, 1979; Biddison *et al.,* 1977; Braciale, 1977). In contrast, the M protein of the rhabdovirus VSV has no hydrophobic domain but may interact electrostatically with the cytoplasmic membrane by means of a highly basic sequence of amino acids near the amino terminus (Rose and Gallione, 1981). The M protein may have two binding sites: one of high specificity for the NC and the other of low specificity for the glycoproteins. Indeed, in cells simultaneously infected with two different budding viruses, envelope glycoproteins can be interchanged on budding virions. This results in pseudotypes containing the NC and M protein of one virus with a mixture of glycoproteins from both viruses or, under certain experimental conditions, with only the foreign glycoprotein (reviewed by Zavada, 1982). For instance, the rhabdovirus VSV can incorporate glycoproteins of alpha-, paramyxo-, orthomyxo-, or retroviruses. In cases where virus glycoproteins are absent, even host glycoproteins can be incorporated into the virion (Lodish and Porter, 1980 c). A "pseudotypic paradox" (Zavada, 1982) has been defined as follows: cell membranes can incorporate both cellular and viral glycoproteins but, in the viral envelope, cellular glycoproteins are excluded under normal circumstances. Enveloped viruses can nonselectively incorporate their own glycoproteins and foreign viral glycoproteins. To explain the paradox, Zavada (1982) postulates that several enveloped RNA viruses must share a common, highly specific mechanism of envelope assembly. Two models have been proposed. First, the cytoplasmic and/ or transmembrane regions of different virus glycoproteins may be structurally related but different from the structure of all host glycoproteins. Among viruses with an M protein, this protein could then specifically recognize the virus glycoprotein, and could aggregate into a crystal-like lattice and push away non-virus glycoproteins. A variation of this model suggests that M proteins alter the properties of the membrane to cause phase partitioning between virus and cellular glycoproteins (Zavada, 1982).

Another type of specific interaction required for virus assembly consists of lateral interactions between the different molecules incorporated into the membrane at the assembly site. Most of these known interactions occur in the extracellular domains of virus glycoproteins or on the internal side of the bilayer between matrix or matrix-like proteins (Compans and Klenk, 1979). Clustering of virus glycoproteins results in the formation of distinct spikes or peplomers at the budding sites. In cells infected with retro- or rhabdoviruses and treated with inter-

Table 1-5. *Properties of M proteins in three virus families*[a]

Virus families (-viridae)	MW ×10⁻³	Chemical properties	Phosphorylated	Relationship to viral envelope[b]	Relationship to nucleocapsid[c]	Regulatory role
Rhabdo- (prototype VSV)	26	No long hydrophobic domain. Basic at amino terminus	+	Does not penetrate deeply inner side of bilayer. May interact with G	Binds to NC only at plasma membrane. Triggers NC coiling	Limits rate of assembly. Inhibits transcription
Paramyxo-	34	Hydrophobic basic. Forms tubular crystalline aggregates *in vitro* (Sendai virus)	+ (intracellular form)	Penetrates inner side of bilayer and binds to lipids or gp? Forms crystalline aggregates on inner leaflet of cell membranes (Sendai virus)	Binds to NC at plasma membrane (measles virus)	Limits rate of assembly. Role in replication?
Orthomyxo-	28	Hydrophobic region in center of molecule	−	Penetrates inner side of bilayer and binds to lipids and gp (may be exposed to cell surface)	Binds NC only at plasma membrane but may interact with genome in nucleus	Limits rate of assembly. Role in replication?

a See individual Chapters 2, 3 and 4 for details and references.
b gp = glycoproteins.
c NC = nucleocapsid.

feron, these clustering events appear to be hampered, perhaps by alteration of membrane fluidity and/or submembraneous structures. Interferon is known to increase the number of actin fibers and decrease the rate of cell locomotion and ruffling (Pfeffer *et al.*, 1980). As a result, interferon may inhibit various steps in final assembly of retro- and rhabdoviruses, including budding and incorporation of virus glycoproteins (Friedman, 1977). The mechanisms of this inhibition are probably multiple, but, in the case of some retroviruses, the apparent inhibition of clustering and release is definitely not related to a defective interaction between NC and viral envelope proteins or to defective cleavage of *gag* proteins (Sen and Pinter, 1983) (see Chapter 9).

Virus Budding

The forces responsible for the formation of the viral buds are not known. Complete (or relatively complete) replacement of host proteins by virus proteins in particular areas of the membrane may induce budding. The cell, recognizing an altered area of the plasma membrane, may try to excise this area. Virus envelopes have indeed increased rigidity which may promote budding (Lenard, 1978). In the budding virus and in the virion, glycoproteins probably restrict lipid motion and lateral diffusion. Alternatively, viral components may have a self-assembly force. The lateral aggregation of envelope glycoproteins and/or M proteins in the membrane may lead to the formation of a non-planar surface and budding. Incorporation of M protein into the plasma membrane increases its rigidity (Compans and Klenk, 1979). The observation that virions which bud from membranes often have a characteristic diameter suggests that viral components are important in deter-mining the diameter and shape of the buds. Virus mutants which yield filamentous or multiploid particles rather than spheres also indicate that budding forms may be determined by specific interactions of viral components.

The progressive coiling and organization of a NC inside a nascent virion may also be a factor contributing to the growth of the bud. In all virus buds, there is a close interaction between NC and envelope. Some enveloped RNA viruses, which do not code for an M protein, possess a solid icosahedral nucleocapsid, while the M protein-containing viruses have helical NCs, as if they require M protein as a structural support for the virion (Lenard, 1978) (Table 1-1). In rhabdoviruses, the M protein appears to control NC coiling and virus shape (see Chapter 2). Viruses with tightly coiled or icosahedral NCs, such as *Rhabdo-, Corona-, Toga-,* and *Retroviridae,* are usually smaller in diameter and more regularly shaped than those with loosely coiled NCs, such as *Paramyxo-* and *Arenaviridae.* The latter also incorporate cell components such as actin and/or ribosomes and may be quite pleomorphic. Inter-mediate situations are observed with the *Orthomyxo-* and *Bunyaviridae.*

Virus Release

Virus can be released from the cell in one or two steps: with viruses budding from the plasma membrane, a narrow neck forms at the base of the mature bud. Pinching off occurs by fusion of the closely apposed membranes in the neck. A similar event takes place with viruses which bud at internal membrane sites, such as corona-, bunya-, and flaviviruses. From there, transport vesicles are used to carry complete viruses to the cell surface (Table 1-3). These vesicles fuse with the plasma membrane, releasing one or several particles into the extracellular space by exocytosis. This is the case for coronaviruses and some togaviruses.

The molecular mechanism of virus release from the infected cell membrane is still a mystery. Possible mechanisms range from a cleavage of virus envelope protein (Green *et al.,* 1981 b) to a contractile process. Since actin has been found in various enveloped viruses (Wang *et al.,* 1976), it was proposed that actin was a contractile element triggering virus release (Damsky *et al.,* 1977). Studies with cytochalasin B, which inhibits actin polymerization (Table 1-4) (Tannenbaum *et al.,* 1978; Lin *et al.,* 1980), do not support this hypothesis, since cytochalasin B did not inhibit maturation of influenza virus or VSV (Griffin and Compans, 1979). However, a different result has been obtained with measles virus. Here, both cytochalasin B and cyto-chalasin D (a similar compound which has no effect on glycosylation) impair the release of virus from the cell (Stallcup *et al.,* 1983). Perhaps actin filaments form a ring at the base of a virus bud. Contraction of this ring might trigger the physical separation of the virion from the cell in the same way that daughter cells are separated by a circumferential filamentous ring of actin at cytokinesis.

Viruses of several families undergo further maturation after release. *Retro-, Paramyxo-, Corona-,* and, perhaps, *Orthomyxoviridae* show retraction of the NC from the envelope. This modification may facilitate virus entry, since NC and genome will be free to move into the cytoplasm after fusion of the virus envelope with the cell surface or endosomal membrane. The enveloped form of rotaviruses is transient and changes into a mature double-shelled virion in the lumen of the ER. This maturation is associated with major changes in the envelope including loss of lipids (Estes *et al.,* 1983).

Defective Assembly

Host cell factors appear to influence the virus assembly process, since the efficiency of virus assembly varies from one cell to another. NCs are often synthesized in excess and accumulate in the cytoplasm or even in the nucleus, as is sometimes the case for paramyxo-, alpha-, and flaviviruses (Table 1.1). There are numerous examples of inefficient or abortive assembly that result in accumulation of NCs, even in cells which are highly productive. NC accumulation may become impressive in chronic infections, where the frequency of defective assembly increases. Noninfectious virions may either have a defective genome, as in defective interfering particles (reviewed by Lazzarini *et al.,* 1981) or a defective envelope, so that they are unable to fuse with the host membrane and deliver their NCs into the cytoplasm. Thus, a large excess of virions and viral components is made in an infected cell.

VI. New Avenues

Until recently, most of the information about assembly of enveloped RNA viruses has been obtained from biochemical and physico-chemical studies of complete virus and biochemical analysis of infected cells. Very few studies have addressed the analysis of virus assembly directly in the intact cell, but recent technical developments may facilitate such studies. First, the preparation of purified polyclonal and monoclonal antibodies against each virus protein now allows sequential analysis of the emergence and transport of each protein at the light and electron microscope level. Incubations with antibodies can be carried out successfully on thin sections (Schwendemann *et al.*, 1982; Griffiths *et al.*, 1983). A promising approach is the development of microinjection techniques (Wehland *et al.*, 1981), allowing introduction of antibodies to specific virus proteins into the cells before infection or during virus replication. With such a method, one can analyze the roles of specific proteins and/or epitopes in viral replication and assembly (Arnheiter *et al.*, 1984). In addition, recent dramatic improvements in the resolution of video intensification (Willingham and Pastan, 1978) and video-Nomarski techniques, combined with computerized image analysis (Allen *et al.*, 1981) may permit detailed analysis of the fate of specific virus proteins within the living cell. Another new approach is to prepare rotary-shadowed surface replicas of the cytoplasmic sides of plasma membranes after fast-freezing and deep-etching to study events of virus assembly by stereo electron microscopy (Büechi and Bächi, 1979; Hirokawa and Heuser, 1982; Odenwald *et al.*, in press). Further information can be obtained by labeling proteins with purified immunoglobulins coupled to 5—10 nm colloidal gold particles prior to the preparation of replicas (Odenwald *et al.*, in press). Finally, a powerful approach to virus morphogenesis is the genetic manipulation of various viral proteins or RNAs by recombinant DNA techniques. Studies on cells transfected with cloned viral genes inserted into suitable eukaryotic expression vectors have already contributed to the understanding of the intracellular transport and processing of viral proteins in the absence of the viral genomic RNA (Rose and Bergmann, 1982; Sveda *et al.*, 1982; Kondor-Koch *et al.*, 1982; Roth *et al.*, 1983 b; Sprague *et al.*, 1983). Thus numerous new approaches are now available for the molecular analysis of the complex processes involved in assembly of enveloped animal viruses in the living cell.

2
Assembly of *Rhabdoviridae*

I. Introduction

Rhabdoviruses are widely distributed among vertebrates and are also found in invertebrates and plants. They are grouped in two genera, the *Lyssavirus* and the *Vesiculovirus*. The prototype *Vesiculovirus* is vesicular stomatitis virus (VSV), of which various strains have been isolated from cattle and horses as well as flies and mosquitoes (Brown and Crick, 1979). VSV produces a vesicular exanthem in bovine species and therefore is of great economic importance. The prototype of *Lyssavirus* is rabies virus, which can fatally infect all warm-blooded animals. Rhabdoviruses contain a bullet-shaped nucleocapsid and are bacilliform when released (Murphy and Harrison, 1979). Virion length varies from 160 and 170 nm and diameter is about 70 nm. The virus envelope is covered with surface projections (spikes) and surrounds a tightly coiled nucleocapsid (NC) containing the virus genome.

We will first briefly review the molecular organization of rhabdoviruses before describing assembly events. We will only review in detail the prototype, VSV, because it has been most extensively studied. However, whenever possible, comparison will be made with the *lyssa virus* genus. The reader is referred to extensive reviews on the subject, expecially in the recent series of three books on rhabdoviruses, edited by D.H.L. Bishop (Boca Raton, Florida: CRC Press, 1979 and 1980) (Etchison and Summers, 1979; McSharry, 1979; Murphy and Harrison, 1979; Ball, 1980; Rose, 1980; Morrison, 1980; Francki and Randles, 1980).

II. Molecular Organization

Rhabdoviruses have a single-strand negative-polarity RNA genome of approximately 11 Kilo-bases (Kb) that contains five genes encoding (from the 3' to the 5' end) the five proteins N, NS, M, G, and L (Fig. 2-1). In addition to the five genes, the 3' end of the VSV genome has a 47 nucleotide sequence coding for plus strand "leader" RNA (Leppert *et al.*, 1979) and an analogous 59 nucleotide long extra-

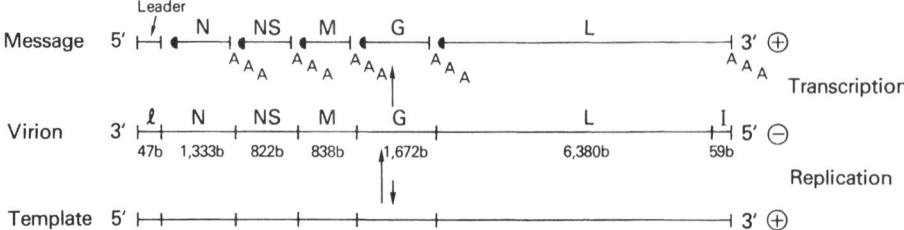

Fig. 2-1. A scheme of transcription and replication of the rhabdovirus prototype vesicular stomatitis virus (VSV). There are five genes in the negative strand genome of this virus. They have all been sequenced and the numbers of bases are indicated on the virion RNA and the mRNAs. In addition, there is a 47 base nucleotide sequence (labeled l) at the 3' end, representing the leader template and a 59 base extracistronic region (labeled I) at the 5' end. Additional nucleotides (not represented in the scheme) exist at the intercistronic regions (three bases between l and N and two bases between adjacent genes). The total number of bases is 11,162. All mRNAs are capped and polyadenylated. The template for the virion genomic RNA is represented in the lower part of this scheme. (Based on personal communication of Drs. R. A. Lazzarini and M. Schubert)

cistronic region at the 5' end (Schubert *et al.*, 1980). Three proteins, N, NS, and L, are associated with the NC and form the transcription complex. N is the major structural component of NC and L and NS participate in viral RNA polymerase activity. There is some antigenic cross-reactivity between the NC proteins of *Vesiculovirus* and *Lyssavirus* genera. The envelope proteins are M and G. M is believed to be located on the inner side of the lipid bilayer (Wagner, 1975) and is a group-specific antigen very similar in both *Lyssa* and *Vesiculo virus* genera (McSharry, 1979; Morrison, 1980). G is a transmembrane glycoprotein with a large portion exposed on the envelope surface and forming the spike. The G glycoproteins of both genera do not show cross-reactivity, although some homologies have been found between rabies virus and VSV by amino acid sequencing, suggesting that both genera have a common ancestry (Rose *et al.*, 1982). The relative order of the *Lyssavirus* appears to be similar to that of the *Vesiculovirus*. The *Lyssavirus* proteins are similar to those of the *Vesiculoviruses:* they have been called N, G, L, M_1, and M_2. M_2 of *Lyssavirus* corresponds to M in the *Vesiculovirus* and M_1 probably to NS. All VSV genes have been cloned and sequenced (Rose and Gallione, 1981; Gallione *et al.*, 1981; Harmison *et al.*, 1984).

Replication and Transcription

Virus entering the cell is uncoated and releases its NC in the cytoplasm. This incoming NC contains a negative polarity RNA strand which serves initially as a template for five subgenomic monocistronic mRNAs, each translating a different protein, and, later in infection, for a full-length positive RNA strand (Fig. 2-1). Primary transcription of the incoming genome generates the proteins necessary for replication (N, NS, L). The polymerase (L + NS) has a single entry site at the 3' end of the genome (Emerson, 1982). In the replication mode, the polymerase proceeds to the 5' end, ignoring all internal termination, polyadenylation and reinitiation signal sites along the RNA strand. The resulting plus strand antigenome then becomes the template for full-length negative strand genomes which, in turn, are used to generate more mRNAs; this is called secondary transcription. Replication requires protein synthesis (Hill *et al.*, 1981) and may be enhanced in the presence of N which encapsidates genomic RNA (Blumberg *et al.*, 1981). Encapsidation of genome by N protein might regulate synthesis of replicative RNA since, when N protein availability decreases, the synthesis of monocistronic mRNAs may be favored (Kingsbury, 1974; reviewed by Lazzarini *et al.*, 1981).

In the transcription mode, the polymerase will first synthesize the leader RNA, terminate, and, without dissociating from the template, reinitiate transcription at the start of the next gene. This results in the sequential synthesis of five mRNAs (Roy and Bishop, 1973; Testa *et al.*, 1980; Emerson, 1982). Transcription complexes can be isolated from infected cells (Murphy and Lazzarini, 1974). Transcription requires a complete NC with its three proteins, but no new protein synthesis (Emerson and Yu, 1975), and may be inhibited by M protein (Clinton *et al.*, 1978). A host cell protein (the LA protein), immunoprecipitated by serum from patients with lupus erythematosus, may favor transcription of VSV. Indeed, La protein binds to plus and minus strand leader RNAs and may serve as a modulator of virus transcription versus replication (Kurilla and Keene, 1983; Wilusz *et al.*, 1983). When N protein accumulates as a result of transcription, N molecules might progressively displace La molecules and favor replication.

Rhabdoviruses are often potent inhibitors of host protein synthesis for two reasons. Cellular and viral transcripts may compete for a limited number of ribosomes (Lodish and Porter, 1980a) and the plus strand leader RNA (found in the nucleus early in VSV infection) may trigger the shut-off of host macromolecular synthesis (McGowan *et al.*, 1982; Kurilla *et al.*, 1982).

Virus Proteins

The properties of the VSV proteins and their presumptive functions are summarized in Table 2-1.

(1) *N Protein* (MW: 47 K). N is the NC structural protein. Approximately 1200 N molecules are noncovalenty bound to each full-length genomic RNA plus or minus strand and protect the genome from ribonucleases. N molecules mediate attachment of NCs to budding envelope. The possible role of N protein as a positive regulator of viral RNA replication has been described above.

(2) *NS* (MW: about 40 K). NS is a minor component of the NC (about 50–150 molecules/genome) and is heavily phosphorylated. Based on its electrophoretic mobility, NS has an apparent MW of 40 K, but nucleotide sequencing of the NS gene reveals a coding capacity for only 222 amino acids and a predicted MW of about 25 K without the phosphates (Gallione *et al.*, 1981). NS has multiple forms (at least 13) with various degrees of phosphorylation (Clinton *et al.*, 1979). The more highly phosphorylated form stimulates transcription (Kingsford and Emerson, 1980; Kingsbury *et al.*, 1981; Hsu *et al.*, 1982). Phosphate residues, located at specific sites in NS and accessible on the outside of NCs, are responsible for the potentia-

Table 2-1. *Proteins of the rhabdovirus VSV*[1]

Species	Amino acids	MW ×10^{-3}	Molecules per virion	Location	Properties	Known or putative function
N	422	47	1200	Genome and its template.	Structural component of NC.	Regulates balance between transcription and replication.
NS	222	25	50–150	Genome and its template.	Phophoprotein.	Required for transcription. Needed for L binding.
M	229	26	2000 to 4000	Inside virion envelope and/or on coiled NC.	Phosphoprotein. Highly basic at amino-terminal domain. No hydrophobic domain.	Assembly protein? Maintains NC coiling. Inhibits transcription.
G	511	57	~1800	Spikes and envelope.	Transmembrane glycoprotein with hydrophobic domain in membrane.	Interacts with virus receptor. Elicits neutralizing antibodies. Role in virulence.
L	~2109	240	30–50	Genome and its template.	Slightly basic.	Transcriptase and replicase methylase? Poly-A-polymerase? Guanylytransferase?

[1] From *Rhabdoviruses* (Bishop D.H.L., ed.), Boca Raton, Florida: CRC Press, 1979 and 1980, and from Drs. R.L. Lazzarini and M. Schubert, personal communication.

tion of transcription by this protein. Thus NS may have different functions regulated by phosphorylation and dephosphorylation (Kingsbury *et al.,* 1981). The L protein will not bind to genome in the absence of NS. The nature of the interaction between L and NS is not clearly understood: NS may be closely associated with L, and thus be a component of the transcriptase. NS may stabilize L, anchor L to the template, or expose the L protein sites which bind to the RNA.

(3) *M Protein* (MW: 26 K, about 2000 to 4000 molecules/virus) (reviewed by McSharry, 1979; Morrison, 1980). M is a very basic and phosphorylated protein (Clinton *et al.,* 1979) apparently associated with the virion inner membrane and also interacting with the NC. The highly basic amino-terminal domain in M may interact with the virus genome (Rose and Gallione, 1981). The conclusion that M is located on the internal side of the envelope is based on its resistance to protease digestion in intact virions, its inability to be labeled by reagents that do not penetrate the envelope, and its strong association with G after digitonin treatment of virions. Interestingly, the M protein sequence, predicted from the nucleotide sequence of its cloned gene, does not contain any long hydrophobic or nonpolar domains that might promote association with the membrane (Rose and Gallione, 1981). Studies with probes of membrane protein-phospholipid interactions have shown that the M protein is closely associated with the virion envelope but does not penetrate it very deeply (Zakowski and Wagner, 1980). Most negatively charged lipids, such as phospholipids, phosphatidic acid, and phosphatidylinositol, can bind to the positively charged M protein (Zakowski *et al.,* 1981). M apparently stabilizes the structure and shape of the virion (see section IV of this chapter) and may favor replication by inhibiting VSV transcription (see above) (Carroll and Wagner, 1979).

(4) *G Protein* (MW: 57 K) is the only protein which constitutes the virion spikes; it is transmembranous and glycosylated (Morrison, 1980). There is a maximum of 1800 G molecules per virion and probably four molecules of G per spike. The nucleotide sequence of the cloned G gene reveals two hydrophobic domains, one in the signal peptide and one in the transmembrane portion (Rose and Gallione, 1981). Since the signal peptide is cleaved after insertion of G in the rough ER membrane (see section III of this chapter), there are three main regions in the G molecule on the infected cell surface and virion. The largest, amino-terminal portion, protruding outside of the membrane is glycosylated and can be cleaved by proteases, leaving the second, hydrophobic portion, anchored in the membrane bilayer. The third, carboxyl-terminal portion, is hydrophilic and exposed to the cytoplasm.

G has multiple functions: it interacts with virus receptors of susceptible cells, it triggers the synthesis of neutralizing antibodies (Volk *et al.,* 1982), and is involved in virulence. Antigenic variants of the G protein of rabies virus have been isolated in the presence of monoclonal antibodies and a change from virulence to avirulence of these variants may occur as a result of a single amino acid substitution (Dietzschold *et al.,* 1983).

(5) *L, the largest protein* (MW: 240 K, 30 to 50 molecules/virion), is probably the viral RNA dependent-RNA polymerase, which is responsible for both replication and transcription (see above). The entire sequence of this very large gene is now

established (Harmison *et al.,* in press). When NCs are purified, *in vitro* transcription requires addition of both L and NS, since NS is necessary for L to function (see above). Since the template for L is NC and not naked RNA (Bishop and Roy, 1971), it is clear that L must interact with the N protein.

III. Intracellular Synthesis of Virus Components

Up to now, we have briefly described replication and transcription of VSV and the general properties of the VSV gene products. Now, we will review how this virus uses the intact cell to reach its final goal: to be assembled and completed, to free itself from the cell, and to infect other cells. Therefore we will now describe when and where virus proteins are first synthesized in the cell, how they are modified post-translationally, and how they are transported to the site of assembly.

The first virus protein to be detected by immunological and biochemical methods is the N protein, about 30 minutes before the other proteins. N protein forms scattered foci throughout the cytoplasm (Fig. 2-2a) that are not closely associated with the rough ER, as shown by electron microscopic studies (Ohno *et al.,* unpublished observations). Later in infection, N protein is found on a large number of linear structures, probably corresponding to the NCs (see section IV of this chapter). This is in agreement with earlier cell fractionation studies which showed that N is synthesized on free ribosomes and is first found as a cytoplasmic soluble protein before being incorporated into the NC (Knipe *et al.,* 1977 a; Hsu *et al.,* 1979 a). NC strands do not appear in cells expressing a cloned N gene but lacking viral genomic RNA (Sprague *et al.,* 1983).

Similar to N, the NS protein is synthesized on free ribosomes and is abundant in the cytoplasm of infected cells, as detected by immunofluorescence (Fig. 2-2c). Most of the NS antigen is colocalized with N antigen. NS and N are almost equimolar in infected cells. However, there are about 20 times fewer NS than N molecules bound to the genome, and how much NS is bound may depend on its degree of phosphorylation. How and where NS is associated with virus genome is unknown. Infected cells contain a large soluble pool of NS, of which the function is obscure.

Fig. 2-2. Intracellular distribution of four proteins of VSV detected by immunofluorescence at 6 hours after infection of Madin-Darby bovine kidney cells. Cells were fixed in formaldehyde and made permeable with Triton X-100. *(a)* The N protein is organized in specks of various sizes throughout the cytoplasm and is concentrated in some areas close to the membrane. *(b)* Same field as (a), labeled with antibody to the G protein. The G protein is mostly present in round and elongated cisternae around the nucleus, probably corresponding to the Golgi apparatus, in transport vesicles and at the plasma membrane. *(c)* The NS protein forms scattered specks and inclusions close to the membranes in a pattern similar to the N protein. Double-label experiments (not shown) indicate that N and NS are colocalized. *(d)* The M protein is diffuse throughout the cytoplasm with a denser lining along the plasma membranes of infected cells. Magnification (a, b) × 560, (c, d) × 300. (Courtesy of Dr. H. Arnheiter)

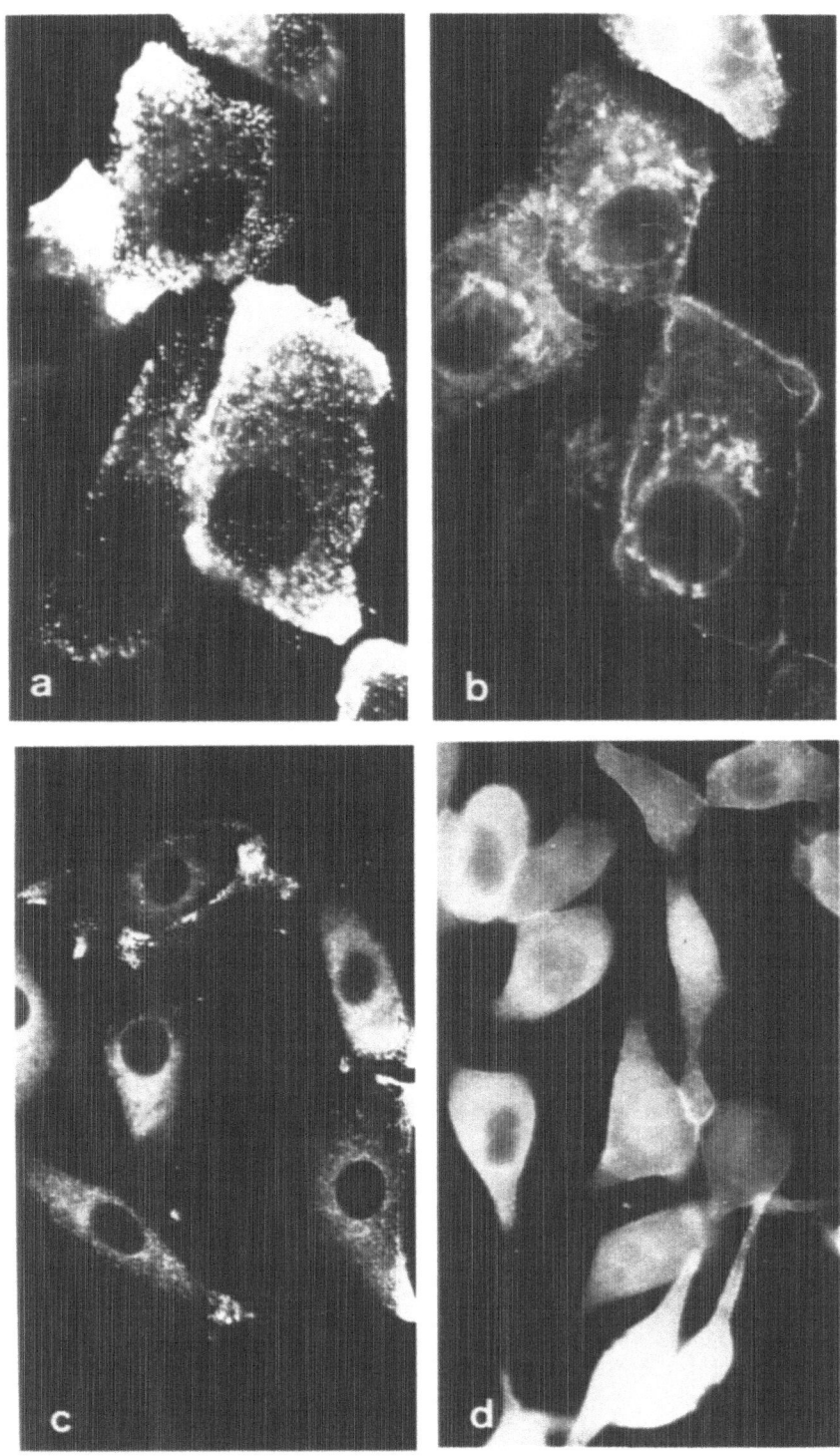

A third virus protein synthesized on free polysomes is the M protein (Knipe *et al.*, 1977 a). M protein is distributed diffusely throughout the cytoplasm of infected cells, as detected by immunofluorescence, and does not colocalize with any other protein early in infection (Fig. 2-2d). However, late in infection, N, NS, and M all show some accumulation close to the membrane when examined by immunofluorescence. Thus, it is likely that M interacts with NC only close to the assembly site at the membrane. Cell fractionation studies have shown that M becomes associated with cellular membranes 5 to 10 minutes after synthesis (Knipe *et al.*, 1977 a). It is intriguing that, *in vitro*, M can accumulate in all classes of membranes, including those of uninfected cells, while in infected cells, M association with cellular membranes seems to be only transient (Morrison, 1980).

G protein is synthesized on membrane-bound ribosomes and its insertion in the rough ER, as well as co- and post-translational glycosylation and transport to the cell surface, have been described in detail (Rothman and Lodish, 1977; Rothman *et al.*, 1980; Lodish and Rothman, 1980; Etchison and Summers, 1979; Morrison, 1980; Bergmann *et al.*, 1981; Rose and Bergmann, 1982; Wehland *et al.*, 1982). The synthesis of virus glycoproteins was reviewed in Chapter 1. It takes about 30 minutes for a G molecule to be processed and arrive at the cell membrane. The hydrophobic signal sequence has 16 amino acids and is cleaved in the rough ER. An immature form, designated G_1, containing mannose-rich oligosaccharide chains, comes out of the rough ER. Mature G_2 comes out of the Golgi with complex oligosaccharide chains. Whether or not the vesicles transporting G to the surface are coated with clathrin is still a matter of controversy (Rothman *et al.*, 1980; Wehland *et al.*, 1982). By immunofluorescence and electron microscopy, G protein is localized in the rough ER and Golgi apparatus, as well as at the cell surface (Figs. 2-2b and 2-3a and b). The movement of G protein has been analyzed with a G temperature sensitive (*ts*) mutant which blocks G protein in the rough ER at nonpermissive temperature (*ts* 045) (Bergmann *et al.*, 1981). Switching to the permissive temperature allowed the movement of G protein successively to the Golgi and the cell surface. With the Indiana strain of VSV, but not the New Jersey strain, additional post-translational modification occurs just before or when G arrives at the plasma membrane. One or two fatty acids are added and may act as a lipophilic anchor of G protein to the membrane (Schmidt and Schlesinger, 1979; Capone *et al.*, 1982).

The transport of G protein occurs normally in cells transfected with a vector containing the cloned G gene (Rose and Bergmann, 1982). Thus, when expressed in the absence of the four other genes and proteins of VSV, the G protein has all the essential signals for its processing and transport. When a gene coding for "anchorless" G (lacking the region coding for 79 amino acids at the carboxyl-terminal end) was expressed, the protein was not inserted into the plasma membrane, but accumulated within the ER and was slowly secreted into the medium (see also Chapter 1, Fig. 1-2b).

Further understanding of the maturation and transport of G comes from studies combining the use of a drug blocking glycosylation (tunicamycin) and *ts* mutants of G (group V), of which all, except one (*ts* L 511), block transport of G between rough ER and the Golgi (Zilberstein *et al.*, 1980; Pringle, 1982; Lodish and Kong, 1983). It appears that the polypeptide conformation is essential for normal

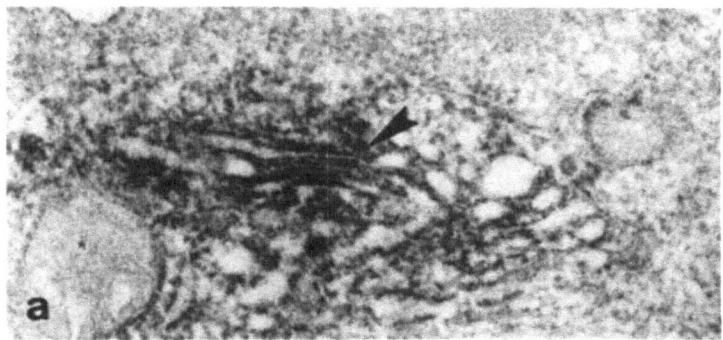

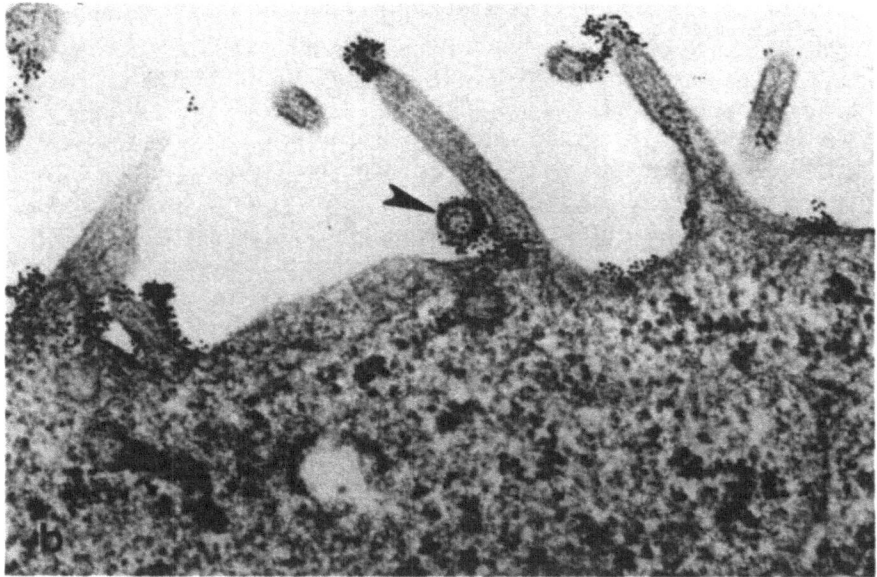

Fig. 2-3. Localization of the G protein in VSV-infected cells by immunoelectron microscopy. In *(a)*, G has been localized by the immunoperoxidase method in cells fixed with glutaraldehyde and made permeable with saponin and is found predominantly in the Golgi apparatus (arrowhead). In *(b)*, living cells were incubated at 20 °C with monoclonal anti-G antibody coupled to 5 nm gold particles. Cells were then fixed and processed for transmission electron microscopy. Groups of gold particles are clustered on the tips of villi as well as in some patches on the cell membrane. The envelope of one virion is also labeled (arrowhead). Magnification: (a) × 44,000; (b) × 57,000

transport to the specific sites where glycosylation events take place and for determining the extent of glycosylation and the oligosaccharide structure. The amino acid sequence of G determines if carbohydrates are needed to form a functional protein which can be inserted in the plasma membrane (Chatis and Morrison, 1981). On the other hand, glycosylation is important to allow achievement of the conformation that facilitates transport of G to the cell surface (Gibson *et al.,* 1979).

IV. Assembly of Virus Components

NC Assembly

The association of virus proteins with the viral genome is called encapsidation. *In vivo* studies of NC assembly suggest that it is a stepwise process starting with the initial condensation of N with viral RNA (Hsu *et al.*, 1979 a). The nature of the interactions between N protein and viral RNA during encapsidation of plus or minus RNA strand have been recently studied *in vitro* (Blumberg and Kolakofsky, 1981; Blumberg *et al.*, 1983). N encapsidates the leader RNA preferentially over other transcripts and over RNA fragments with different sequences. It appears that the aggregating property of the N protein is essential for the formation of a helical NC. In fact, purified N protein in solution can self-assemble and form doughnut-like structures or rings closely resembling in size the rings (or turns) of a coiled isolated NC (Fig. 2-4a and b). Each ring of self-assembled N is made of small disks which have the same diameter as the extended NC. When the leader RNA of positive or negative polarity encapsidated by N was isolated, identical circular rings were also observed which resemble tobacco mosaic virus (TMV) coat protein disks. Not only are the leader NCs structurally similar to genome NCs but they exhibit resistance to RNAse and have the same density in cesium chloride. Assembly of N with the genome appears to be a highly cooperative process in which there is linear addition of N proteins starting at the leader RNA which prevents the establishment of a secondary structure which could hinder assembly (Blumberg *et al.*, 1983). In addition, the synthesis of RNA is probably closely coordinated with N binding to the nascent product molecule, preventing the polymerase from moving very far ahead of the assembling NC. This process is very different from TMV assembly, which occurs bidirectionally from an internal nucleation site (Butler and Klug, 1980).

Recently, NC assembly has been obtained *in vitro* in the presence of newly synthesized or pre-formed L, NS, and N (Patton *et al.*, 1983; Peluso and Moyer, 1983). However, recent evidence suggests that newly synthesized N assembles rapidly with genome to form NC *in vitro* (Davis *et al.*, personal communication). Such *in vitro* systems should facilitate further study of NC assembly.

When NCs are isolated from virions, purified on cesium chloride gradients and negatively stained, they appear as loose, undulating strands (5 to 8 nm in diameter) which, in some places, form helical structures 20 nm in diameter with a periodicity of 10 nm spacing (Fig. 2-4c). From studies on isolated NCs, it appears that at least two factors, salt concentration and M protein, can influence the NC helical organization and, perhaps, the transcription process. In the absence of M, NC can form a helix in the presence of 1 M NaCl (Heggeness *et al.*, 1980). This helix, however, has half the diameter (20 nm) and twice the periodicity (10 to 12 nm) of that observed in the budding and complete virion (40 nm wide, 5 to 6 nm spacing). Other studies have shown that addition of M to purified NCs *in vitro* can increase NC compaction at low ionic strength (De *et al.*, 1982). As a result of M binding to NC, the movement of the polymerase might be restricted and transcription inhibited. Conversely, moderate ionic strength (0.1 to 0.2 M NaCl) dissociates M protein, unfolds the NC, and probably favors RNA chain elongation, reinitiation and transcription *in vitro*.

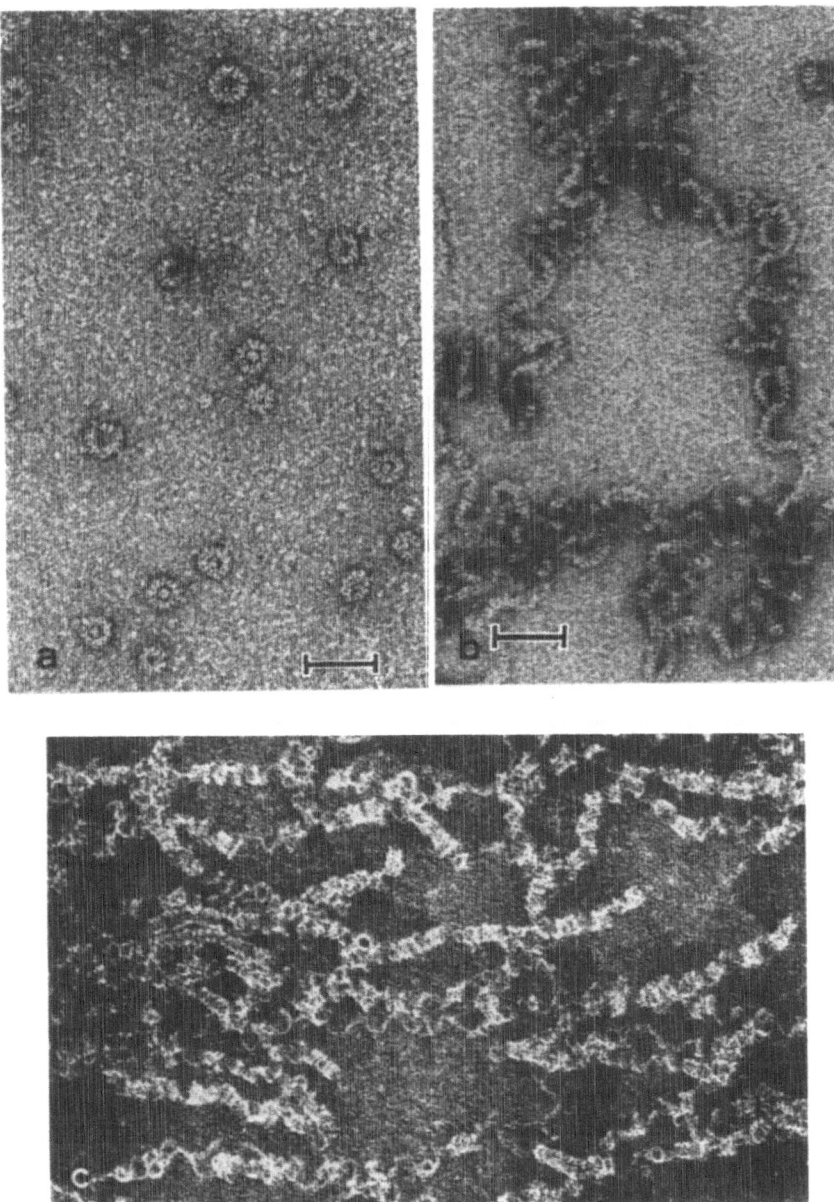

Fig. 2-4. A negatively stained preparation of isolated VSV N protein (a) and isolated VSV nucleo-capsids (b and c). *(a)* N protein has self-assembled in rings made of small disks stacked face to face toroidally. These disks within the rings have the same diameter as the nucleocapsids shown in *(b)*. The nucleocapsids are almost completely uncoiled and subunits of the N protein associated with the genome are visible. The bar corresponds to 14 nm. *(c)* A group of purified nucleocapsids which are either uncoiled or coiled in a helix 20 nm in diameter. The degree of coiling depends on salt concentration. Magnification: (c) × 125,000. Fig. (a) and (b): courtesy of Dr. D. Kolakofsky, from Blumberg *et al.*, 1983. (Reproduced with permission of MIT Press)

Inside the cytoplasm of infected cells, VSV NCs are loosely coiled and not easy to discern, because they are dispersed, of low electron density, and measure only 10 nm (Fig. 2-5a). In acute infection with VSV, NCs are incorporated rapidly into the budding virion. In contrast, with *Lyssaviruses,* replication is slower and, therefore, numerous and distinct NCs accumulate in the cytoplasm of the infected cell (Fig. 2-5b).

Sites of Budding

How the NCs move to their site of packaging in the viral envelope and how budding works is poorly understood. The most frequent site of rhabdovirus budding is the plasma membrane, but other sites, such as internal membranes of ER cisternae, have been observed, especially in the *Lyssavirus* (Fig. 2-6) (Murphy and Harrison, 1979; Fekadu *et al.,* 1982). Plant rhabdoviruses sometimes bud exclusively in nuclear cisternae (Francki and Randles, 1980).

Depending on the cell type, G protein may play a more or less important role in attracting NCs to the infected cell membrane. In many cells, the presence of G protein on the cell surface is not a prerequisite for virion maturation but it may facilitate the extent of budding (Lodish and Porter, 1980 b). When VSV-infected cells are treated with monensin, which blocks the transport of G to the cell surface, virus buds are not detected at the cell surface (Johnson and Schlesinger, 1980). In polarized epithelial cells, the preferential incorporation of G at the basal pole will determine where NCs will be enveloped, although NCs are scattered throughout the cell (Rodriguez-Boulan and Pendergast, 1980). The viral glycoproteins might share with some surface proteins of epithelial cell common informational features (or sorting-out signals) that determine their compartmentalization. Even when epithelial cells do not make tight junctions, VSV buds preferentially from the cell side adherent to substrate (Rodriguez-Boulan *et al.,* 1983). The interaction with substrate or with another cell may serve as a sufficient signal to trigger the segregation of plasma membrane components. The polarity of VSV budding is maintained after tunicamycin treatment, indicating that glycosylation of G protein is not a determinant of the virus budding site in this system (Roth *et al.,* 1979; Roth and Compans, 1981). Rather, the carboxyl-terminal portion of G may be the essential element allowing selective incorporation of the virus glycoprotein in a specific membrane site.

It seems clear from the study of some *ts* mutants of G that, even when G protein is not at the assembly site, NCs can select certain cellular and normal plasma mem-

---

Fig. 2-5. Cytoplasmic inclusions in neurons infected with VSV (a) or with Lagos bat rabies virus (b). In *(a),* delicate nucleocapsid strands accumulate in inclusions *(INC)* under the neuron plasma membrane (upper part of the picture) and a few viruses are budding (small arrowheads) while many extracellular viruses are detected (white arrow). One virus is entering a coated pit (large arrowhead at right). *(b)* An early inclusion body in the rabies virus-infected cell contains nucleocapsid strands in a rather dispersed array. The coiling of individual strands can be detected (arrowhead). Compare to Fig. 2-4b. Magnification: (a) × 36,000; (b) × 87,100. (Courtesy of Drs. F. A. Murphy and A. K. Harrison, from Murphy and Harrison, 1979; reproduced with permission of CRC Press: Boca Raton, Florida)

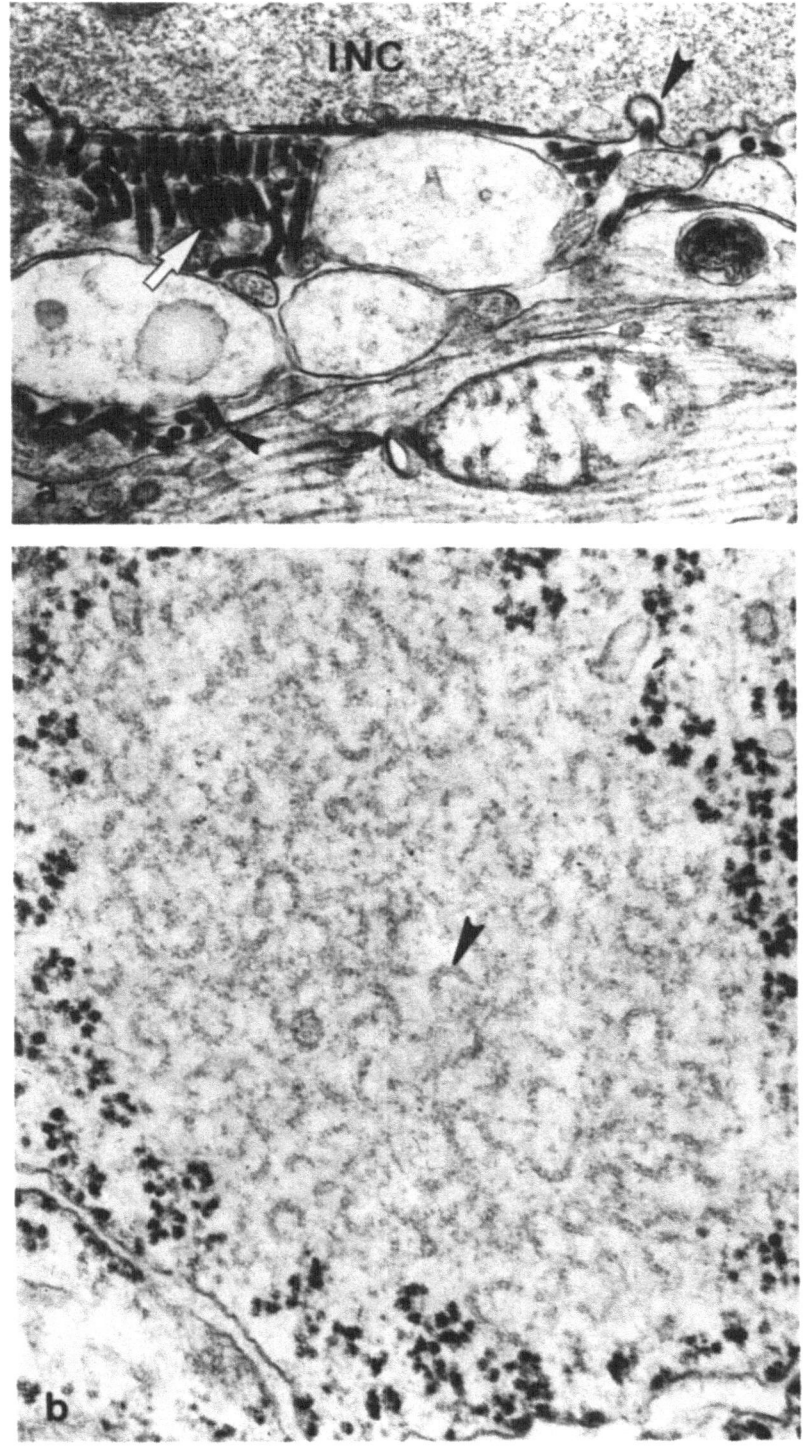

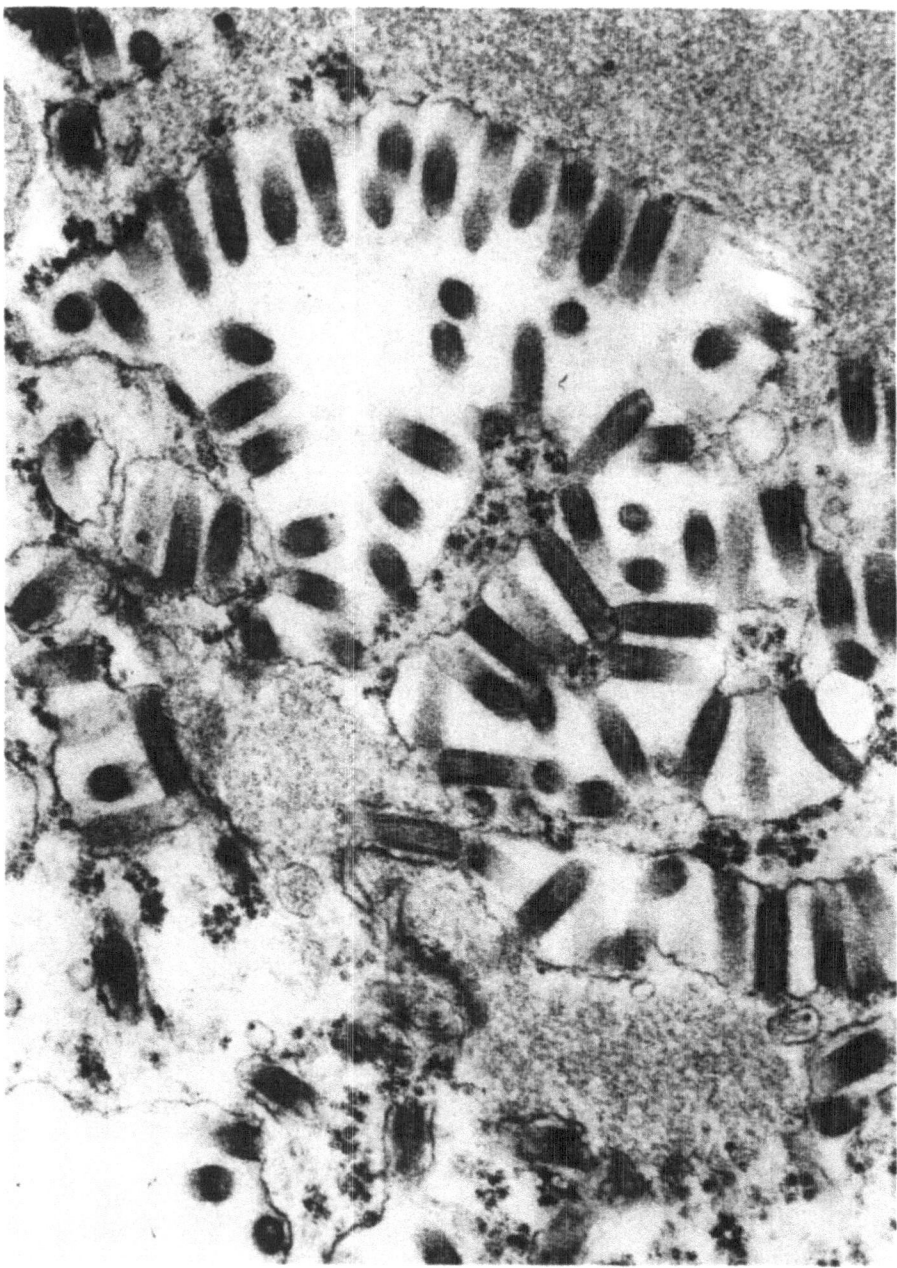

Fig. 2-6. Intracellular budding of an Ethiopian canine rabies virus strain. The virus is mostly budding from membranes of the endoplasmic reticulum. Magnification × 77,200. [Courtesy of Dr. A. K. Harrison, from Fekadu *et al.*, 1982. (Reproduced with permission of Springer-Verlag, Wien)]

brane proteins for incorporation into the virion. Such virions are spikeless, as in *ts* 045 infection (Schnitzer and Lodish, 1979). A spikeless G-lacking virus can even incorporate a glycoprotein of another virus, such as a retrovirus (Weiss, 1980). For instance, VSV selectively assembles proteins of human cells into progeny virions (Zavada *et al.,* 1983). These host proteins may share with virus glycoproteins some structural properties critical to assembly in a viral envelope; some may even correspond to uncleaved proteins coded for by *env* genes of incomplete retrovirus genomes (see Chapter 9). Thus, an interaction between a precise number of surface G proteins with NCs is not a prerequisite for budding, but a small number of G proteins may play an essential role in the first transmembrane events which trigger final virion assembly.

Interactions Between NC and Envelope Proteins

There must be a dual recognition event before maturation and budding occur: NCs must recognize factors in the membrane (G or M plus G) and, conversely, the membrane at the budding site must recognize encapsidated negative strand genomes to the virtual exclusion of encapsidated positive strands. Whatever mechanisms of recognition are active, they must accomodate the budding of so-called defective interfering (DI) particles. DI particles are virus particles that contain only a portion of the genetic information of the parental virus (Lazzarini *et al.,* 1981). DI particle genomes are efficiently replicated by enzymes provided by the complete standard virus and therefore must share with the complete virus the minimum structural requirement for replication, a competent replication initiation site at the 3' end of the RNA and the complement of an initiation site at the 5' end (see section II of this chapter).

The events triggering final virus assembly involve the specific association of NC, M, and G at the budding site. Isolated plasma membranes from infected cells do indeed contain the three proteins, G, M, and N (Scheefers *et al.,* 1980). Several assembly scenarios have been proposed, but the matter is still unsettled. Here are three of these scenarios:

(1) M may first bind to NC, forming a M-NC complex. This binding is specific, since M of a rhabdovirus cannot bind to NC of a paramyxovirus (Weiss, 1980). M binding to a NC probably occurs at the membrane, since (a) M and N proteins are not colocalized by immunofluorescence in the cytoplasm of infected cells and (b) in infection with a *ts* mutant defective in N protein, M remains diffuse in the cytoplasm without associating with the membrane (Knipe *et al.,* 1977 b). The cytoplasm of infected cells may contain a factor which inhibits M binding to NC or lack a factor (such as the presence of G in the membrane) triggering this binding. Thus, the M-NC complex would only form at the membrane and would recognize the carboxyl-terminal end of G and induce clustering of G molecules and the formation of spikes. Thus, there would be two binding sites on M; one to NC and one to G, but the latter binding site would be less specific, since certain host membrane proteins can enter the viral envelope when G is absent. Studies with *ts* mutants of M are in favor of this scenario (Knipe *et al.,* 1977 b, c). G protein goes to the cell surface and

stays diffuse, whereas a small amount of N protein is incorporated into the few virions that are produced. In such mutants, the association of M with the membrane is increased, whereas M association with the NC is decreased (Lenard *et al.*, 1981). Consequently, only 10% of the small number of virus particles formed contain NC, whereas M and G are present in all particles (Schnitzer and Lodish, 1979).

(2) M first forms a patch in the membrane by self-aggregation and binds to G molecules and/or cellular membrane proteins. Subsequently, a NC recognizes this patch of associated M-G and induces further clustering (Jacobs and Penhoet, 1982). In cells infected with a *ts* mutant defective in viral NC synthesis, there is no clustering of G molecules; they remain diffuse over the cell surface at restrictive temperature (Jacobs and Penhoet, 1982). The lateral mobility of host and viral surface glycoproteins within the plane of the membrane has to be preserved for clustering events to occur. Photobleaching recovery experiments have shown that a small fraction of G becomes immobilized at sites where budding occurs (Johnson *et al.*, 1981). The majority of G molecules, however, remains mobile throughout the infectious cycle. G mobility can be altered by cross-linking of its external glycosylated portion by antibodies or lectins such as concanavalin A (Con A) (Cartwright, 1977; Alsteil and Landsberger, 1981). During Con A treatment, all virus proteins are synthesized and transported normally, but no maturation event takes place. As soon as the drug is removed by α-methylglucoside treatment, virus can be released within a few minutes. From these results, one can postulate that, in normal infection, M protein binding to proteins on the inner side of the virus envelope produces increased rigidity of the bilayer during budding (Lodish and Porter, 1980 c).

(3) NC recognizes and binds directly to the carboxyl-terminal end of G, and M only plays a role in coiling of NC (Odenwald *et al.*, 1984) (see below).

Virus Budding

After these first transmembrane assembly events, what triggers and governs the growth of the virus bud in coordination with NC coiling? During this event, an increasing number of NC coils are incorporated into the bud, and the virus envelope is growing, tightly packed with virus protein molecules and excluding most host cell proteins. Changes in molecule conformation might occur at that time. When the infected cell membrane is split by freeze-fracture, the inner leaflet

---

Fig. 2-7. Membrane modifications during budding and assembly of VSV at the surface of Vero cells 6 hours post infection. The main figure is a high resolution scanning electron micrograph showing many budding viruses and occasional free virus lying on the surface of the cell. Knobs corresponding to groups of spikes are better resolved in surface replicas of infected cells (lower inset). The upper inset demonstrates the inner membrane leaflet of virus particles (between arrowheads) seen after freeze-fracture. The virus membrane is mostly devoid of intramembrane particles seen on the infected cell plasmalemma on the left *(P)* and is covered with fine granular material. Magnification: ×180,000; upper inset ×180,000; lower inset ×90,000

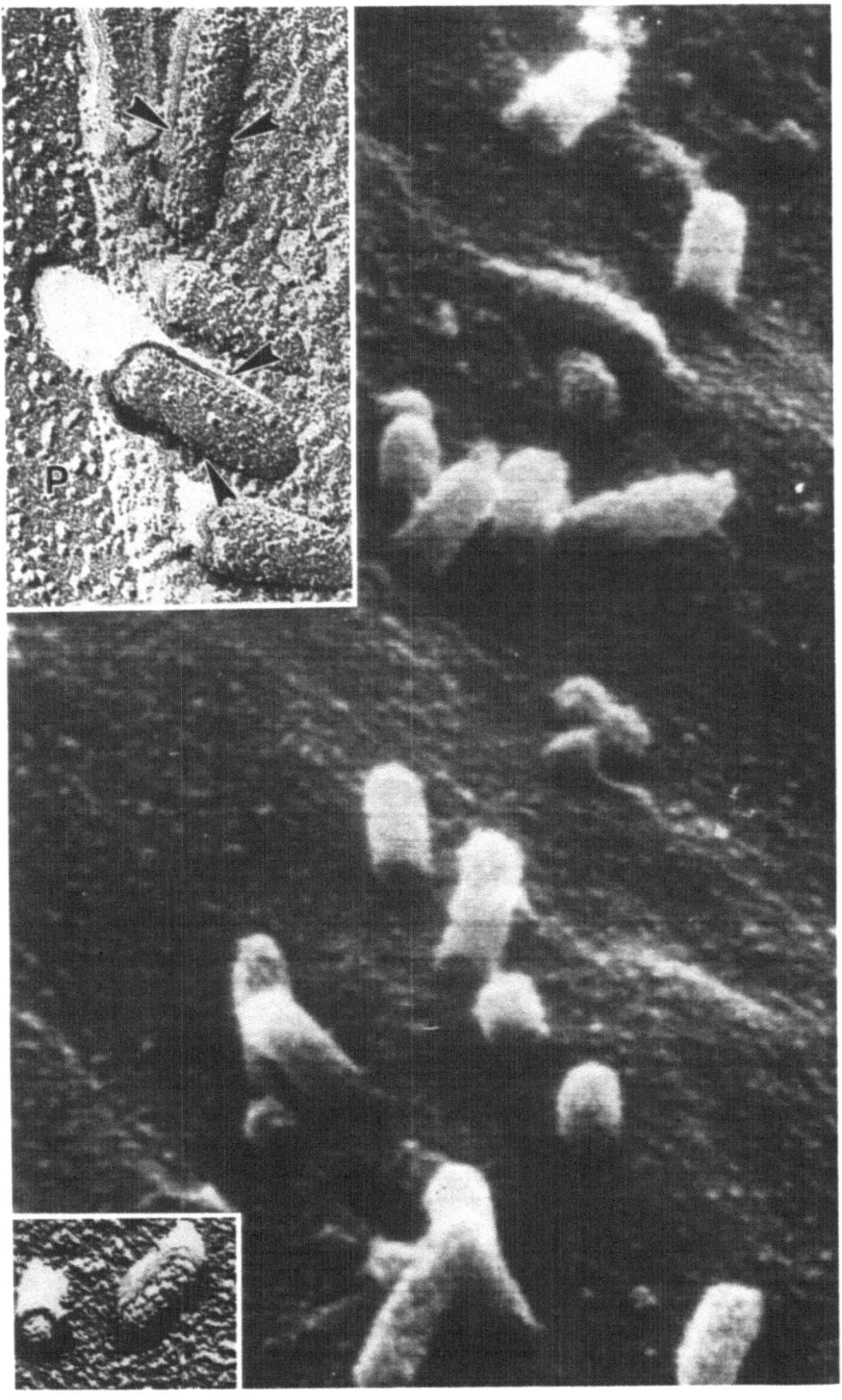

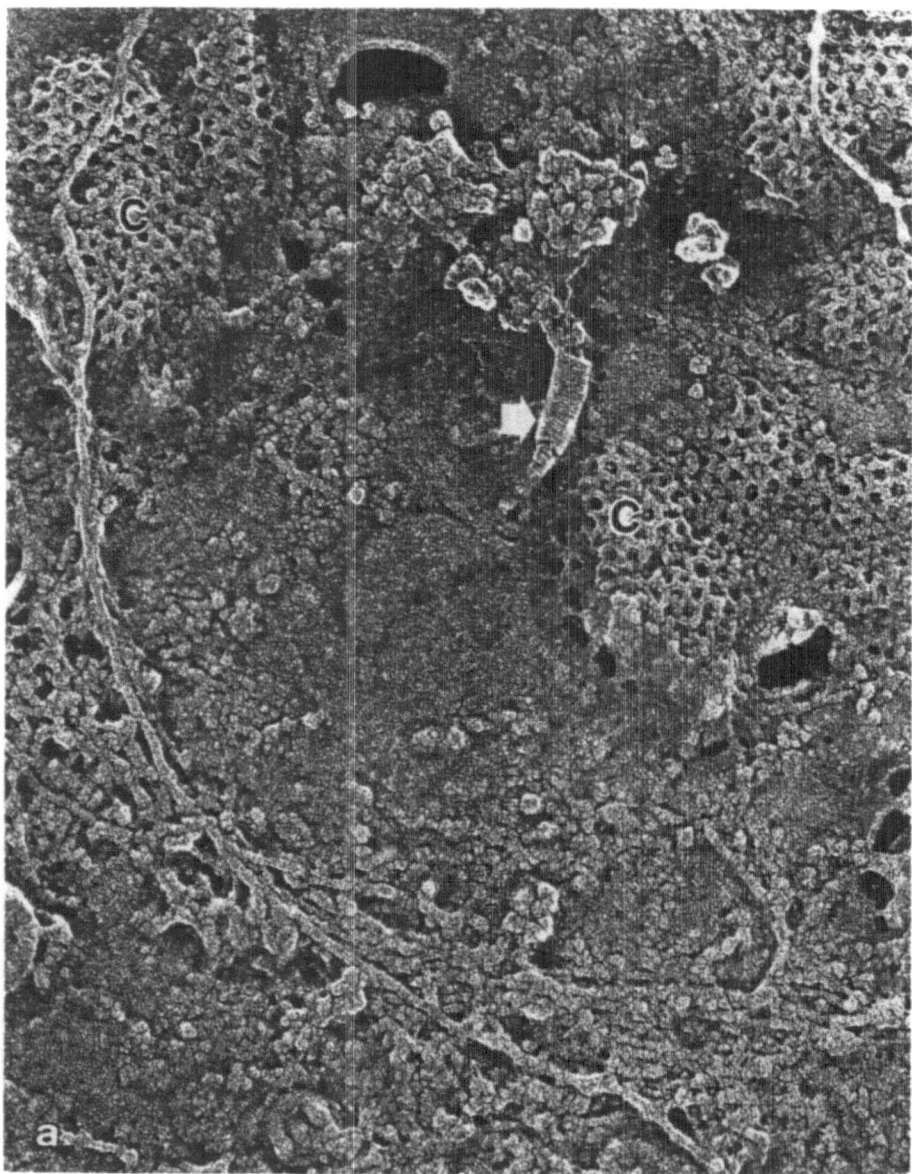

Fig. 2-8. Comparison between the coiled nucleocapsid of VSV attached to the cytoplasmic face of the membrane of infected cells (a and b) and in isolation after treatment of the virion with octylglucoside (c). In *(a)*, infected cells were mechanically opened, fixed, fast frozen and deep-etched. A replica of the cytoplasmic face of the membrane reveals clathrin baskets *(C)* and one tightly coiled nucleocapsid (white arrow). *(b)* A stereo view of two almost completely coiled NCs and one NC of which only one part is tightly coiled and forms a nose cone. Tortuous NC is attached to the base of this cone (white short arrow). *(c)* Tightly coiled nucleocapsids form skeletons identical to those isolated from complete virions. The coils in both pictures have a diameter of 45 nm and a periodicity of 5 nm. Magnification: (a) × 120,000, (b and c) × 200,000. [Courtesy of Ward Odenwald (a and b), and Drs. W. W. Newcomb and J. Brown, 1982 (c)]

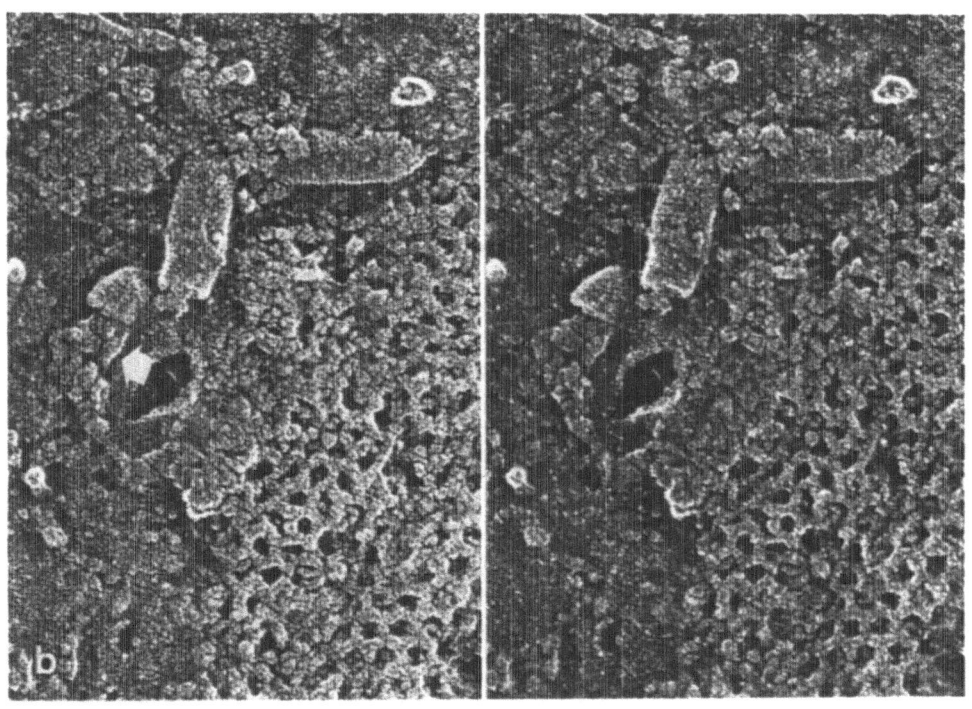

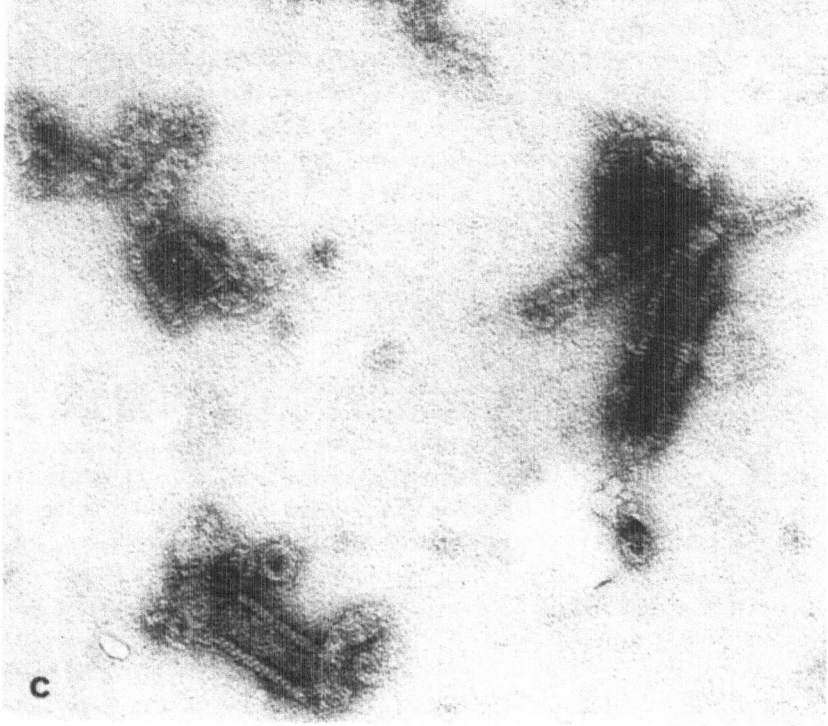

of the budding virus envelope shows a fine granular material instead of the usual intramembrane particles seen on the rest of the membrane (Fig. 2-7) (Brown and Riedel, 1977; Dubois-Dalcq *et al.*, 1979). Such morphological changes probably reflect exclusion of host proteins in the virus buds.

The dynamics of NC coiling may be the determining force in the formation of the virus bud. Inside the bud, the NC changes from the loose configuration seen in the cytoplasm to a tightly helical one 40 nm in diameter with a 5–6 nm periodicity. Whether this coiling occurs at once all along the NC or progresses from one end to another during budding is not known, but recent images obtained from the cytoplasmic face of infected cell membranes favor the latter hypothesis (Odenwald *et al.*, in press) (Fig. 2-8a and b). Tightly coiled NCs are detected on the inner side of the infected cell membrane and, in some cases, are in continuity with a tangle of loose NC strands (Fig. 2-8a and b). The M protein can be localized immunologically only on this entangled NC, suggesting that M interaction with NC induces the formation of the tight coil (Odenwald *et al.*, 1984).

Studies on complete VSV virions further support the role of M in coiling NC inside the bud (Newcomb and Brown, 1981). Nonionic detergent (Triton-X or octylglucoside) treatment of virions yields a particle devoid of G (and of spikes) but still containing M in association with the NC. These particles have been termed "skeletons" (Fig. 2-8c), and contain a coiled NC identical to that seen in intact virions (Fig. 2-8b) (Newcomb and Brown, 1981; Newcomb *et al.*, 1982). At higher salt concentration (0.5 m NaCl) M falls off and the NC becomes extended. If salt is removed by dialysis, M reassociates with NC, which then recoils in a tight helix, as seen in native virions (Heggeness *et al.*, 1980).

The role of cytoskeletal elements, especially actin, in the budding process has been debated. Does actin participate in virus budding? Or is actin depolymerized during viral infection and not involved in the budding process? The latter hypothesis is favored by studies of cytochalasin B effects on normal and VSV-infected cells (Griffin and Compans, 1979; Genty and Bussereau, 1980). This drug, which inhibits actin polymerization, does not inhibit production of infectious particles.

Defective Assembly

Besides the defects in assembly observed with *ts* mutants mentioned above, abnormal budding and viral maturation have been related to two other factors, DI particles and interferon. In mouse neuron cultures infected with a mixture of wild type virus and DIs, internal and surface virus proteins were detected, but virus budding and production were strongly inhibited and the neurons survived for two weeks (Faulkner *et al.*, 1979). In VSV-infected cells treated with interferon, infectivity may be reduced, because reduced amounts of G and M protein may be incorporated into virions (Jay *et al.*, 1983). This lower infectivity seems to be related to a reduced amount of G protein glycosylation (Maheshwary *et al.*, 1980; Friedmann, personal communication). However, these observations were not confirmed by others (Olden *et al.*, 1982).

V. Virus Release and Organization of the Virion

Once the rhabdovirus bud has reached its final shape and contains the entire coiled NC (there are about 35 coils in VSV), it is released from the cells as a bacilliform particle with a bullet-shaped NC. The end, where the virus has detached from the cell, is often deformed or partly collapsed; the other end, where the viral bud started to grow, contains four or five NC turns of increasing diameter. These first turns of the NC helix are followed by 30 to 35 turns of equal diameter (40 nm) separated by a regular 5 nm space. The unwound NC of VSV is 3.6 μm in length and is infectious (Cartwright *et al.*, 1970). The NC length, number of turns, and virus shape varies somewhat with the type of rhabdovirus. The shorter the genome, the fewer turns and the shorter the virus: DI particles can be as short as 65 nm, in the case of VSV, and as short as 70 to 100 nm, in the case of rabies virus. Among rabies virus strains, Canyon virus has fewer NC turns than VSV and is shorter, while Mount Elgon bat virus is much longer. Plant viruses are especially long, measuring up to 360 nm

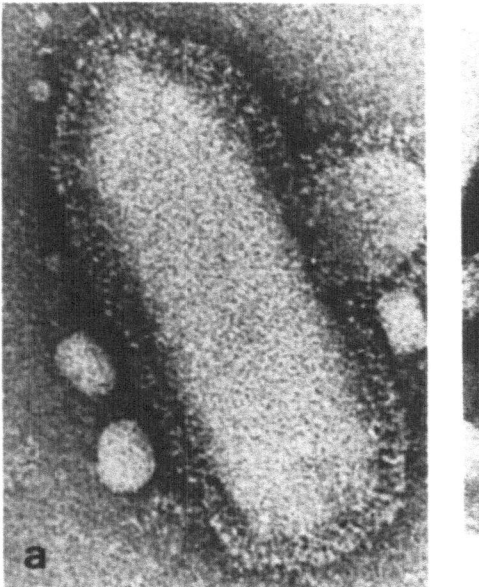
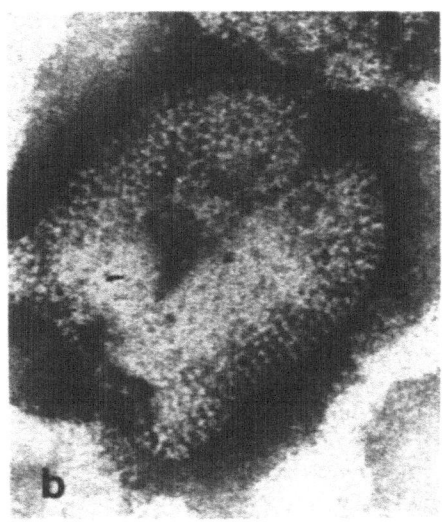

Fig. 2-9. Negatively stained mature rhabdoviruses. *(a)* VSV. Spikes surround the entire virion. *(b)* Rabies virus Klamath strain. The virus is partially open. A sixfold or honeycomb symmetry indicating the placement of spikes is seen. Magnification: (a) ×400,000; (b) ×218,000. (Courtesy of Drs. F. A. Murphy and A. K. Harrison, from Murphy and Harrison, 1979; reproduced with permission of CRC Press, Boca Raton, Florida)

(twice the length of VSV), as if two particles were attached end to end (Francki and Randles, 1980). Other lyssa strains are not cylindrical but conical.

Between the NC and its envelope is a space of a few nm. The envelope is covered with a maximum of 500 spikes (Fig. 2-9a), each of which probably contains

three to four G molecules; however, the number of G molecules and spikes per
virion may vary considerably. A honeycomb arrangement of the spikes is clearly
seen on some rabies virions by negative staining (Fig. 2-9b). These surface projections are 6–9 nm long and spaced 4–5 nm from each other. As in many other
enveloped viruses, the spikes have a knob at their distal end and a narrow stalk.

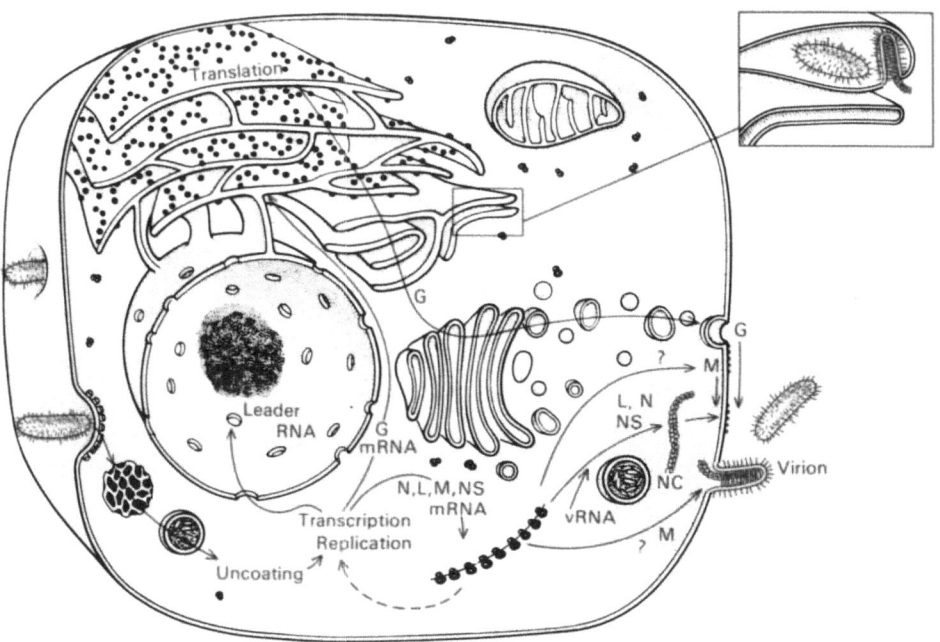

Fig. 2-10. Representation of the events of replication, transcription and assembly of the prototype
rhabdovirus VSV in the cell as schematized in Fig. 1-1. VSV enters cells via coated pits. After uncoating,
primary transcription yields N, NS, M, G, and L mRNA, as well as the leader RNA which goes to the
nucleus and may be involved in shuting off host protein synthesis. After translation of the L and NS
messenger RNAs, the L/NS polymerase complex is involved in replication and secondary transcription. N, NS, and M mRNAs are translated on free ribosomes, and the N and NS proteins associate with
the nucleocapsid and move towards the cell surface while M moves to the cell surface without association with nucleocapsids. Meanwhile, G mRNA is translated on membrane-bound ribosomes and the
transmembrane G protein, after glycosylation in the Golgi apparatus, moves to the cell surface in
transport vesicles. Interaction between the different components through the membrane results in the
budding process. The M protein apparently can only interact with the nucleocapsids with the
membrane at the budding sites. It appears that M triggers nucleocapsid coiling. The inset shows that
virus can also bud in smooth endoplasmic reticulum cisternae is some cell types

We have summarized in Fig. 2-10 the replication and assembly of rhabdoviruses. Several models of the structure of mature, free rhabdoviruses have been
proposed (Cartwright *et al.*, 1972; Wagner, 1975; Francki and Randles, 1980). Fig. 2-11
is a slight modification of models of others. Since the exact location of the M
protein within the virion is still unclear, three possibilities are shown in Fig. 2-11.

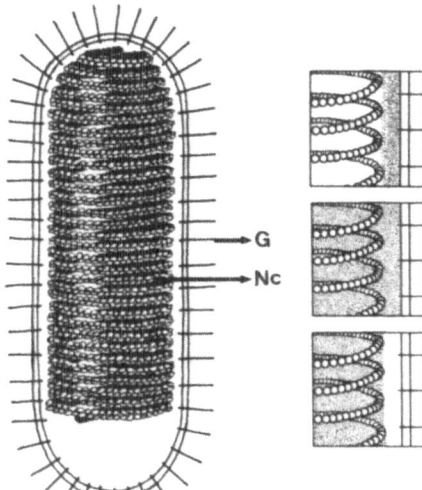

Fig. 2-11. Model of the organization of VSV. The proteins N, NS, and L are all associated with the nucleocapsid and the G protein constitutes the spikes and is a transmembrane protein. The three boxes depict the three hypothetical locations of the M protein, represented by dots. This intriguing protein is undoubtedly essential to allow correct assembly. In the upper box, the M protein is located between the tight helix and the inner membrane layer of the envelope. In the second box, the M protein is located around the nucleocapsid and in between the nucleocapsid turns, but also interacts with the membrane inner layer. In the bottom box, the M protein is located only around and in between NC turns

3

Assembly of *Paramyxoviridae*

I. Introduction

The *Paramyxoviridae* family is a large group of viruses divided into three genera according to their biological and biochemical characteristics (Kingsbury *et al.*, 1978). *Paramyxoviridae* replicate in a wide range of animal species, tissues and organs and readily establish persistent infections *in vivo* and *in vitro*. They are responsible for a large variety of diseases, including respiratory illnesses, measles, mumps, and several acute or chronic neurological disorders.

(1) The *Paramyxovirus* genus includes mumps virus, four species of human parainfluenza virus, Sendai virus of the mouse (HVJ), simian virus 5 (SV5) of primates and Newcastle disease virus (NDV) of birds, and many other species infecting birds and mammals. Members of this group have both neuraminidase and hemagglutinating activities (Scheid *et al.*, 1972). (2) The *Morbillivirus* genus consists of human measles virus, canine distemper virus, rinderpest virus of cattle, as well as other bovine viruses. *Morbilliviruses* differ from *Paramyxoviruses* in the lack of a neuraminidase activity. (3) The *Pneumovirus* genus was named after pneumonia virus of mice (PVM) and also comprises human and bovine respiratory syncytial (RS) viruses. The *Pneumoviruses* differ from the two other genera in the absence of both neuraminidase and hemagglutinating activities and in other respects (Kingsbury *et al.*, 1978).

The majority of data gathered on *Paramyxoviridae* has been obtained with *Paramyxoviruses*, mostly with SV5, Sendai virus and NDV. Parallel information on members of the other genera is similar where it has been studied, with a few specific differences which we will mention.

Paramyxoviridae virions have a characteristic morphology (reviewed by Choppin and Compans, 1975). They are large, somewhat pleiomorphic virions with an overall spherical shape ranging from 150 (SV5) to 400 (measles) nm in diameter. The envelope is covered with spikes 8 to 12 nm long. The RNA molecule is contained in a helical NC.

Many excellent reviews of the replication and biological properties of *Paramyxoviridae* should be consulted for details (Okada, 1969; Kingsbury, 1973, 1974, 1977; Klenk, 1974, 1977; Choppin and Compans, 1975; Rott and Klenk, 1977; Hosaka and Shimizu, 1977; Ishida and Homma, 1978; Compans and Klenk, 1979; Matsumoto, 1982).

II. Molecular Organization

The genome of *Paramyxoviridae* (Fig. 3-1) is a single stranded RNA molecule of negative polarity, 15 to 18 kb long (Kingsbury, 1974; Kolakofsky *et al.*, 1974) which contains at least six genes. These genes code for the six virion proteins: NP (sometimes designated N*), L, and P are elements of the viral NC and are thus associated

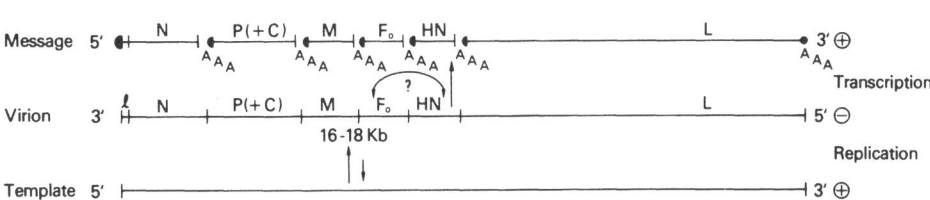

Fig. 3-1. Schematic representation of replication and transcription of the negative stranded genome of *Paramyxoviridae* such as Sendai and measles viruses. The gene order is based on the most recent evidence; the exact length is not yet known for each gene. The L gene, coding for a protein involved in RNA polymerase activity, may represent about 50% of the genome. Lower-case l marks the leader template sequence on the (−)strand. C is a non-structural virus protein found in Sendai and measles virus infected cells. It is encoded in the bicistronic P gene (see text). The genes coding for the fusion factor, F_0, and for the HN (Sendai virus) or HA (measles virus) are located between the M and the L genes, but their respective order is not known. All mRNAs are capped at their 5′ ends and polyadenylated at their 3′ ends. Based on Collins *et al.*, 1980; Dowling *et al.*, 1983; Bellini *et al.*, 1984

with the genome; HN, F, and M are associated with the envelope. In Sendai virus-infected cells, there is also a small non-structural, virus-specific protein C (Table 3-1). As in all negative-strand viruses, isolated virion RNA is unable to initiate replication. Within the infected cell, the RNA-dependent RNA polymerase from an infecting virion directs the synthesis of a full length 50S complementary positive strand RNA that becomes encapsidated and serves as a template for the synthesis of virion RNA (Kingsbury, 1977).

In the *Paramyxovirus* genus, primary transcription can occur in the presence of inhibitors of DNA or protein synthesis, because the infecting virus carries its own RNA polymerase. Primary transcription yields at least six mRNAs, which are capped at their 5′ ends and polyadenylated at their 3′ ends. The mRNAs are of two size classes: 35S and 18S. The 35S mRNA is translated into the L protein. In NDV infection, the 18S RNA consists of 5 mRNA species of similar size, each of which

---

* For the sake of uniformity, the protein designated "NP" for N protein, the major structural protein of NC in *Paramyxo-, Orthomyxo-,* and *Arenaviridae,* will be referred to as N protein in Chapters 3, 4, and 6.

codes for a different protein (Spanier and Bratt, 1977; Collins *et al.*, 1981). In the case of mumps virus, 50S RNA is the major RNA species in NCs (East and Kingsbury, 1971; McCarthy, 1981). Genome replication and secondary transcription require synthesis of virus proteins.

The gene order of *Paramyxoviruses* is not yet completely determined. However, information on NDV (Collins *et al.*, 1980), Sendai virus (Glazier *et al.*, 1977; Dowling *et al.*, 1983) and measles virus (Rozenblatt *et al.*, 1982; Bellini *et al.*, 1984) suggests the sequence of genes on the negative stranded genomic RNA shown in Fig. 3-1. In Sendai and measles virus, the P gene appears to be bicistronic and codes for P and C. P and C appear to be translated alternatively from overlapping reading frames in the same RNA species (Dowling *et al.*, 1983; Giorqi *et al.*, 1983; Bellini *et al.*, personal communication).

Virus Proteins

The structural and functional properties of *Paramyxoviridae* proteins are summarized in Table 3-1.

Table 3-1. *Proteins of* Paramyxoviridae[1]

Species	Apparent MW × 10⁻³	Molecules per virion	Location	Known or putative function
N(NP)	56– 60	2600	NC	Main structural component of NC
P	70– 79	300	NC	In transcriptase complex
C(V)*	22	–	Cell	Regulation of replication and transcription?
M	34– 40	?	Inner side of virion envelope	Viron assembly. Possible inhibitor of transcription?
HN (or HA)	65– 74	?	Spike	Hemagglutination and neuraminidase (HN) or hemagglutination only (HA), or no activity in pneumoviruses
F_0 (F_1) (F_2)	65– 68 (47– 56) (10– 16)	?	Spike	Fusion of cells, hemolysis, virus entry
L	150–200	30	NC	RNA synthesis

[1] From Matsumoto (1982); Strauss and Strauss (1983).
* Identified and characterized in Sendai virus.

(1) *N Protein* (NP) (MW: 56 K to 60 K). N protein binds tightly to the virion RNA, protecting it against nucleases. Phosphorylation of N protein may play a role in the regulation of transcription, perhaps by induction of a conformational change in the protein (Lamb and Choppin, 1977 b). N is the most abundant virus protein in the infected cell.

(2) *P Protein* (MW: 70 K to 79 K). P is a phosphorylated protein more abundant in cells than in the virion. P associates with the NC (Stone *et al.*, 1972; Portner and Kingsbury, 1976).

(3) *C Protein* (MW: 22 K). At least one non-structural protein, C (sometimes called V), is found in Sendai and measles virus-infected cells and not in virions (Lamb *et al.*, 1976; Bellini *et al.*, in preparatum). It may play a role in regulation of viral replication and transcription (Lamb and Choppin, 1978; Etkind *et al.*, 1980). Other non-structural proteins have been identified recently in RSV-infected cells (Huang *et al.*, in press) as well as in NDV, mumpsvirus, SV5 and CDV (discussed in Georgi *et al.*, 1983).

(4) *M Protein* (MW: 34 K to 40 K). M is an envelope-associated protein with hydrophobic properties. It has a high affinity for lipids and for viral NCs, and is therefore involved in virus assembly. The amino acid sequence deduced from the sequence of the M gene of measles virus does not reveal a long hydrophobic sequence (Bellini *et al.*, personal communication). M is an important structural component of the virion and imparts stability to the envelope (Matsumoto, 1982). In the infected cell, approximately half of the M polypeptides are phosphorylated, while the non-phosphorylated forms predominantly in the virion. However, Hsu and Kingsbury (1982 b) have detected phosphate residues in domains of the M protein located on the inside of the virion envelope. The phosphorylated form of M may regulate virus replication and/or transcription (Choppin and Compans, 1975; Yoshida *et al.*, 1976, 1979), while the non-phosphorylated form may be involved in virion formation (Lamb and Choppin, 1977 a).

(5) *HN Protein* (MW: 65 K to 72 K). HN is the largest glycoprotein molecule in the *Paramyxovirus* genus and it is transmembranous (Bowen and Lyles, 1981). It carries both hemagglutinating and neuraminidase activities (Scheid *et al.*, 1972; Portner, 1981; Sugawara *et al.*, 1982; Örvell and Grandien, 1982). HN plays a role in virus attachment to cells (Tozawa *et al.*, 1973; Shimizu *et al.*, 1974; Scheid and Choppin, 1974). Since the *Morbilliviruses* have no neuraminidase activity, the corresponding glycoprotein is called H or HA. In the spike, HA is a dimer in which the two molecules are linked by disulfide bonds (Hardwick and Bussell, 1978; Lund and Salmi, 1981; Shapshak *et al.*, 1982). The *Pneumoviruses* have a similar glycoprotein but are devoid of both hemagglutinatnig and neuraminidase activities.

(6) F_0 *Protein* (MW: 65 K to 68 K). F_0 is the precursor of the F_1 and F_2 subunits of the surface glycoprotein F (Samson and Fox, 1973; Scheid and Choppin, 1974). F_2 is smaller than F_1 but contains a higher concentration of carbohydrates (Scheid and Choppin, 1977). These two polypeptide chains are linked by disulfide bonds and form the active F protein. F is responsible for the fusion of the virus envelope with the plasma membrane of its target cell, hence F is required for virus penetration. The amino terminus of F_1, which is highly conserved among *Paramyxoviridae*, appears to be responsible for virus penetration, since a synthetic peptide mimicking this sequence can inhibit virus entry (Richardson *et al.*, 1980). F also causes both lysis of red blood cells (hemolysis) and fusion of cells (Homma and Ohuchi, 1973; Scheid and Choppin, 1974).

(7) *L Protein* (MW: 200 K). L is present in small amounts in the virion and is associated with the NC transcription complex. L is involved in RNA-dependent

RNA polymerase activity and probably also in polyadenylation and capping of the viral mRNAs (Choppin and Compans, 1975).

In addition to virus-coded proteins, cellular actin is found in virions, as in some other virus families (Lamb *et al.*, 1976; Wang *et al.*, 1976; Tyrrell and Norrby, 1978). The possible role of actin in virus assembly will be discussed below.

III. Intracellular Synthesis of Virus Components

N protein is synthesized on free ribosomes and travels rapidly to the plasma membrane (Nagai *et al.*, 1976; Lamb and Choppin, 1977 a). The N protein of Sendai virus is detected in infected cells as early as 4 hours after infection. At that time, N protein represents 50% of the total virus protein synthesized. A large amount of N protein is made in Sendai virus-infected cells, but only a small fraction is chased into virions (Lamb *et al.*, 1976). Using immunofluorescence and monoclonal antibodies, Sendai virus N protein is detected 7 to 8 hours after infection in large perinuclear inclusions (Örvell and Grandien, 1982). The N protein of measles virus is detected in large cytoplasmic inclusions (Fig. 3-2a) and in the nuclei of cells infected with the Lec strain (Norrby *et al.*, 1982). In contrast, P protein of measles virus is not detected in the nucleus but is limited to perinuclear cytoplasmic inclusions, which contain N protein (Hogan *et al.*, in press).

M is synthesized on free ribosomes, but very quickly reaches the plasma membrane in areas where viral glycoproteins are present (Nagai *et al.*, 1976). M may bind to smooth cytoplasmic membranes which transport virus glycoproteins to the cell surface or onto the plasma membrane. Örvell and Grandien (1982) have shown by immunofluorescence studies using monoclonal anti-M antibodies that only a small portion of Sendai virus M becomes associated with the plasma membrane and that a pool of M remains associated with intracytoplasmic membranes (ER). In measles virus-infected cells, some M staining is membrane-associated in a pattern like that of the HA on the membrane (Fig. 3-2b, compare with 3-2d) (Norrby *et al.*, 1982; Bohn *et al.*, 1982; Hogan *et al.*, 1984). Whether M is also associated with cytoplasmic NCs is presently unclear.

Fig. 3-2. Immunofluorescence staining of measles virus proteins using monoclonal antibodies to viral proteins. *(a)* Multinucleated measles virus-infected monkey cell 30 hours after inoculation were microinjected with monoclonal antibody to N protein, fixed, permeabilized with Triton X-100 and stained with a rhodamine-conjugated antibody recognizing the monoclonal antibody. Large perinuclear inclusions of NCs are intensely stained (in uninfected control cells, the injected antibodies stay diffusely distributed). *(b)* Fixed and permeabilized cells stained with a monoclonal antibody against M protein (courtesy of Dr. W. Bohn) 30 hours after infection with measles virus. M protein is found in association with the cell membrane in small dots and large patches. *(c)* Cells stained with antibodies to the hemagglutinin of measles virus 6 hours after infection. The hemagglutinin is localized in the perinuclear area and in small cytoplasmic vesicles. *(d)* Localization of the hemagglutinin of measles virus stained after fixation and permeabilization of a giant cell 20 hours after infection. The hemagglutinin is localized in intracellular vesicles and at the cell membrane in large patches. This localization resembles that observed with M protein [compare with (b)]. Magnification: (a) and (c) × 400; (b) and (d) × 330. [(a) and (b) courtesy of Dr. R. N. Hogan, and (c) and (d) courtesy of Dr. F. Rickaert]

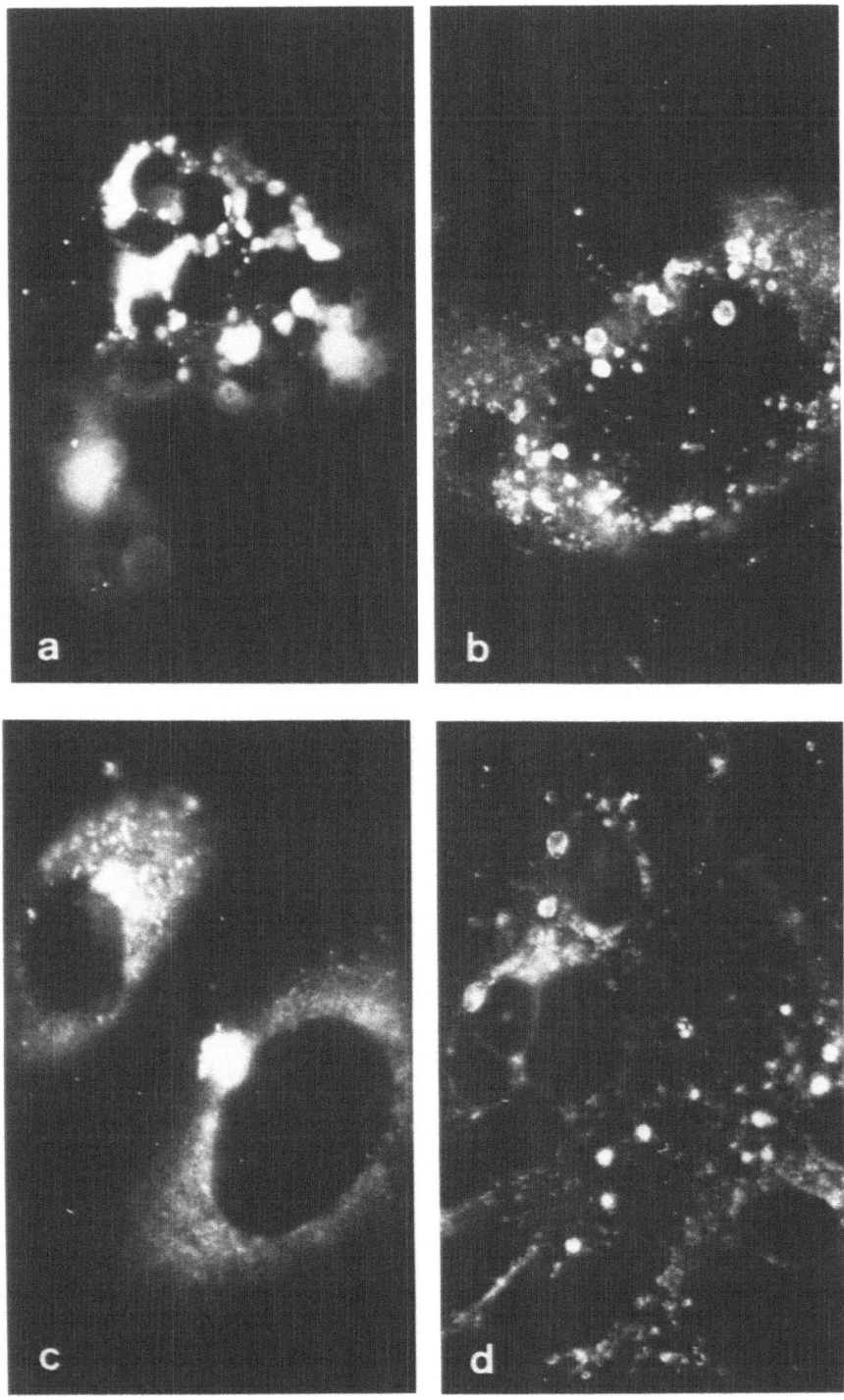

The glycoproteins HN and F of *Paramyxoviruses*, as well as HA of *Morbilliviruses*, are synthesized on membrane-bound ribosomes, inserted into the rough ER and transported through the Golgi to the plasma membrane (see Chapter 1) (Nagai *et al.*, 1976; Lamb and Choppin, 1977 a; Bellini *et al.*, in press). By immunofluorescence, measles virus HA can be detected as early as 6 h post-inoculation in the perinuclear area of Vero cells and in small scattered cytoplasmic granules (Fig. 3-2c) (Rickaert, personal communication). Later in infection, when giant cells have formed, numerous dots and patches are stained on the membrane (Fig. 3-2d). A similar localization at the membrane is seen with F_0 protein (Norrby *et al.*, 1982).

As deduced from partial amino acid sequence data, F_1 of Sendai virus and of SV5 contains a hydrophobic segment at its amino terminus (Gething *et al.*, 1978). Both F and HN span the membrane so that large glycosylated amino-terminal domains are outside the membrane and small carboxyl-terminal domains protrude into the cytoplasm. The carboxyl-terminal domain, although short, can be experimentally cleaved from the rest of the molecule. After Sendai virus has fused with erythrocytes, it is possible to form inside-out vesicles of these erythrocyte membranes and submit them to protease treatment. This results in the removal of a 1 K to 2 K fragment (Lyles, 1979). Similarly, protease treatment of inverted plasma membranes from measles virus-infected cells removes a 3 K fragment from the HA molecule (Bellini *et al.*, in press).

Paramyxoviridae glycoproteins undergo post-translational glycosylation, cleavage and acylation. Glycosylation (see Chapter 1) of HA and F_0 occurs co-translationally in the rough ER and further trimming of the oligosaccharides occurs following transport to the Golgi (see Chapter 1). Inhibitors of glycosylation such as *D*-glucosamine or 2-deoxy-*D*-glucose inhibit virus growth to some extent but do not completely block glycosylation of HA or F_0 (Scholtissek *et al.*, 1974, 1975; Hodes *et al.*, 1975; Schnitzer *et al.*, 1975). Tunicamycin inhibits glycosylation of NDV glycoproteins and changes their activity and conformation, but the unglycosylated HN and F_0 are still transported to the cell surface where they are incorporated into virions (Morrison and Simpson, 1980; Schwalbe and Hightower, 1982; Morrison *et al.*, 1981). Similarly, in the presence of tunicamycin, glycosylation of measles virus HA and F_0 glycoproteins is blocked but non-glycosylated HA and F_0 are inserted into rough ER membranes and transported through smooth ER and Golgi apparatus to the plasma membrane (Stallcup and Fields, 1981; Bellini *et al.*, 1983). Glycosylation thus is not a prerequisite for membrane insertion and intracellular transport.

Cleavage by a host cell protease of the F_0 glycoprotein into F_1 and F_2 subunits, which form the functional F protein (Nagai *et al.*, 1976; Scheid and Choppin, 1976), occurs when the glycoprotein is at or near the plasma membrane. Sendai virus and NDV are apparently not cleaved at the same cellular site F_0 of NDV being cleaved in the rough ER lumen, while Sendai F_0 is cleaved at the plasma membrane or extracellularly (Seto *et al.*, 1981). A lack of cleavage does not interfere with virus maturation, but virions with uncleaved F_0 are non-infectious and are not able to cause cell fusion and hemolysis. The host cell protease is an important determinant of paramyxovirus virulence (Homma and Ohuchi, 1973; Scheid and Choppin, 1974). In some strains of NDV, cleavage of an inactive HN into an active HN form, with loss of a polypeptide fragment, is also triggered by a host protease, but this may not be a general phenomenon in *Paramyxoviridae* (Nagai and Klenk, 1977; Garten *et al.*, 1980; Nagai *et al.*, 1980).

IV. Assembly of Virus Components

NC Assembly

The NC is formed by a single strand (5×10^6 MW) of genomic RNA, with about 2600 molecules of N protein. In addition, about 300 molecules of P and about 30 molecules of L are associated with the virion NC and constitute the replicative complex. Genome-length positive strands are also encapsidated (Kolakofsky and Bruschi, 1975). Affinity between genomic RNA and N protein is strong since their interaction is preserved whether the NC is tightly coiled or not. The flexibility and helical structure of NC are dependent on ionic strength: the higher the salt concentration, the tighter the helix (Heggeness *et al.*, 1980). Within the infected cell, numerous NCs assemble in the cytoplasm and form inclusions. In very permissive cells, assembly of NCs is balanced with budding into virions, so few NCs accumulate in the cytoplasm. In less permissive cells, or late in infection, large aggregates of NCs may accumulate in the cytoplasm (Compans *et al.*, 1966). It is not clear how NCs are transported to the membrane for virus budding.

NCs are found in both the nucleus and the cytoplasm of cells infected with measles virus (Fig. 3-3) (Raine *et al.*, 1974). Nuclear and cytoplasmic NCs are different: cytoplasmic NCs are thicker (about 24 nm) and covered by a "fuzzy" coat thought to be made of N, P, and, possibly, M proteins (Fig. 3-3a). Nuclear NCs are "smooth" and might contain only RNA and N protein (Fig. 3-3a, b). This is suggested by the immunofluorescence detection of N protein, but neither M nor P, in nuclei of some measles virus-infected cells (Norrby *et al.*, 1982). Canine distemper virus, another *Morbillivirus*, may also form nuclear inclusions, but NC strands in the nucleus resemble those in the cytoplasm (Fig. 3-4) (Raine *et al.*, 1976).

Association of Envelope Proteins

The HN and F_0 glycoproteins of *Paramyxoviridae* are first inserted into the plasma membrane in a random fashion and then aggregate into small patches (Nagai *et al.*, 1975; Dubois-Dalcq and Reese, 1975). The clustering of virus proteins in the plasma membrane may require the presence of M protein to exert a transmembrane effect. Indeed, in artificial membranes containing only HN glycoprotein, the density of HN molecules is significantly different from the spike density of viral envelopes (Hsu *et al.*, 1979 b). Thus, contact between glycoproteins and M may occur at intramembrane domains of the molecules. In the plasma membrane, clusters of spikes apparently become visible only in regions where M is present. This suggests that M binds the glycoproteins into visible structures (Nagai *et al.*, 1976; Scheid *et al.*, 1978; Markwell and Fox, 1980). Indeed, evidence for a direct interaction between M and

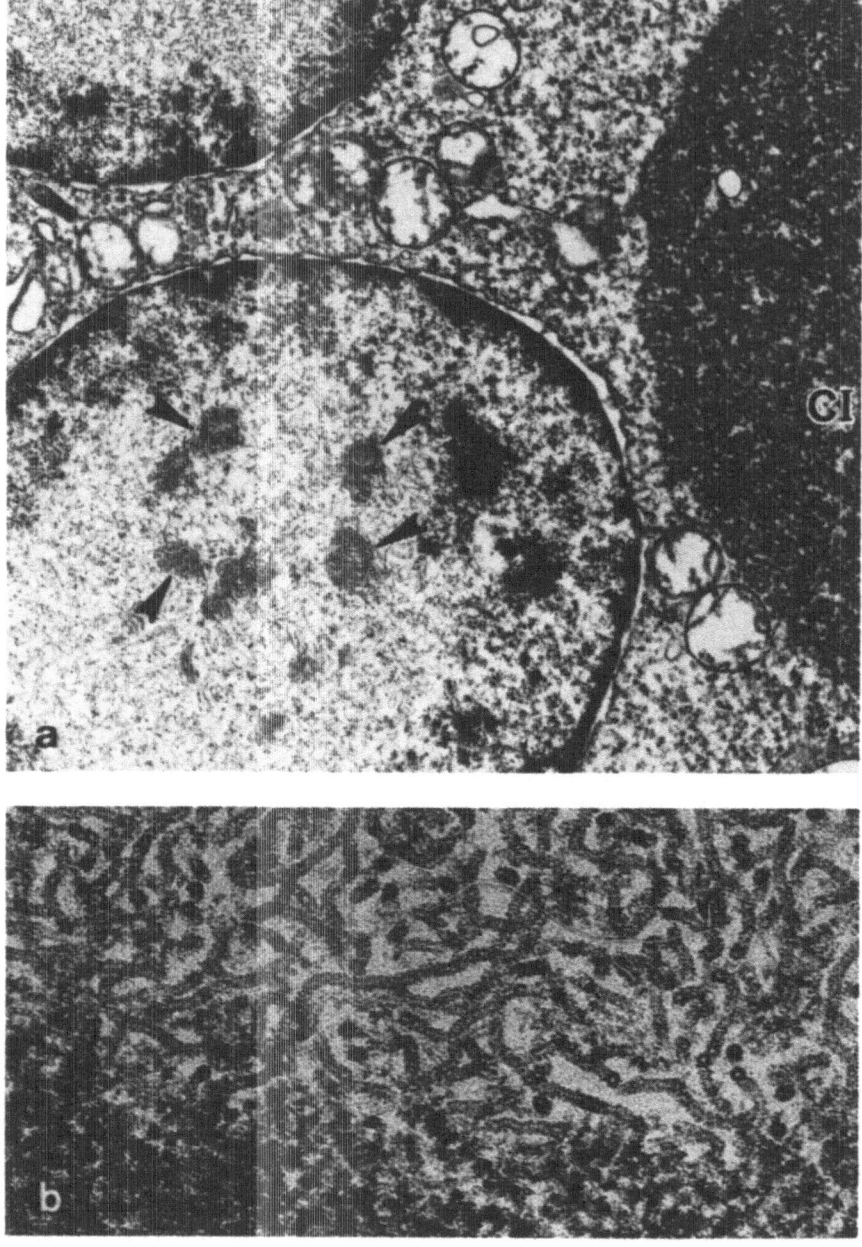

Fig. 3-3. Nucleocapsid inclusions in giant cells produced in the hamster brain by inoculation of a measles virus SSPE strain. *(a)* Large cytoplasmic inclusion *(CI)* on the right is made of fuzzy NCs while the two nuclei on the left contain free or aggregated smooth NCs (arrowheads).
(b) Detailed fine structure of these nuclear NCs. The helix of this *Morbillivirus* nucleocapsid has a 5 nm periodicity. Magnification: (a)×13,500 and (b)×110,000. [(a) courtesy of Dr. C. S. Raine]

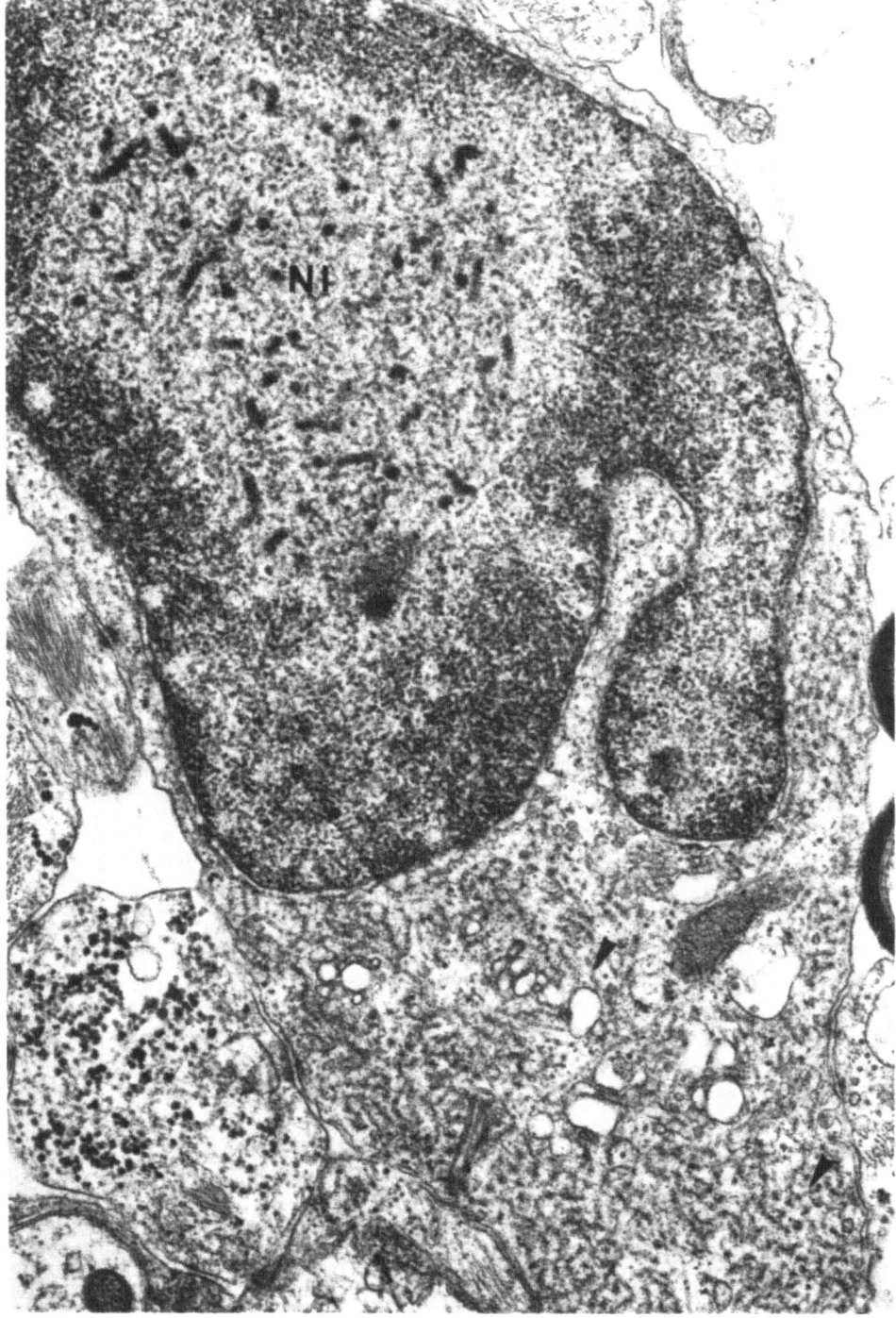

Fig. 3-4. NCs of canine distemper virus, a *Morbillivirus,* in brain cells. Scattered, fuzzy NCs (arrow-heads) can be detected in the cytoplasm and in the nuclear inclusion *(NI).* Magnification: ✕ 28,000. (Courtesy of Dr. C. S. Raine)

virus glycoproteins has been reported (Nagai *et al.*, 1975; Yoshida *et al.*, 1976). Yoshida *et al.* (1979) have shown that on living cells infected with a *ts* mutant of Sendai virus with a defect in M protein, antibody-induced redistribution of glycoprotein is faster at non-permissive temperature, when no M is present in the plasma membrane, than at permissive temperature, when M is bound to the plasma membrane. This strengthens the idea that M at the inner surface of the membrane causes glycoprotein clustering.

Why would M bind to the other viral components only at the plasma membrane, and not while glycoproteins are transported in vesicles with their carboxyl termini protruding in the cytoplasm? Is there a recognition signal that appears only at the plasma membrane? In measles virus-infected cells, M is found very rapidly after synthesis in microsomal fractions where HA is detected (Bellini *et al.*, in press). At that time, HA is not yet fully glycosylated, indicating that M could associate with glycoproteins when they are still in cytoplasmic vesicles (W. J. Bellini, personal communication).

Interactions Between NC and Envelope Proteins

Where patches of viral glycoprotein and M are formed in the plasmalemma, NC adheres to the internal face of the modified membrane (Berkaloff, 1963; Compans *et al.*, 1966). Glycoproteins align in twisted ribbons at the surface of the cell membrane, exactly over nucleocapsids (see below) (Dubois-Dalcq and Reese, 1975). The mechanism by which virus glycoproteins, M, and NC recognize each other and assemble is still controversial. Transmembrane interaction between NC, M, and glycoproteins is demonstrated by co-capping of these components by antibodies against the glycoproteins (Tyrrell and Ehrnst, 1979). How deeply M penetrates into the lipid bilayer is not known, but it probably spans at least the inner leaflet, as suggested by freeze-fracture studies of Sendai virus-infected cells (Bächi, 1980; Büechi and Bächi, 1982). When the inner leaflet of the plasma membrane of Sendai virus-infected cells is observed, areas of interaction between viral components are devoid of normal cellular intramembrane particles. This morphological phenomenon corresponds to the exclusion of cellular membrane proteins from the viral bud. However, within the viral buds, a fine crystalline-like array of very small particles (about 5 nm), with a four-fold orthogonal symmetry, is observed (Fig. 3-5). This intramembranous array may be formed by M protein molecules protruding into the inner leaflet (Bächi, 1980). When the outer surface of the plasma membrane is visible in the same field, spikes are present, but their arrangement is clearly different from that of the intramembranous array. Observation of the internal (cytoplasmic) face of the plasma membrane reveals the presence of NCs adhering to the crystalline arrays (Fig. 3-5, inset). Thus, M appears to form the crystalline arrays which span at least the internal layer of the plasma membrane and interact with NCs (Fig. 3-5, Büechi and Bächi, 1982).

In measles virus-infected cells, intramembranous patches devoid of normal cellular intramembrane particles have also been shown at ridges where NCs lie under the membrane and at budding sites. Arrays of extremely small intra-

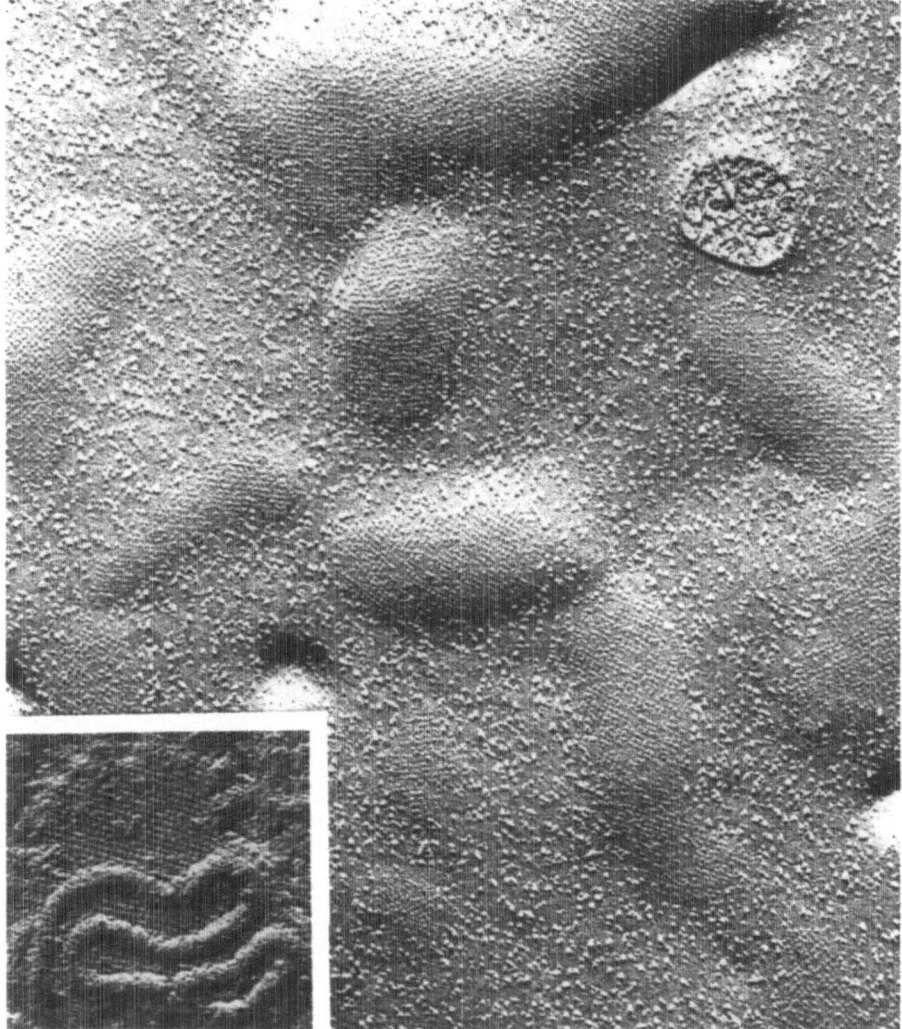

Fig. 3-5. Platinum replica showing intramembrane modifications after freeze-fracture of the plasma-lemma of cells infected for 48 hours with Sendai virus. Distinct crystalline formations of small particles are seen on bulging areas of the inner leaflet of the membrane. These particles measure only 4 to 5 nm, whereas the normal cellular intramembrane particles measure 8 to 13 nm. The inset shows a view of the cytoplasmic face of the inner membrane leaflet. A crystalline array of particles similar to those seen on the other side of the membrane is detected in contact with a loosely folded nucleocapsid. Magnification: × 80,000. Inset: × 130,000. (Courtesy of Dr. T. Bächi, from Bächi, 1980, and Büechi and Bächi, 1982; reproduced with permission of Academic Press, New York)

membranous particles are observed in the membrane at these regions, but no crystalline organization is detected (Dubois-Dalcq and Reese, 1975). Similarly, when the cytoplasmic face of the membrane is exposed, nucleocapsids are detected on a granular background, but crystalline arrays have not been found yet (Bächi, personal communication; N. Hogan, personal communication) (Fig. 3-6a). The

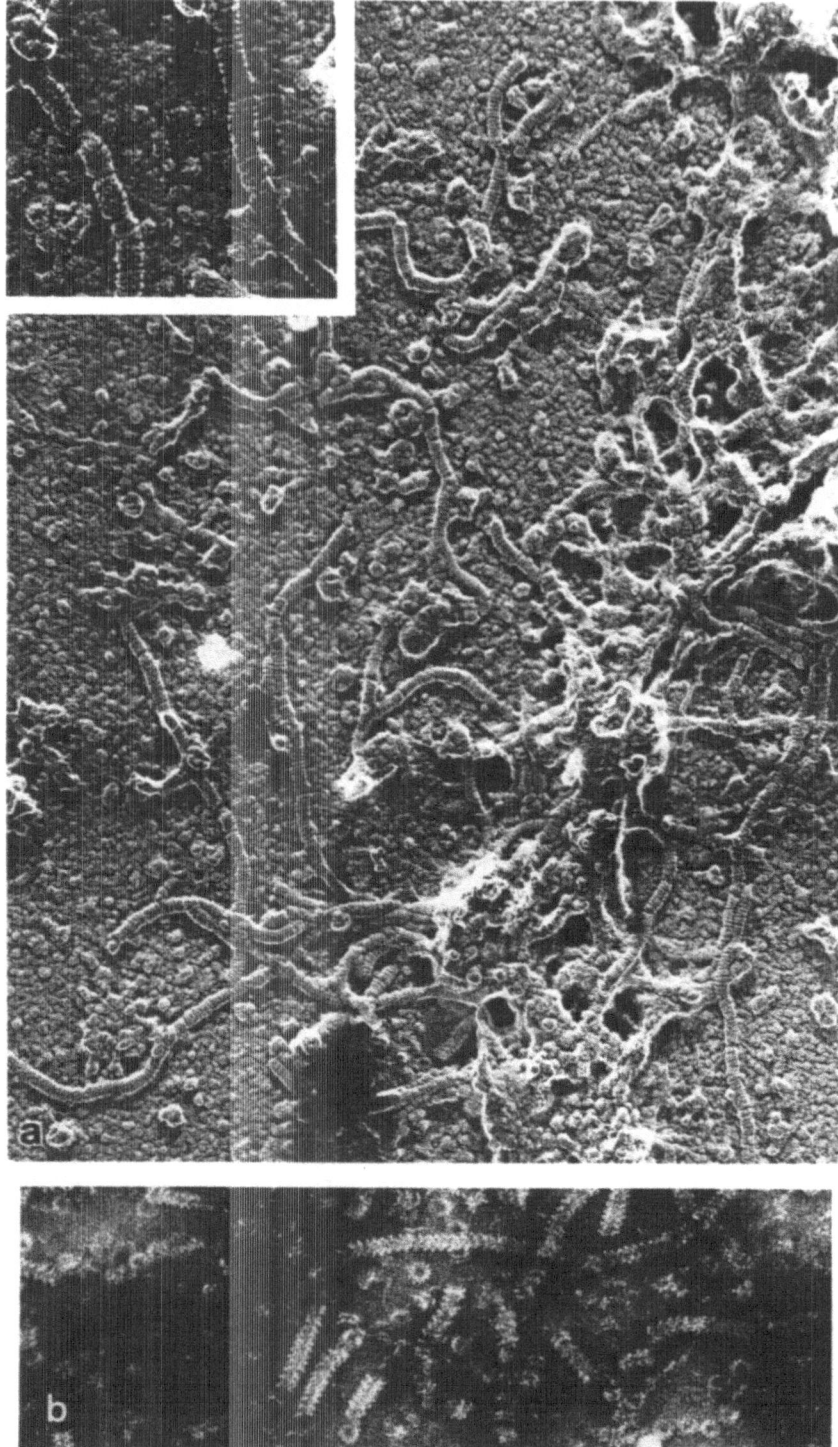

measles virus NC shows its typical periodicity and, in high resolution replicas obtained after fast-freezing and rotary shadowing, the subunits of the N protein may be visualized (Fig. 3-6a, inset). NCs identified in replicas of infected cells are a few nm larger than those seen after purification and negative staining (Fig. 3-6b). This is probably due to the preservation of all NC-associated proteins and the presence of the platinum layer in the replica of the fast-frozen cells.

Experiments performed *in vitro* on isolated proteins show attachment of glycoproteins to NC only if M is present, indicating that M acts as a bridge between glycoproteins and NC (Yoshida *et al.,* 1976, 1979). Interactions between glycoproteins and M are not strictly specific for molecules of the same virus species, since pseudotypes can form which contain the NC and M of one virus and glycoproteins of another (see Chapter 1), such as those obtained during mixed infection with the rhabdovirus VSV and the paramyxovirus SV5 (Choppin and Compans, 1970; McSharry *et al.,* 1971). In contrast, affinity between M and N proteins is virus-specific (McSharry *et al.,* 1975; Shimizu and Ishida, 1975; Yoshida *et al.,* 1976; Hewitt and Nermut, 1977; Markwell and Fox, 1980).

It is not known whether M attaches to cytoplasmic NCs early in infection. Although measles virus M protein becomes associated with HA in microsomal membranes early in infection, no N protein is ever found in these intracellular fractions. Therefore, attachment of N protein to envelope proteins is probably a late event which occurs only at the plasma membrane (W. J. Bellini, personal communication). The incorporation of M into the plasma membrane appears to be a rate-limiting step in virus assembly (Famulari and Fleissner, 1976; Nagai *et al.,* 1976).

Virus Budding

Virus budding occurs as a reverse pinocytosis process, forming a protruding sphere or cylinder that pinches off, freeing the virion. Among the several mechanisms of budding which have been proposed (Georges and Guedenet, 1974), it has often been suggested that M plays a crucial role in this process (Shimizu and Ishida, 1975; Blough and Tiffany, 1975). Isolated and purified M proteins of Sendai virus may form regular sheets or helical tubes made by the juxtaposition of annular subunits, 6 nm in diameter with a central hole (Hewitt and Nermut, 1977; Heggeness *et al.,* 1982 a). M thus spontaneously forms ordered arrays, an observation which fits well with the concept of M serving as a scaffolding structure for assembly of other viral components. This self-aggregating property of M could be related to the *in vivo* ability of M to impose curvature to the plasma membrane during budding (Hewitt, 1977). The crystalline aggregates observed by freeze-fracture (see

Fig. 3-6. *(a)* Platinum replica of measles virus-infected cells, which have been mechanically opened. Cells were fast-frozen, deep-etched, and rotary-shadowed. Numerous NCs have spilled over the granular glass surface. The NCs are closely intermingled in some points. Inset shows the fine structure of NCs in the replica. *(b)* Fine structure of negatively stained purified measles virus NCs. Both (a) and (b) reveal the helical organization of the NC with its 5 nm periodicity. However, in the replica of infected cells, the NCs are larger [24 nm versus 16 nm in (b)]. The herringbone structure is only seen after negative staining. Magnification: (a) × 92,000, (b) × 202,000. Inset: × 184,000. (Courtesy of Dr. R. N. Hogan)

above) in the membrane at the early stages of budding strengthen this hypothesis (Fig. 3-5a).

Alignment of NCs under the virus envelope is a typical feature of the growing virus buds in paramyxoviruses. Alignment means that the surface organization of virus glycoproteins mimics exactly the shapes of NCs lying under the membrane (Figs. 3-7 and 3-8). Surface replicas of measles virus-infected cells reveal numerous ridges of glycoproteins on the cell surface and on the viral buds (Fig. 3-7). Interestingly, these ridges are more closely apposed to each other in the viral bud, reflecting the progressive packing of the nucleocapsid during budding (Dubois-Dalcq and Reese, 1975; Mannweiler *et al.,* 1980) (Fig. 3-7a). In virus-induced giant cells, large areas of the cell surface may have glycoprotein ridges, but no budding virions (Fig. 3-7b). In measles virus-infected HeLa cells, virus budding can be inhibited by phenothiazine and virus-specific strands accumulate at the plasma membrane. As soon as the drug is removed, strands become incorporated into virus buds (Bohn *et al.,* 1983). Thin sections also reveal the regular alignment of NCs along the plasmalemma and in virions of nerve cells infected with mumps virus (Fig. 3-8a and b) (Wolinski *et al.,* 1974). Detection of HN and N proteins by immunoelectron microscopy also shows the interaction of viral components at the level of the viral envelope (Wolinski *et al.,* 1982) (Fig. 3-8c and d).

It has been proposed that cytoskeletal elements play a role in virus budding (see Chapter 1). Actin, the essential component of cellular microfilaments, is present in the virion. However, actin may interact with many hydrophobic proteins. In measles virus-infected cells, actin co-caps with the glycoproteins in the presence of anti-HA antibodies (Tyrrell and Ehrnst, 1979). In NDV and Sendai virus, actin has been shown to interact with M (Giuffre *et al.,* 1982). These observations on the role of actin microfilaments in budding are similar to those made with the retrovirus MMTV, but they are different from observations on the rhabdovirus VSV or orthomyxoviruses (Damsky *et al.,* 1977; Griffin and Compans, 1979; Genty and Bussereau, 1980). The involvement of cytoskeletal elements in paramyxovirus budding has also been inferred from data on the effect of inhibitors of cytoskeleton activity (Chapter 1). However, the effects of inhibitors should always be regarded with caution, since they often affect many cellular functions, including glycosylation and intracellular transport (Griffin and Compans, 1979). Stallcup *et al.* (1983) have shown an inhibition of measles virus maturation by cytochalasin B. This inhibition appears to be due to an effect of the drug on actin microfilaments and not on glycosylation of virus proteins. With cytochalasin B, cell morphology is drastically changed and measles virus budding is restricted to aggregates on the polar regions of cells (Fig. 3-9). Such virus can only be released by freeze-thawing. This suggests that actin may play a role in virus release.

Thus, several aspects of paramyxovirus budding remain unsolved. It is still unclear how the multiple elements of the virion recognize each other, how they

Fig. 3-7. Surface replica of Vero cells infected with measles virus. *(a)* Numerous ridges formed by viral glycoproteins are scattered on the cell surface and are more closely apposed to each other in the spherical buds. *(b)* Numerous ridges formed by glycoproteins are seen without any budding virions on the surface of a giant cell produced by measles virus infection. Magnification: × 30,000

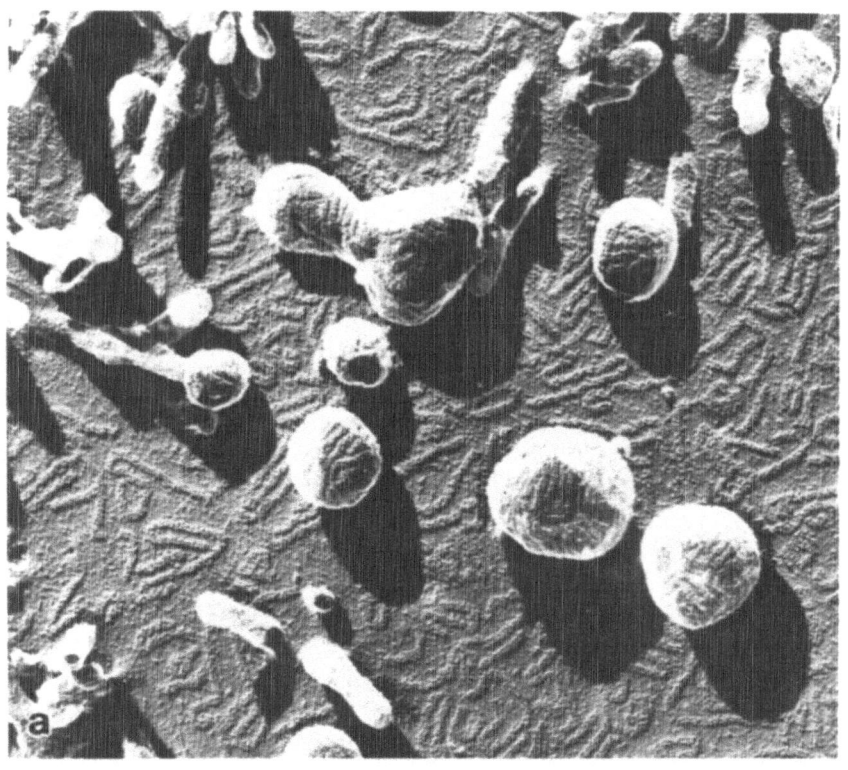

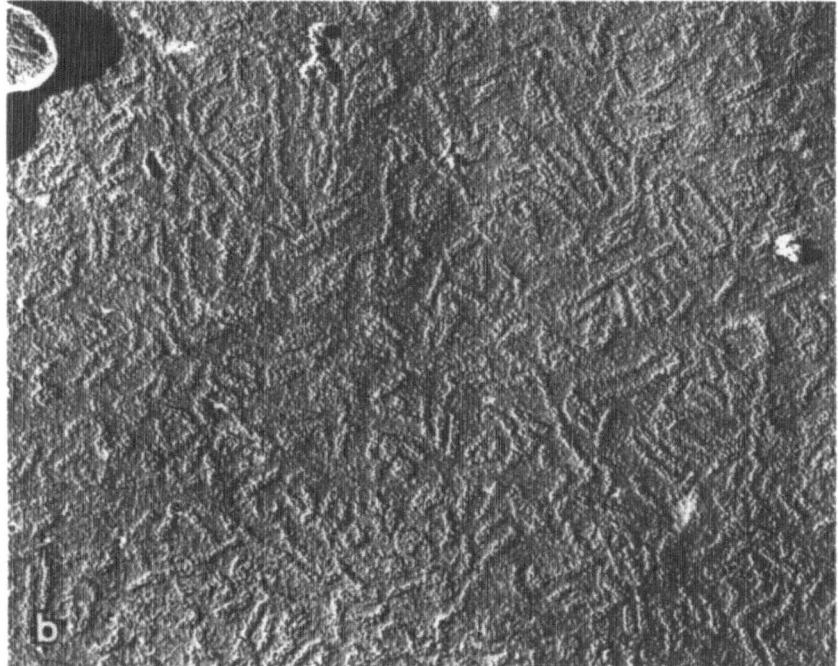

induce changes in each other and become assembled, and, finally, how the final incorporation of NC in the spherical or pleomorphic bud occurs. The selection of encapsidated genomic RNA for incorporation into the virion is not very selective since encapsidated positive strands have been found also in complete virions (Kolakofsky and Bruschi, 1975).

Defective Assembly

Incomplete or aberrant virus production, as well as defective budding, are frequent events in *Paramyxoviridae* infections, and provide interesting clues to the mechanisms of virus maturation. Incorporation of several NCs into a virion may occur (Granoff, 1962; Roman and Simon, 1976 a). *ts* mutants of Sendai virus, which have no HN or no active HN, are able to form non-infectious virions at the non-permissive temperature. Thus, the HN polypeptide is not necessary for the budding process (Portner *et al.*, 1975). Several *Paramyxoviridae* such as Sendai virus (Kingsbury and Porter, 1970), mumps virus (East and Kingsbury, 1971) or NDV (Roman and Simon, 1976 b) can generate defective interfering particles, indicating that a deletion in viral RNA does not impair particle formation, provided wild type virus is also present. An interesting block of virus release was observed in a human cell line chronically infected with a subacute sclerosing panencephalitis (SSPE) strain of measles virus (Dubois-Dalcq *et al.*, 1976 a). Apparently, the NC is unable to coil in this strain. This rigid NC becomes associated with the membrane and forms elongated buds which do not detach from the cell. Thus a flexible, coiled NC appears to be a prerequisite for virus release (Dubois-Dalcq *et al.*, 1976 a).

Persistence of *Paramyxoviridae* can arise from several mechanisms of defective budding. Accumulation of large numbers of ribbons of virus glycoproteins on the surface of measles virus-induced syncytia (Fig. 3-7b) are apparently due to a defect in virus maturation and assembly since giant cells do not produce much virus (Rentier *et al.*, 1978). During a persistent Sendai virus infection of BHK-21 cells, M protein could not be detected, suggesting that it was either absent or at least antigenically modified (Roux and Waldvogel, 1982). The authors concluded that absence of M protein was due to an instability of the molecule rather than a lack of synthesis. However, it is difficult to establish cause and effect in the relationship between this observation and induction of virus persistence. The abnormal properties of M may be either the cause or the consequence of viral persistence.

Fig. 3-8. NC structure and assembly of mumps virus. *(a)* Mumps virus NCs (Kilham strain) are seen in the cytoplasm of a neuron of a newborn hamster brain. On the right, several NCs are aligned under the membrane in a periodic manner (arrowheads). *(b)* A complete virion is squeezed in between two nerve cells. Regularly aligned NCs are detected close to the envelope and NCs are seen in the middle of the virion. *(c)* Vero cells infected with the O'Take strain of mumps virus 4 days after inoculation. The HN glycoprotein of mumps is labeled by the immunoperoxidase technique using monoclonal antibody. Intense labeling is present on the viral envelope while the reaction product of the peroxidase is less abundant on the rest of the membrane and the microvilli. *(d)* The N protein of mumps has been immunolabeled with a monoclonal antibody by the immunoperoxidase technique. The granular reaction product of the peroxidase is exclusively localized inside the virus particles and along the virion envelope. Magnifications: (a) ×23,000, (b) ×60,000, (c) and (d) ×50,000. (All pictures courtesy of Dr. J. Wolinsky)

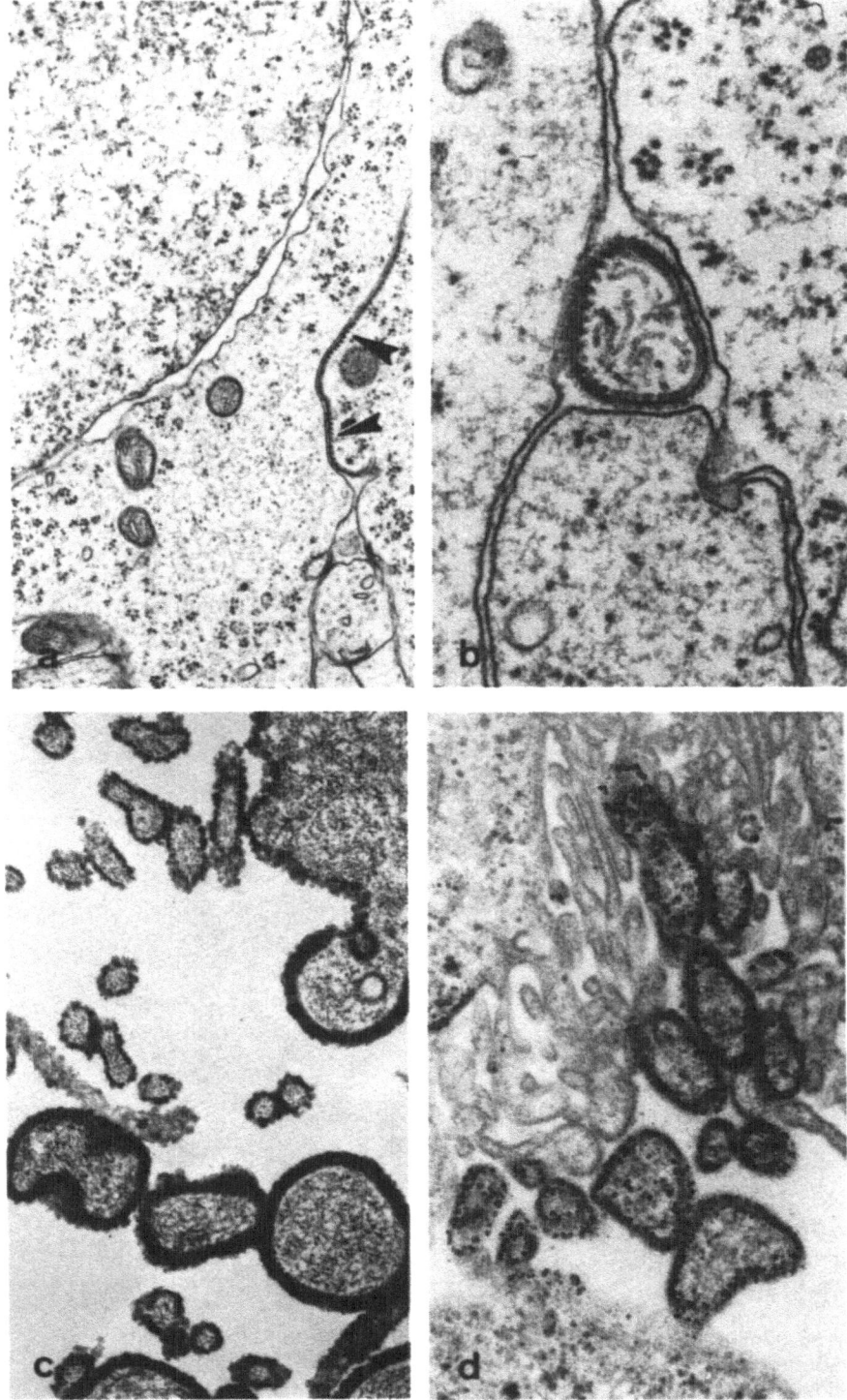

In persistent infections with measles virus, various modifications of the M protein have been observed. M protein was not found in infected brain cells of patients with SSPE, a slow degenerative disease of the nervous system caused by measles virus (Hall and Choppin, 1979) or in tissue culture cells infected with some SSPE isolates (Lin and Thormar, 1980). These observations correlate with the

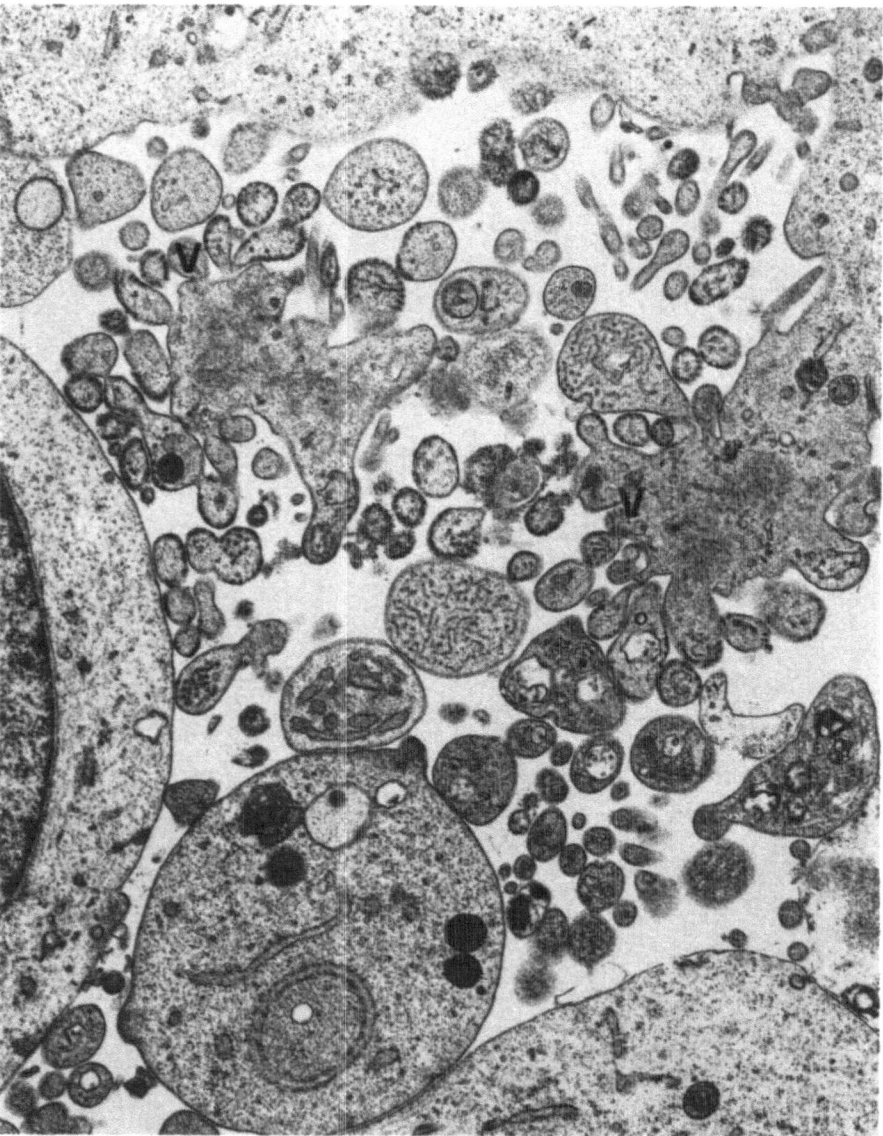

Fig. 3-9. Monkey kidney cells infected with measles virus were treated with cytochalasin B for 24 hours. Most of the cell surface is free of virus and virions are budding only in aggregates at the polar regions. Groups of virions are indicated at V. Magnification: ×14,000. (Courtesy of Dr. C. S. Raine, from Stallcup *et al.*, 1983; reproduced with permission of Academic Press, New York)

complete absence of budding and free infectious virions in these cells (Dubois-Dalcq *et al.*, 1974). However, M protein and all other virus proteins were found in several SSPE-derived viruses (Hall *et al.*, 1979), in HeLa cells persistently infected with measles virus (Wechsler *et al.*, 1979) and in cultured mouse neurons persistently infected with measles virus (Rentier *et al.*, 1981). A modified M protein with an altered electrophoretic mobility has been reported in some viruses rescued from an SSPE infection (Wechsler and Fields, 1978; Schleuderberg *et al.*, 1974). However, the significance of this finding is not clear, since considerable variation in the apparent size of several viral polypeptides is observed in different measles virus strains (Mountcastle and Choppin, 1977). Absence of M in infected cells can be due to inhibition of synthesis or instability of the molecule. Degradation of M could depend on host cell proteases and/or intracellular transport which may vary with the stage of differentiation of the cells. When differentiation is induced in transformed neural cells, M protein disappears from infected cells as shown by immunofluorescence. The effect is reversible when the differentiating agent is removed (Miller and Carrigan, 1982). The synthesis of other proteins, such as HA and F_0, is sometimes decreased in persistent infections *in vivo* and *in vitro* (Johnson *et al.*, 1982; Norrby *et al.*, 1982). The relationship between these observations and virus persistence is still speculative.

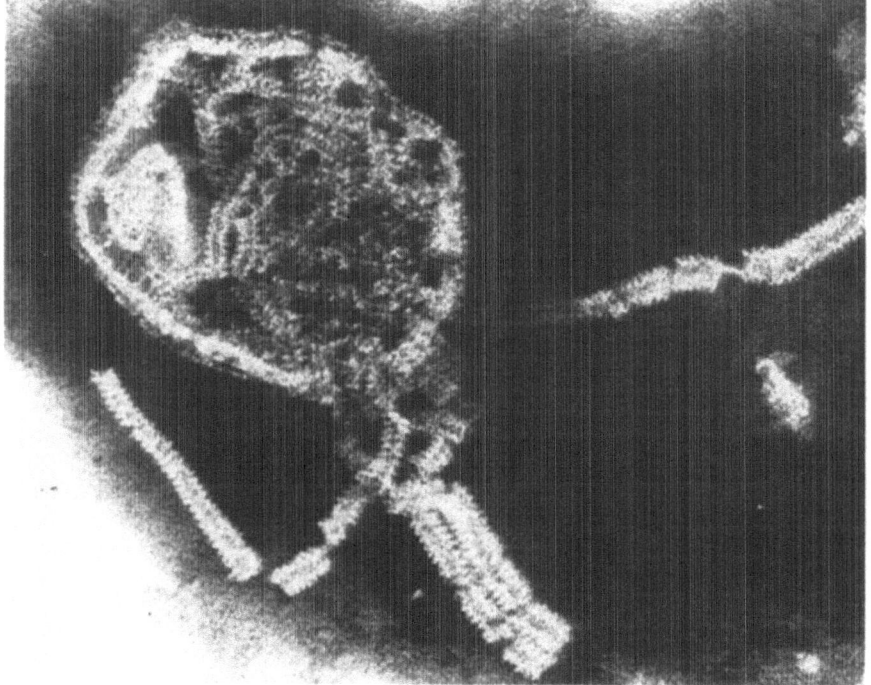

Fig. 3-10. Negative staining of a disrupted Newcastle disease virion. Stain has penetrated inside the disrupted virion. The internal NC with its typical herringbone aspect is partially extruded. Note the regular arrangement of spikes on the envelope. Magnification: ×165,000. (Courtesy of Dr. C. M. Calberg-Bacq)

V. Post-Release Maturation and Organization of the Virion

A complete virus with its distinct spikes is shown in Fig. 3-10. While the NC in the viral bud is in close contact with the envelope, NC is loosely folded and not associated with the envelope when the particle is detached. This post-budding modification could be related to the morphological changes which occur within the envelope of mature virions. The crystalline array of fine intramembranous particles, still present in the freshly released virion envelope, disappears some time after virus release (Kim *et al.*, 1979; Bächi, 1980; Markwell and Fox, 1980). Interestingly, this loss of intramembrane array in viral envelopes also correlates with the appearance of hemolytic and cell fusing activities of the virus. This suggests that a reorganization occurs in the viral envelope following virus release. The bonds between NC, M, and glycoproteins, that were necessary for assembly and budding, appear to be lost in the mature virion. This modification may allow glycoproteins to

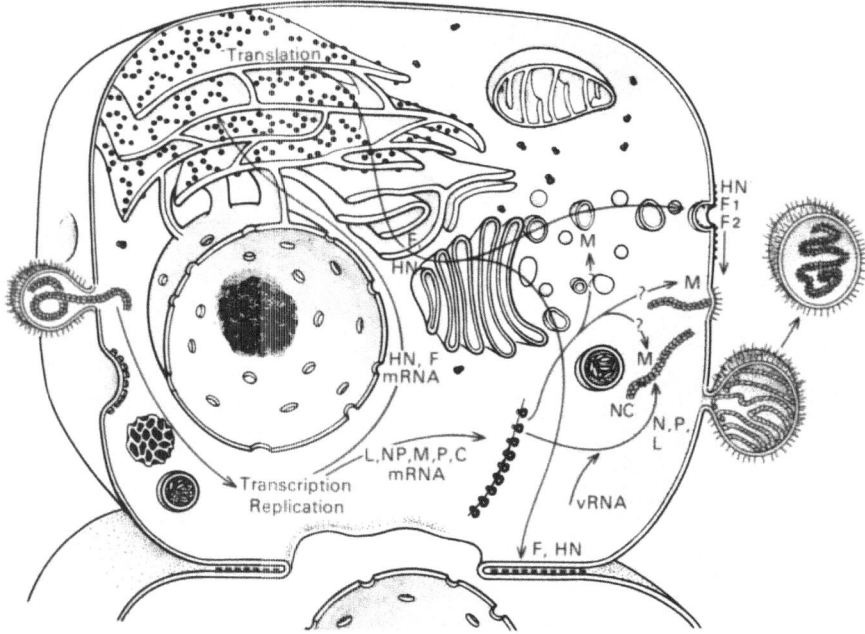

Fig. 3-11. Representation of the replication and assembly of paramyxoviruses. These viruses enter the cell by fusion of the virus envelope with the plasma membrane and not *via* clathrin-coated pits. Transcription results in the synthesis of messenger RNA for L, N, M, P (and C) proteins which are translated on free polysomes. The resulting proteins N, P, and L associate with the genome to form the NC. Transcription also results in the synthesis of mRNA for HN (Sendai virus and others) or HA (measles virus) and F_0 glycoproteins which are translated on membrane-bound polysomes. The resulting F_0 and HN (or HA) glycoproteins move through the Golgi to the cell surface inside transport vesicles. Cleavage of F_0 into the two subunits F_1 and F_2 may occur at the membrane or in the virion depending on virus and host cell. The M protein may interact with the cytoplasmic domain of the viral glycoproteins exposed on transport vesicles or on the plasmalemma. M probably also binds to NCs. NCs become aligned under the membrane and spikes are clustered in ridges over these NCs. Progressive coiling of the helical NCs occurs in the budding virion. The F_1 protein is responsible for membrane alterations resulting in cell fusion between neighboring cells

mediate more efficiently virus adsorption and fusion with target cells. At the same time, a NC detached from the viral envelope may be more easily released into the cell cytoplasm during virus penetration.

The known events of replication and assembly of paramyxoviruses are summarized in Fig. 3-11. The mature virus is schematized in Fig. 3-12. The loosely folded NC is organized in a tight helix 16 to 18 nm in diameter and with a 5 nm periodicity. NCs usually are 1 μm long and consist of a left-handed helix (Finch and Gibbs, 1970; Compans *et al.*, 1972 b). Several hypotheses for the possible localization of the M protein have been proposed (reviewed in Matsumoto, 1982), and are presented in Fig. 3-12.

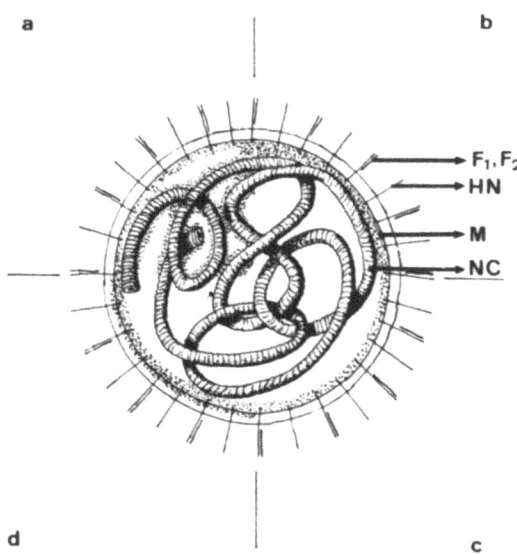

Fig. 3-12. Representation of a mature paramyxovirus. A one micrometer long NC organized in a helix of 5 nm periodicity is loosely folded inside the virion. The two subunits of the F protein, F_1 and F_2, are represented in the spikes and may be located in a different spike from the HN protein. Different possible locations of the M protein have been proposed and are represented in the four quadrants a, b, c, and d. In *(a)*, M is forming a matrix around the NC and between the NC and the envelope. In *(b)*, M protein closely associated with the internal side of the lipid bilayer binds to the NC when it is interacting with the envelope. In *(c)*, there is no interaction between M and the nucleocapsid, and the M protein is exclusively located on the inner side of the envelope. In *(d)*, the M protein is shown more deeply embedded into the bilayer than in a, b, and c

4

Assembly of *Orthomyxoviridae*

I. Introduction

The representatives of this family of viruses are the three serologically unrelated types of influenza viruses A, B, and C, which are widely distributed in humans, other mammals and birds (Melnick, 1980). Type A is responsible for influenza pandemics and the regular epidemics. Type B has been found only in man so far and causes more localized epidemics. Type C is, in many respects, somewhat different from the two other types and causes rare sporadic infections in humans. Very recently, it has been reported that orthomyxo-like viruses have been isolated from arthropods (Clerx *et al.*, 1983). *Orthomyxoviridae* are spherical, measure 80 to 100 nm, contain helical NCs and are surrounded by an envelope covered with radial spikes.

Influenza viruses may evolve in an animal reservoir between epidemics but the mechanism of this evolution is not clearly established yet. Birds and mammals are candidate reservoirs, since genetic information similar to that in human strains is present in representatives of both groups. Influenza virus may also evolve in humans, since occasional influenza virus isolations are made from humans at all times of the year.

Type A and B influenza viruses are able to escape the host's immune response by modifying their glycoproteins slightly. This frequent phenomenon is termed antigenic drift. In addition, the antigenicity of type A influenza virus sometimes changes dramatically. This modification is called antigenic shift and corresponds to the rapid spread of pandemics.

II. Molecular Organization

The structure of the influenza virus genes and proteins has been recently reviewed in detail (Scholtissek, 1979; Lamb and Choppin, 1983; Lamb, 1983). The genomes of influenza A and B viruses consist of eight segments of negative strand RNA with a total length of 13.5 Kb, wrapped in eight distinct nucleocapsids (Fig. 4-1). Influenza

C virus probably contains only seven genomic segments (Cox and Kendal, 1976), although one report describes nine segments (Petri et al., 1979). It would be logical to have one segment less in influenza C virus, since it does not have a neuraminidase. Thus, influenza C virus appears to be an outsider among the influenza viruses (Herrler et al., 1981). In influenza A and B viruses, each of the eight segments encodes at least one specific protein and can thus be considered as an individual gene (McGeogh et al., 1976). In influenza A virus, there are ten known virus poly-peptides (seven structural and three non-structural) and two genes are bicistronic. The coding assignments of influenza A virus genes, numbered by order of decreasing size are: $1=PB_2$, $2=PB_1$, $3=PA$, $4=HA$, $5=N$ protein, $6=NA$, $7=M_1$ and M_2, $8=NS_1$ and NS_2 (Table 4-1 and Fig. 4-1). In influenza B, the NA gene is also bicistronic (Shaw et al., 1983).

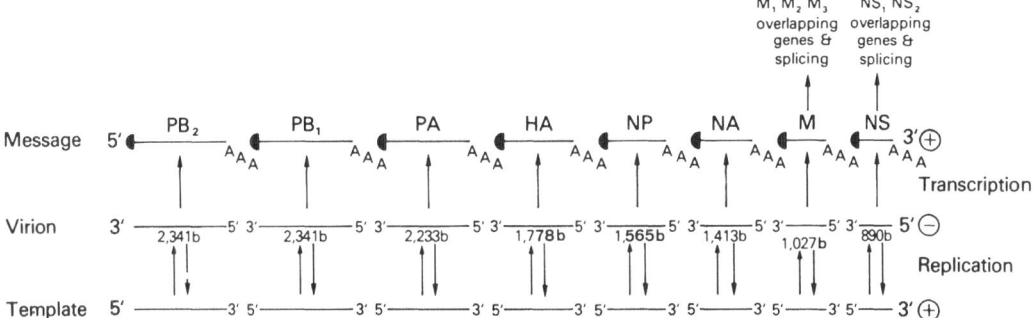

Fig. 4-1. Scheme of influenza A virus transcription and replication. The eight genome segments are shown in order of decreasing size from left to right. The encoded proteins are indicated. The numbers of bases, as determined by nucleotide sequencing, are indicated on the negative strands. All messages are capped at their 5′ ends and polyadenylated at their 3′ ends. The M segment codes fo M_1, M_2 and possibly a small peptide, M_3. The NS segment codes for NS_1 and NS_2. Both M and NS segments have overlapping coding regions using different reading frames, and splicing occurs to generate the different mRNAs. From Lamb and Choppin, 1983

Naked virus RNA is unable to initiate its own replication and transcription. The presence of the complete NC is required for these functions since the RNA-depen-dent RNA polymerase activity is associated with the NC complex. Virion RNA seg-ments all have identical conserved nucleotide sequences at their termini, which, in addition, are partially complementary (Skehel and Hay, 1978; Robertson, 1979; Desselberger et al., 1980).

Two kinds of RNA species are specified by the negative strand virion RNA in the cell (see Fig. 4-1). (1) Incomplete positive transcripts, capped at their 5′ ends by a classical methylated cap I structure (m^7GpppA_mpX) followed by a heterogeneous oligonucleotide (9 to 15 nucleotide long), and polyadenylated at their 3′ ends (Krug et al., 1976; Plotch and Krug, 1977). Poly (A) tracts are added approximately 20 nucleotides before the 3′ end of the template, but the transcripts contain the entire protein-coding sequence. The positive transcripts play the role of mRNAs (Plotch and Krug, 1978; Bouloy et al., 1978). The 5′ cap and oligonucleotide are taken by the virus from the host cell mRNAs in a "transcapping" event, a unique feature of influenza virus replication (Plotch et al., 1979). (2) Complete positive strand copies of the virion RNA segments contain triphosphates at their 5′ termini and are not polyadenylated. These RNA species are the templates used for synthesis of new virion RNA and are called A(−)cRNAs or A(−)template RNAs (Skehel and Hay, 1978; Smith and Hay, 1981; Hay *et al.,* 1982).

A number of observations indicates that influenza virus RNA synthesis and encapsidation occur in the nucleus of the host cell. The presence of an active nucleus is indeed necessary for replication (Kelly *et al.,* 1974). Replication is sensitive to UV irradiation and to actinomycin D, at least early in the replication cycle (reviewed by Nayak, 1977). Moreover, cell fractionation techniques have also indicated that RNA synthesis is achieved in the nucleus, at discrete fixed sites (Herz *et al.,* 1981; Jackson *et al.,* 1982). Virus replication requires concommitent host cell DNA transcription (Mahy *et al.,* 1980). Influenza virus RNA transcripts acquire their caps in the nucleus of the infected cells from newly synthesized heterogeneous nuclear RNA species. Replication is blocked by α-amanitin, an inhibitor of DNA-dependent RNA polymerase II (Mahy *et al.,* 1972). In cells that have an RNA polymerase II resistant to α-amanitin, virus replication occurs normally (Lamb and Choppin, 1976). Since α-amanitin inhibits primary transcription effectively even when added to cells just prior to infection, it is thought that priming can only occur with newly synthesized capped host RNA species and not with preexisting nuclear RNA (Lamb and Choppin, 1983).

Virus Proteins

The structural and functional properties of influenza virus proteins are summarized in Table 4-1.

(1) *N Protein* (MW: 55 K to 57 K). N protein is the main protein of the NC. As opposed to what is observed in other families, such as *Paramyxoviridae,* this protein does not protect RNA from ribonuclease (Pons *et al.,* 1969). N protein is phosphory-lated and this appears to enhance virus transcription (Kamata and Watanabe, 1977). N protein is a type-specific antigen of influenza viruses. Monoclonal antibodies have been produced against different epitopes on the N protein (Van Wyke *et al.,* 1980). One of them inhibits *in vitro* transcription (Van Wyke *et al.,* 1981). However, this observation does not allow attribution of a transcriptase activity to N protein itself because these bound antibodies may inhibit sterically the function of other proteins involved in transcription.

(2) *Polymerase* (MW: 82 K to 95 K). The RNA-dependent RNA polymerase activity is carried by viral components, formerly called P_1, P_2, and P_3, now defined as PB_2, PB_1, and PA, two of them being basic and one acidic (Horisberger, 1980). PB_2 binds the cap structure of heterologous RNAs while PB_1 is involved in initiation of transcription and mRNA elongation (Braam *et al.,* 1983). The role of PA in transcription and RNA replication is unclear.

(3) M_1 *protein* (MW: 28 K). This envelope-associated membrane protein is the most abundant component of the virion. It is present in such a large amount that it has been considered as a structural matrix of the particle (Compans *et al.,* 1970 a) or

Table 4-1. Proteins of influenza A virus[1]

Species	Gene segment	MW × 10^{-3}	Amino acids[2]	Molecules per virion	Location	Known or putative function
N(P)	5	55–57	498	500–1000	NC	Main structural component of NC.
PB$_2$	1	86	759	30– 60	NC	Cap binding, mRNA synthesis, endonuclease.
PB$_1$	2	86.5	757	30– 60	NC	Initiation of transcription, mRNA elongation
PA	3	82	716	30– 60	NC	vRNA synthesis?
M$_1$	7	28	252	3000	virion envelope	Virion assembly. Regulation of replication?
HA (HA$_1$) (HA$_2$)	4	77 (50) (27)	566 (328) (221)	500	spike	Surface glycoprotein, adsorption to target cells, penetration.
NA	6	50–56	454	100	spike	Surface glycoprotein, neuraminidase activity. Role in virus release?
M$_2$	7	15	97	—	cell	?
NS$_1$	8	26	230–237	—	cell	Cellular protein synthesis shut-off?
NS$_2$	8	14	121	—	cell	?

[1] From Lamb and Choppin (1983).
[2] Deduced from nucleotide sequence.

of the envelope (Schulze, 1973). Since it is encoded by gene 7, which also codes for at least one other product (see below), it is now designated M_1. The antigenicity of M_1 is different from that of N protein, but it is another type-specific antigen. The amino acid sequence of M_1, deduced from the nucleotide sequence of gene 7, reveals a strongly hydrophobic region in the middle of the molecule (Lamb and Choppin, 1983), as expected from its physico-chemical behavior (Gregoriades, 1973). It is thus through this region that M_1 may bind to envelope lipids (Gregoriades, 1980). Because of this property, M_1 can be easily inserted into liposome membranes and proteolytic digestion of these liposomes hydrolyzes the molecule except for a 5 K polypeptide corresponding probably to the hydrophobic region (Bucher *et al.*, 1980). Indications that M_1 is deeply inserted in the membrane and can even protrude on the outer surface come also from direct analysis of protease-treated, sucrose gradient-banded, intact virions (Reginster *et al.*, 1976) and from immune recognition of type-specific M protein on the surface of living infected cells and virions (Braciale, 1977; Biddison *et al.*, 1977; Ada and Yap, 1979; Reginster *et al.*, 1979). Thus the matter is still unsettled whether both ends of M_1, separated by the central hydrophobic region, are located on the same side or on opposite sides of the lipid layer. M_1 apparently interacts with genome RNA in the nucleus of infected cells (Gregoriades, 1977; Rees and Dimmock, 1981). Thus M_1 may have two functions, in the regulation of replication and in the assembly of the virus.

(4) *HA Protein* (MW: 77 K). The hemagglutinin (HA) is a transmembrane glycoprotein spanning the virus envelope and protruding from the virion. It is responsible for the hemagglutinating activity of the virus and carries the main antigenic specificity. It is involved in virus attachment to and penetration into cells (Klenk *et al.*, 1975; Lazarowitz and Choppin, 1975). Bromelain-cleaved HA, lacking the hydrophobic C terminus of the native molecule, has been purified and crystallized (Brand and Skehel, 1972). It has a trimeric structure (Wiley *et al.*, 1977). Its three-dimensional configuration has been determined to 3 Å resolution by X-ray crystallography (Wilson *et al.*, 1981). The trimer is an elongated cylinder 13.5 nm long, with each monomer consisting of a fibrous stem ($HA_1 + HA_2$) and a globular region (HA_1) which is connected to its apex.

(5) *NA Protein* (MW: 50 K to 56 K). The neuraminidase (NA) molecule is the second glycoprotein of *Orthomyxoviridae*. Its 3Å resolution structure has also been determined: isolated NA is a tetramer with 6 β sheets in each subunit, arranged like the blades of a propeller (Varghese *et al.*, 1983). The role of the virus NA, an enzyme that cleaves sialic acid residues, is not clear. It may trigger virus release (Palese and Schulman, 1974; Palese *et al.*, 1974), render HA accessible to its cleaving enzyme (Sugiura and Ueda, 1980) or help the virus penetrate through the barrier of sialic acid-containing glycoproteins in the respiratory tract (Lamb and Choppin, 1983).

(6) *Non-structural proteins: M_2, NS_1, and NS_2* (MW: 15 K, 26 K, and 14 K). These proteins are not incorporated in the virion, being found only in infected cells. M_2 is a small polypeptide of unknown function (Lamb and Choppin, 1981). It is coded by gene 7, the gene that codes also for M_1. Its coding region on gene 7 overlaps the M_1 coding region in a different reading frame (Lamb and Lai, 1981). Gene 7 contains an additional open reading frame, suggesting the existence of another peptide, M_3,

which has not yet been detected. Such a peptide would be only 9 amino acids long and would be identical to the carboxyl terminus of M_1 (Lamb *et al.*, 1981).

NS_1 is a phosphorylated protein which might be involved in the shut-off of cellular protein synthesis (Lamb and Choppin, 1983). The role of NS_2 is unknown.

III. Intracellular Synthesis of Virus Components

Many observations suggest that the host cell can influence the level of influenza virus protein synthesis and/or its normal assembly (Skehel, 1973; Meier-Ewert and Compans, 1974; Inglis *et al.*, 1976; Lamb and Choppin, 1976; Dimmock *et al.*, 1981). N protein is synthesized on free ribosomes in the cytoplasm. Later it seems to be transported to the nucleus and then returns to the cytoplasm (Taylor *et al.*, 1970; Flawith and Dimmock, 1979; Narmanbetova and Burkrinskaia, 1980; Briedis *et al.*, 1981). Earlier immunofluorescence studies revealed N protein in the nuclei of infected cells (Liu, 1955; Watson and Coons, 1954; Breitenfeld and Schafer, 1975). N protein can be detected as early as 3 hours after infection in cells infected with an influenza A virus (Fig. 4-2a). The traffic of N protein towards the nucleus is likely to be a functional property of this protein since an identical localization is seen with recombinant N protein obtained after transfection with an SV_{40} vector containing a cloned N gene (Lin and Lai, 1983). N protein traffic from the nucleus to the cytoplasm is blocked in certain non-permissive cells (Kelly and Dimmock, 1974).

As with N protein, M_1 protein is apparently produced on free ribosomes (Compans, 1973; Hay, 1974). It has no signal peptide and is probably transported to the membrane following an intracytoplasmic membrane-independent pathway. It is also detected in the nucleus (nucleoplasm) of infected cells (Gregoriades, 1977). In pulse-chase experiments, M_1 is incorporated into the plasma membrane within minutes (Lenard, 1978) and does not bind to NC as long as it is not membrane-bound (Hay and Skehel, 1974). Thus, M_1 is in low amounts in the cytoplasm of infected cells while it is a main component of virions (Lazarowitz *et al.*, 1971). As mentioned earlier, type-specific M_1 has been detected on the surface of virus-infected cells using monospecific antibodies and cytotoxic T cells (Braciale, 1977; Biddison *et al.*, 1977; Ada and Yap, 1979). On the other hand, several monoclonal antibodies could not detect M_1 on the surface of infected cells (Hackett *et al.*, 1980; Sveda and Lai, personal communication). Also, cells transfected with cloned M gene in a SV40 vector did not show fluorescence on their surface (Sveda, personal communication). However, it is possible that epitopes recognized by these monoclonal antibodies can either be masked by glycoproteins, buried in the membrane or localized on the internal region of the M_1 molecule. Thus, it is still possible that M_1 spans the lipid bilayer.

The polypeptide backbone of virus *glycoproteins* is synthesized on rough ER-associated ribosomes (reviewed by Klenk, 1977), and is inserted in the rough ER membrane by vectorial discharge (see Chapter 1) (Fig. 4-3a). A signal peptidase cleaves off the hydrophobic signal peptide of HA, freeing the amino terminus of

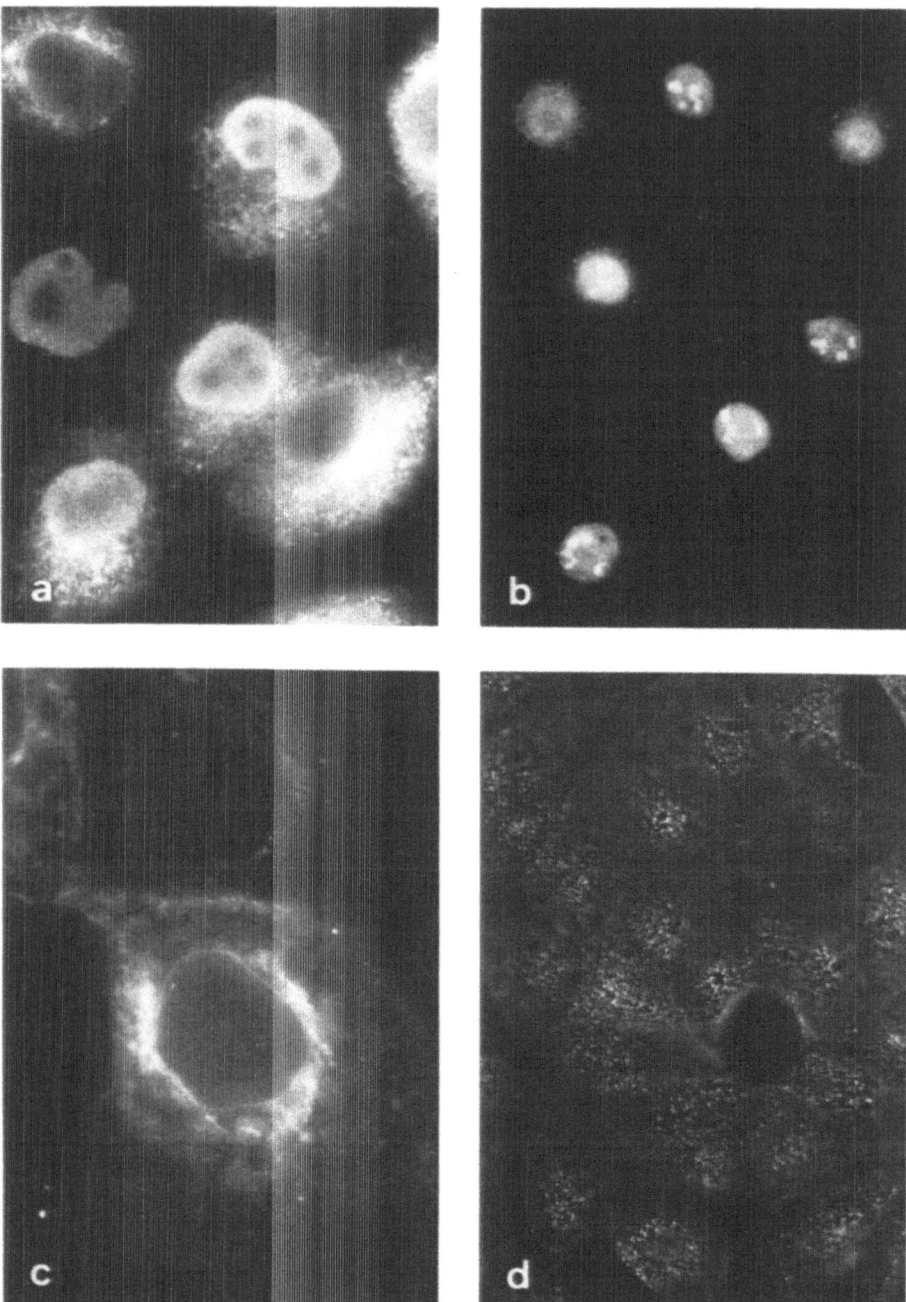

Fig. 4-2. Localization of influenza A virus proteins by immunofluorescence using monoclonal antibodies. *(a)* The N protein is localized mostly in the nucleus of MDCK cells 3 hours after infection with the PR8 strain. Dots of stain are also found in the cytoplasm of some cells. *(b)* The NS protein is predominantly localized in the nucleolus. *(c)* The HA protein is localized mostly in the perinuclear area (ER and Golgi?) at 3 hours after infection. *(d)* The HA protein is localized in small dots on the surface of living infected cells at 6 hours (labeling was performed at 4 °C before fixation). Magnifications: (b) and (d) × 360; (a) and (c) × 600. [Courtesy of Dr. T. Bächi (a), (c), and (d) and Dr. H. Arnheiter (b)]

the molecule (Air, 1979). At the carboxyl terminus, a halt transfer signal anchors the molecule in the membrane (Elder *et al.*, 1979). The very end of the molecule on the cytoplasmic side is hydrophilic (Minjou *et al.*, 1980). Alteration, by genetic manipulation, of the halt transfer signal of HA yields a molecule which is secreted into the medium (Gething and Sambrook, 1981; Sveda *et al.*, 1982) (Chapter 1). Alteration of the amino-terminal insertion signal may cause intracytoplasmic retention of the HA, resulting in its synthesis on free ribosomes (Fig. 4-3b); Sekikawa and Lai, 1983). It has been shown that a single base substitution, inducing a single amino acid change at the signal peptide cleavage site, prevented signal peptide cleavage, and the HA produced did not reach the cell surface (C.-J. Lai, personal communication).

When NA is inserted into the rough ER, the amino-terminal region is not removed. There is also no halt peptide (Fields *et al.*, 1981; Markoff and Lai, 1982). Therefore, the extended signal sequence both transfers the protein across the membrane and anchors the protein to the bilayer, attaching the molecule by its amino terminus to the membrane (Blok *et al.*, 1982) (Fig. 4-3c). How the amino-terminal insertion signal remains anchored in the membrane is not known but three mechanisms have been proposed: the amino-terminal portion could loop back after having protruded in the rough ER lumen, it could remain attached to the membrane and become inverted relative to its insertion orientation (Fig. 4-3c), or it could be completely excreted into the rough ER lumen and become anchored later. Complete or partial deletion of the signal peptide gives rise to a NA molecule that remains intracytoplasmic (L. Markoff, personal communication). The mechanism of transport of the NA molecule to the site of assembly and budding has not been clarified but is of particular interest because of its unusual insertion mechanism. Interestingly, NA transport is 2 to 3 times slower than HA transport (Hay, 1974) and this may be related to its amino-terminal insertion in the membrane.

Glycosylation of HA and NA is performed stepwise in a manner identical to cellular glycoproteins (Nakamura and Compans, 1978aa, b; see Chapter 1). Inhibition of glycosylation has several revealing effects on virus production. D-glucosamine, 2-deoxy-D-glucose and tunicamycin prevent HA and NA glycosylation and production of infectious virus (Kilbourne, 1959; Kaluza *et al.*, 1972; Klenk *et al.*, 1972; Schwartz and Klenk, 1974). However, in the presence of these agents, the non-glycosylated HA polypeptide, called HA_0, migrates normally from rough ER to smooth ER and to the plasma membrane. Virus maturation takes place with non-glycosylated proteins, but the virus produced is not infectious and is sometimes devoid of spikes, probably because spikes are cleaved off by proteases (Nakamura and Compans, 1978 a). Thus, sugar residues on virus glycoproteins might protect spikes against external proteases (Schwartz and Klenk, 1974) and provide specific active sites on the virus glycoproteins necessary for adsorption and penetration. The effect of glycosylation inhibitors may vary with virus strain and host cell. Nevertheless, full glycosylation is not required for transport and assembly.

The HA and NA molecules are both sulfated (Compans and Pinter, 1975). Partial sulfation is already achieved in rough ER membranes and is completed progressively in the smooth ER, plasma membrane and virions. Sulfation is independent of polypeptide synthesis (Nakamura and Compans, 1977).

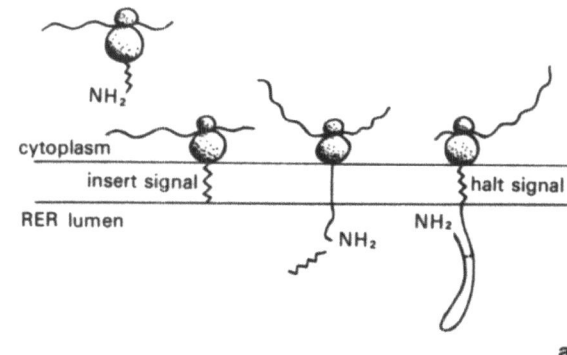

a

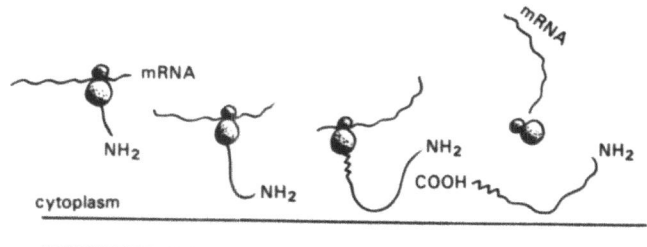

b

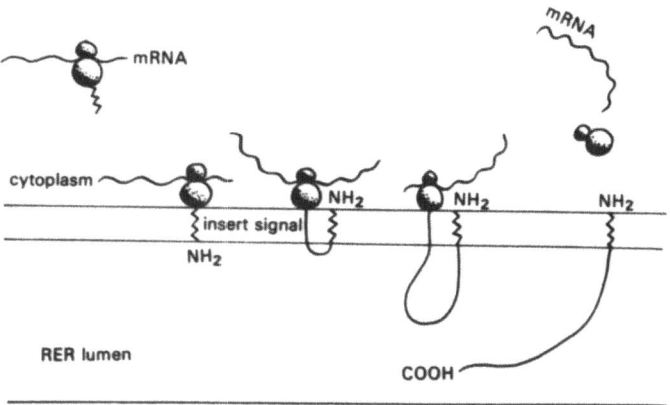

c

Proteolytic cleavage of HA into HA_1 and HA_2 subunits is an important post-translational processing of the hemagglutinin molecule and it results in a large conformational change (Skehel *et al.*, 1982). Virus with cleaved or uncleaved HA can bind equally well to cells but membrane fusion occurs only after cleavage (Lazarowitz *et al.*, 1973 a, b). Disulfide bonds and noncovalent bonds between HA monomers prevent dissociation of HA_1 from HA_2 after cleavage. Cleavage generates a new amino terminus (HA_2) which is involved in the fusion of the virus envelope with the host plasma membrane and thus determines the infectivity of the virion (Klenk *et al.*, 1975; Lazarowitz and Choppin, 1975; Huang *et al.*, 1981). The highly conserved region in the amino terminus of HA_2 that is responsible for virus penetration is homologous to the amino terminus of the F_1 protein of Sendai virus and other paramyxoviruses (Scheid *et al.*, 1978; Richardson *et al.*, 1980). Synthetic peptides mimicking this sequence can inhibit penetration of both influenza and Sendai viruses. HA cleavage occurs at the plasma membrane and is host-dependent: some cells are unable to provide the correct cleaving enzyme (Klenk *et al.*, 1975). Cleaved HA molecules are almost never found in these virions unless they are grown in the presence of trypsin-like proteases.

Travel of HA from rough ER through the Golgi to the plasma membrane (Fig. 4-2c and d) takes approximately 10 minutes (Lazarowitz *et al.*, 1971; Stanley *et al.*, 1973), while NA travels more slowly (Hay, 1974). HA is then found in discrete regions of the plasma membrane (Fig. 4-2d), as opposed to NA, which shows a more diffuse distribution before bud formation (Hay, 1974). The insertion of glycoproteins in the plasma membrane is not random. In epithelial cells, influenza virus buds from the apical pole of the cell and it is the integration of HA in the apical membrane that determines the location of budding (Roth *et al.*, 1983 b; Rodriguez-Boulan, in press; Chapter 1).

NS_1 molecules are first found in polysomes by pulse-chase experiments, then migrate to the nucleus in less than 5 minutes (Taylor *et al.*, 1970). As opposed to M_1 which is found associated with the nucleoplasm, NS_1 accumulates in the nucleolus (Becht, 1971) (Fig. 4-2b). Synthesis continues for approximately 6 hours and then

---

Fig. 4-3. Scheme of the synthesis of the influenza virus hemaglutinin (HA) and neuraminidase (NA) in the rough ER. *(a)* Synthesis of HA. The signal peptide is inserted into the membrane and will later be cleaved by a signal peptidase. The protein is folded so that intramolecular disulfide bonds link the amino-terminal portion of each molecule to its carboxyl-terminal portion. These disulfide bonds and noncovalent bonds prevent dissociation of the molecule after proteolytic cleavage into HA_1 and HA_2. A halt transfer signal maintains the protein inside the membrane. The carboxyl-terminal portion is exposed towards the cytoplasm. Panel *(b)* shows what happens when the HA signal peptide is altered by genetic manipulation. The polypeptide chain cannot attach to the rough ER membrane, resulting in intracytoplasmic retention of the HA. Under these conditions, it is synthesized on free ribosomes and is never exposed on the cell surface. *(c)* Scheme of the synthesis of the NA glycoprotein of influenza virus. Here, the amino-terminal region is inserted into the membrane, but is not cleaved. Rather it stays in the membrane and becomes an anchorage point. After its synthesis, the NA protein is in the inverted orientation compared to the HA, since it has its carboxyl terminus free in the ER lumen. In this scheme, it is proposed that the amino-terminal portion of NA remains anchored in the membrane throughout the synthesis of the polypeptide, and that it may invert its orientation during synthesis. As a result, the amino-terminal portion is turned towards the cytoplasm. However, other mechanisms may be operating during synthesis and insertion of NA (see text)

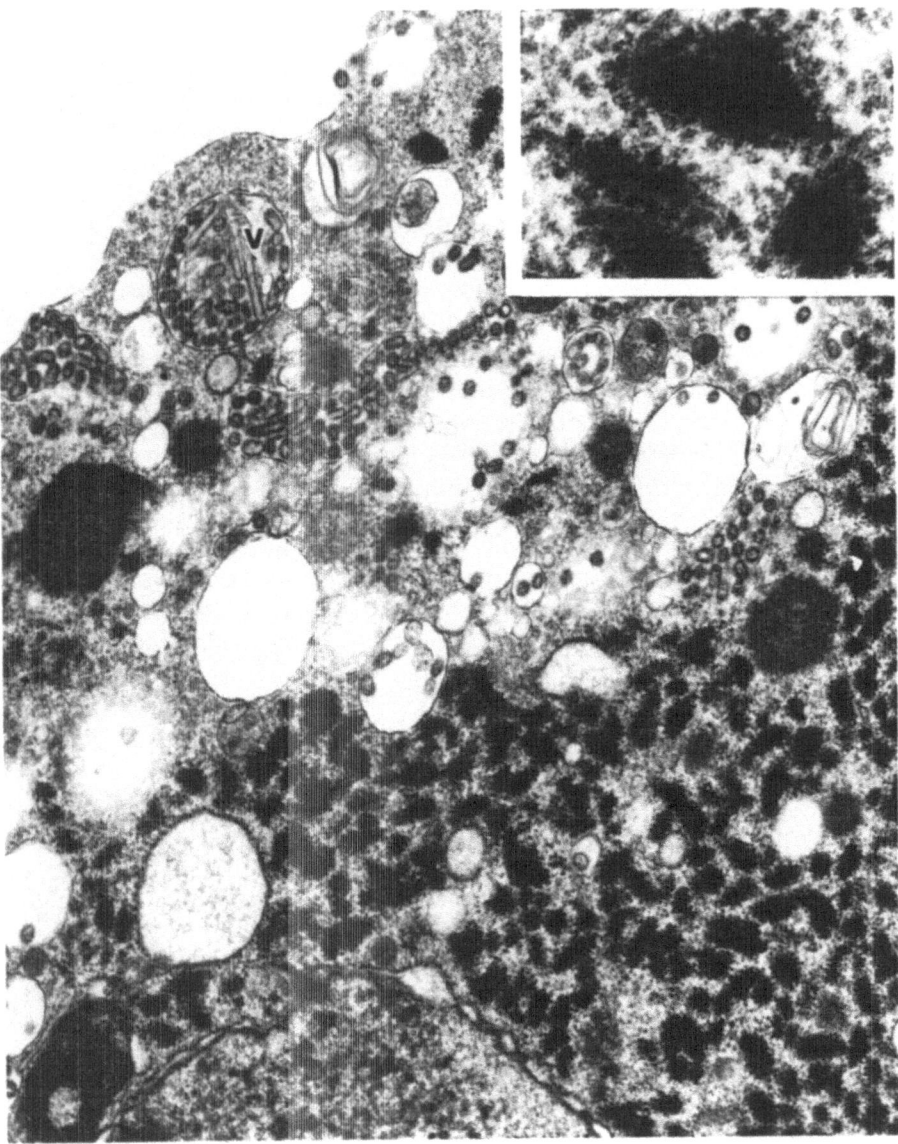

Fig. 4-4. Crystalline inclusions in HeLa cells infected for 24 hours with the influenza virus A/WSN. In these non-permissive cells, influenza virus is produced inside vacuoles *(v)* instead of at the plasma membrane. In addition, numerous cytoplasmic inclusions are present in these cells and are surrounded by ribosome granules (inset). These crystalline inclusions show fine parallel striations in their matrix and are thought to consist mainly of the NS_1 protein of influenza virus. Magnification: ×32,000. Inset: ×120,000. From Caliguiri and Holmes, 1979. (Reproduced with permission of Academic Press, New York)

decreases gradually. Occasionally, late in infection, NS_1 forms crystalline-like cytoplasmic inclusions which also contain cellular RNA species (Morrongiello and Dales, 1977; Shaw and Compans, 1978; Caliguiri and Holmes, 1979) (Fig. 4-4). Little is known about the synthesis or functions of NS_2 and M_2.

IV. Assembly of Virus Components

NC Assembly

From the evidence presented above, it appears that N protein binds to the virus genome in the nucleus. When the three virus P proteins attach to the encapsidated genome, the NC becomes transcriptionally active. However, NCs are not clearly detected in the nucleus by electron microscopy, perhaps because short NC strands may not be easily distinguished from chromatin. How NCs move to the cell surface is mysterious. Clusters of cytoplasmic NC have been observed but are difficult to identify because of the small size of these clusters (Compans and Dimmock, 1969; Compans *et al.*, 1970 b; Compans and Caliguiri, 1973).

Interaction Between NC and Envelope Proteins

Interaction between N protein bound to genome, M_1 and the glycoproteins inserted in the membrane appears to be coordinated. Recognition between glycoproteins and M_1 occurs early in virus assembly. However, there is no stoichiometric equivalence between M_1 and glycoprotein molecules, the ratio being approximately 3 to 1 (Schulze, 1973). Moreover, this ratio can change by a factor of two according to the host cell type (Lenard, 1978). It is still not clear how glycoproteins and M_1 make hydrophobic contact with each other, but M_1 is essential to virion assembly. Indeed, in some abortive influenza infections or when a defect is introduced in M_1 by substitution of an arginine analog, neither M_1 nor N protein accumulate at the cell surface, although HA and NA find their way there (Lohmeyer *et al.*, 1979; Maeno *et al.*, 1979; Bukrinskaya *et al.*, 1981; Aoki *et al.*, 1981). This suggests that a lack of glycoprotein recognition by a deficient M_1 prevents virus budding. However, virions can sometimes be produced in the presence of low amounts of M_1 (Kendal *et al.*, 1977).

Virus Budding

In the early stage of budding, several short NC strands can be seen converging towards the budding site (Fig. 4-5a and b). Transverse sections of virions frequently reveal cross-sections of filamentous strands likely to be NCs (H. Frank, personal communication). These NCs often align under the membrane and their highest number is 8 (Fig. 4-5b). These observations suggest that several NCs enter the viral bud parallel to each other and to the axis of the bud.

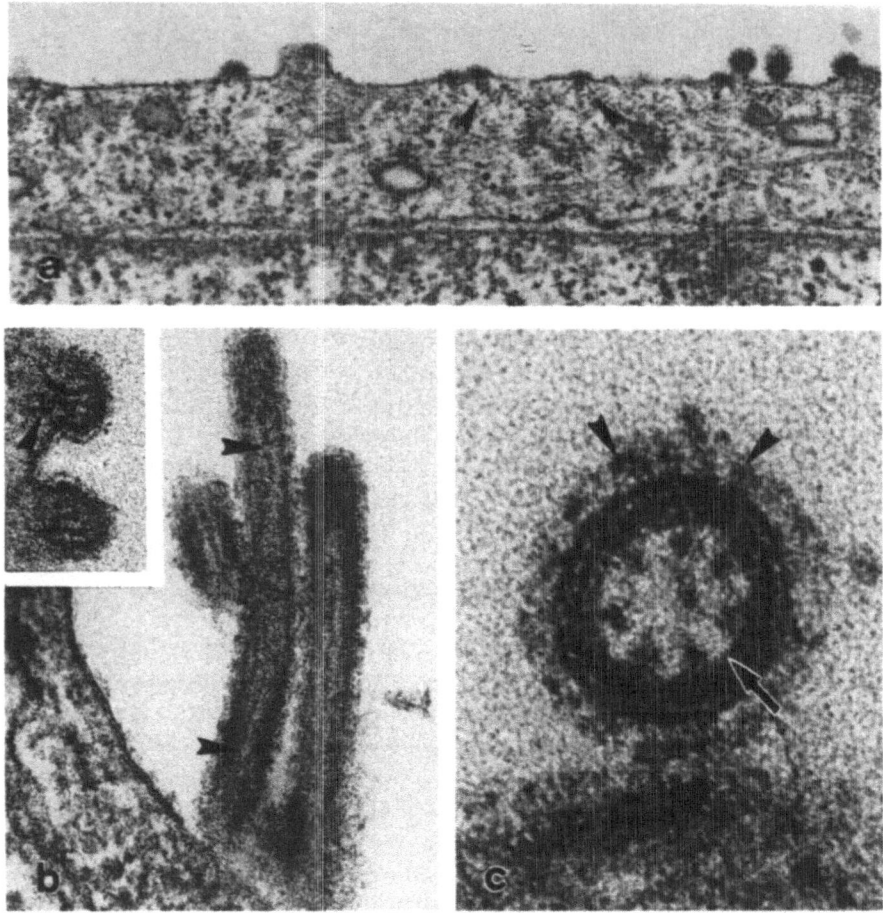

Fig. 4-5. Maturation of influenza virus at the cell membrane. *(a)* Several particles budding from the apical surface of MDCK cells. Small strands, probably representing nucleocapsids, appear at arrowheads and cluster at the early sites of budding where spikes become prominent. (b) and (c) Chick embryo fibroblasts infected with influenza A2 (H2N2) virus, which forms predominantly filamentous virions. In *(b)*, elongated viruses contain NCs running along the membranes. Inset shows spherical viruses budding from the cell. Three to four NC strands parallel to each other can be detected entering the virus bud perpendicular to the tip of the bud surface (arrowhead). *(c)* Cross-section of a filamentous influenza virus bud shown at a very high magnification. Spikes inserted into the lipid bilayer are very distinct (arrowheads). The thick inner coat (arrow) may correspond to the M protein. Eight filamentous strands probably corresponding to NCs can be counted in the transverse section of this virus. Seven are organized in a circle under the membrane and one is present in the middle of the virion. Magnifications: (a) × 40,000; (b) × 110,000; (c) × 450,000. [Courtesy of Dr. F. V. Alonso-Caplen, from Alonso *et al.*, 1982 (a), and of Dr. H. Frank (b) and (c)]

M_1 protein appears to play a critical role in budding. By its polymerization in association with the plasma membrane, M_1 could cross-link the carboxyl-terminal domain of glycoproteins and induce budding. Cross-linking molecules can indeed induce a curvature in membranes, with the concave face towards the side to which they are bound (Hewitt, 1977). On the other hand, anti-NA antibodies (which block enzymatic activity but not virion infectivity), inhibit viral budding, probably because they cross-link NA molecules and directly oppose any curvature-inducing effects of M_1 (Dowdle *et al.*, 1974). Budding is also prevented by anti-HA antibodies and multivalent Con A, but not by Fab fragments nor by monovalent succinyl-Con A (Becht *et al.*, 1971; Poste *et al.*, 1974). Formation of spikeless virus particles in the presence of glycosylation inhibitors (see above) does not necessarily indicate that glycoproteins are not involved in the budding process. Since the effect of these inhibitors is not to prevent HA and NA insertion, but to render them susceptible to proteases, the internal (cytoplasmic) terminus of these molecules is still buried in the envelope and can still be bound by M.

One of the most mysterious aspects of *Orthomyxoviridae* assembly is the mechanism by which the correct set of encapsidated genes or NCs is incorporated into a particle. Even if NCs are in excess in each virion, one would expect that a recognition process operates to insure a complete gene assortment. Apparently, the genome remains segmented at all times during the infectious cycle, despite early reports that RNA molecules or NCs were joined together temporarily (Li and Seto, 1971; Skehel, 1971). All RNA segments having identical terminal nucleotide sequences (Skehel and Hay, 1978; Robertson, 1979; Desselberger *et al.*, 1980), the existence of a differential terminal signal is thus unlikely and N protein may cover other RNA sites along each strand. Lamb and Choppin (1983) have considered the theoretical probability of obtaining an infectious particle when NCs are randomly introduced in virions. This probability is low when only 8 segments are packaged but increases when more segments are incorporated, reaching 10% when 12 segments are randomly incorporated in the virion. An average packaging of twelve NCs per particle would thus yield 10% infectious virions, which is commonly observed with these viruses. This interpretation is strengthened by the observation that fresh isolates, where filamentous particles are frequent and which contain more RNA than suspensions of spherical virions, are more infectious (Choppin and Compans, 1975). A filamentous virion containing a large amount of NCs is shown in Fig. 4-6b.

Defective Assembly

Influenza virus can occasionally bud into intracytoplasmic vesicles (Compans and Dimmock, 1969). Intracellular budding is also seen when extracellular budding is hindered by lectins (Stitz and Becht, 1977) or when the cell is poorly permissive (Caliguiri and Holmes, 1979). Influenza NCs do not accumulate in the cytoplasm as in *Paramyxoviridae* infections.

A high frequency of phenotypic mixing is observed when the same cell is infected by two different strains of influenza viruses, due to gene reassortment

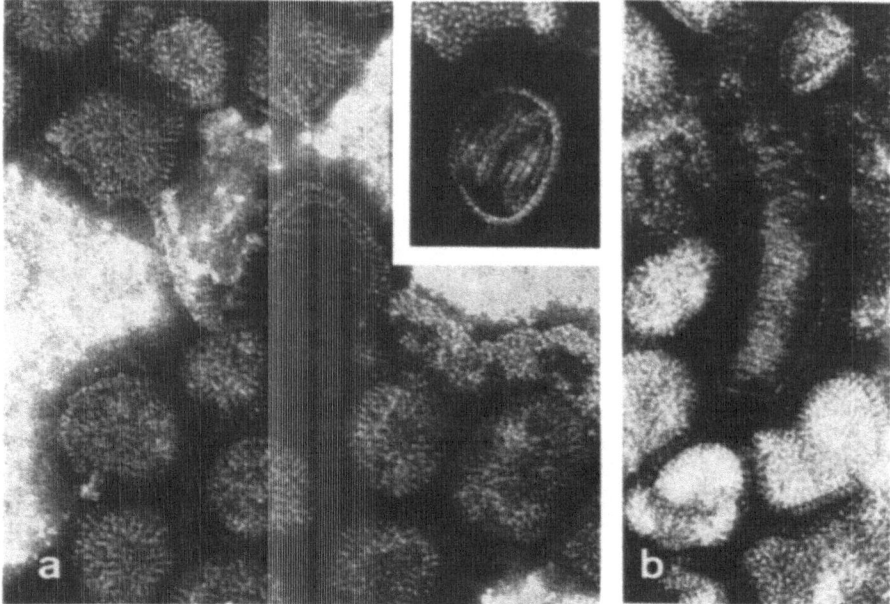

Fig. 4-6. Influenza virus A2 (H2N2) as seen after negative staining. The spikes are very prominent, approximately 10 nm long, and some are club-shaped. Two spherical particles in A and one elongated particle in B have been penetrated by the stain, revealing the internal NC. The NCs are 8 to 10 nm in diameter. The total length of the NC material visible in the inset of *(a)* is approximately 700 nm, a length which may account for the entire genome. The elongated particle shown in *(b)* contains more NCs, stacked on each other. Such packaging of NCs in the released virion is different from the arrangement seen in the budding virion in Fig. 4-5b and c. Magnification: ×130,000. (Courtesy of Dr. H. Frank)

(Hirst, 1971). In addition, defective viruses lacking a segment can complement each other (Hirst and Pons, 1973).

V. Virus Release and Organization of the Virion

Release occurs when the viral membrane at the base of the mature bud is pinched off from the cell and fuses to make a closed particle. The structure of the complete virus is best seen after negative staining (Fig. 4-6). The radially arranged spikes are 10 nm long and some are club-shaped. When the stain penetrates inside the virion, NCs are detected and form stacks or turns (Fig. 4-6a and b), a disposition rarely seen in the budding virus. The length of purified influenza virus NC is heterogeneous and averages 60 nm (Choppin and Compans, 1975). Helical NC segments are tightly coiled in low salt and unwound in high salt, in contrast to the situation seen with paramyxovirus NCs (Heggeness *et al.*, 1982). It appears that influenza NC is a right-handed helix formed by two antiparallel strands, in contrast to the left-handed single helix formed by the paramyxoviruses (Heggeness *et al.*, 1982 b). A model for the NC structure has been proposed by Compans *et al.* (1972 a) and extended by Robertson *et al.* (in press). The RNA must be coiled. The coil folds back on itself and is twisted into a double helix with a pitch of 13 nm. Inside the influenza virion, NCs can be arranged either longitudinally or in a coiled form. Since the first arrangement is mostly seen in thin sections of budding viruses and the second one is most frequently observed in released virions, the most likely explanation is that NC interaction with the envelope during budding requires a close alignment as seen with *Paramyxoviridae*. After release, the NCs may detach from the envelope and rearrange themselves.

The replication and assembly of influenza viruses are summarized in Fig. 4-7. Fig. 4-8 represents the organization of the mature virion (modified from Lamb and Choppin, 1983). Although the gene and protein structures of theses viruses have been extensively studied, their mechanism of assembly, in particular the way by which multiple NCs are incorporated into the virions, are still very mysterious.

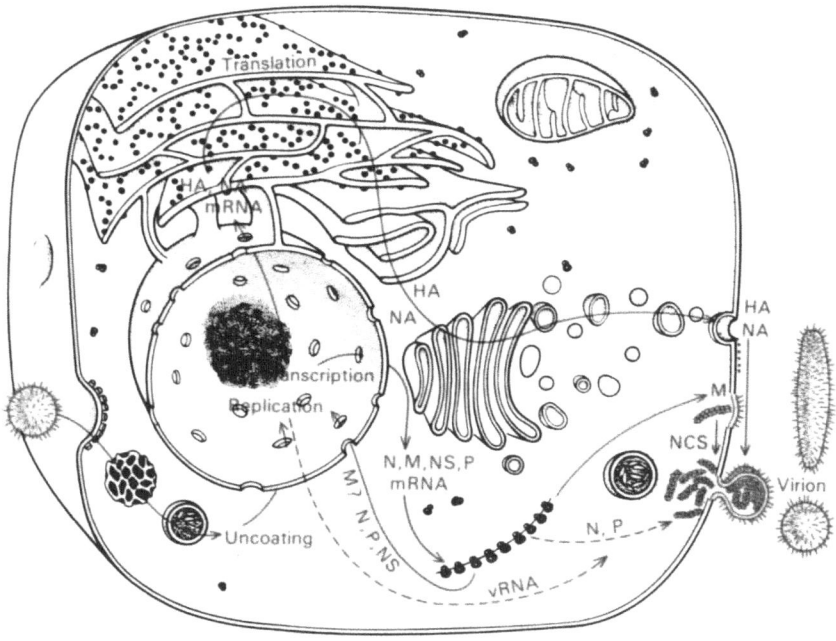

Fig. 4-7. Scheme summarizing the events of influenza virus replication and assembly. Virus entering cells through clathrin-coated pits releases the genome which apparently moves towards the nucleus where replication and primary transcription start. The mRNAs of genome-associated proteins (N and the three P proteins) as well as M and NS are translated on polysomes and the proteins move back to the nucleus. NS may trigger the shut-off of host cell protein synthesis while M may have a regulatory role in replication. It is not known whether the NCs assemble in the nucleus or in the cytoplasm. The genomic RNA, N protein and polymerases assemble into NC complexes which are then seen in association with the plasma membrane, where budding occurs. Messenger RNAs for the glycoproteins HA and NA are translated at the rough ER membrane. Glycosylation occurs in the Golgi apparatus. The glycoproteins probably move in transport vesicles to the cell membrane. NCs and virus-modified membranes with their inserted glycoproteins then interact with each other to form spherical virions. Elongated forms are more rare and vary with the virus strain

a   b

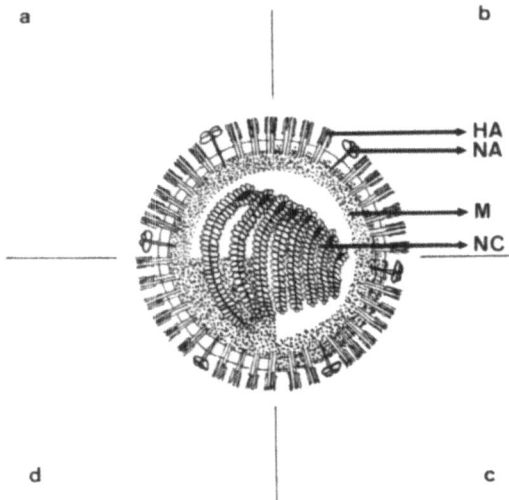

d   c

Fig. 4-8. Representation of an influenza virion showing the general structure of the envelope and the internal nucleocapsids. The trimeric structure of the HA and the tetrameric structure of the NA are represented in the spikes. There are approximately 5 to 6 times more HA molecules than NA molecules in the envelope. Eight NCs of decreasing size are represented but their exact organization in the virion is unclear (see Figs. 4-5 and 4-6). The M protein is the major component of the virion. Its localization in the virion is still debated. Four possibilities are presented in quadrants a, b, c, and d, where the localization of the M protein is indicated by black dots. In *(a)*, M is closely associated with the lipid bilayer on the internal side. In *(b)*, M penetrates inside the lipid bilayer. This is very likely, because M has a long hydrophobic segment. In *(c)*, M is transmembraneous and is partly exposed on the outer surface of the lipid bilayer (see text). In *(d)*, M is the structural matrix of the virion localized on the inner side of the bilayer and also holding the NCs together

5

Assembly of *Bunyaviridae*

I. Introduction

The *Bunyaviridae* family is a large group of over 200 distinct viruses sharing morphological, biochemical and serological properties and formerly included in the now obsolete *Arbovirus* group. They are transmitted by arthropods and occur most frequently in tropical countries. Many of them may cause lethal encephalitis in man and other mammals. Like many other viruses, including togaviruses, they replicate in insects without causing disease. The family has been subdivided in four genera: *Bunyavirus, Nairovirus, Phlebovirus,* and *Uukuvirus.* Clear differences between genera have been reported, such as genetic organization, NC shape, ability to shut off host protein synthesis, glycoprotein spike arrangement, etc. Mature virions have a spherical or slightly oval shape and are 80 to 110 nm in diameter (Holmes, 1971) with a central core formed by a helical NC (von Bonsdorff *et al.*, 1969).

Today, more and more information is being generated on the molecular aspects of *Bunyaviridae* replication and maturation, although only a few representative viruses of this large family have been successfully grown in culture (Bishop *et al.*, 1980). Since there are more dissimilarities between genera than was previously thought, we will specify for which virus or group of viruses information has been obtained.

II. Molecular Organization

The *Bunyavirus* genome (Fig. 5-1) consists of three unique segments of single-stranded RNA of negative polarity, coding for three major structural proteins, a virion-associated RNA polymerase catalyzing the synthesis of a positive mRNA strand (Ranki and Pettersson, 1975; Bouloy and Hannoun, 1976) and probably some non-structural viral polypeptides (Ulmanen *et al.*, 1981; Ushijima *et al.*, 1981; Fuller and Bishop, 1982; Smith and Pifat, 1982). So far, an RNA polymerase has not been formally demonstrated in every member of the family, and non-structural polypeptides have not been detected in cells infected with the prototype *Bunyavirus* itself, Bunyamwera virus (Lazdins and Holmes, 1979).

The total length of the virus genome is between 14 and 18 K-bases. The three segments, which are neither polyadenylated nor capped (Obijeski *et al.,* 1976; Obijeski and Murphy, 1977 and 1980; Gentsch *et al.,* 1977; Pettersson *et al.,* 1977; Abraham and Pattnaik, 1983), are designated according to their respective lengths: L for large (7.5 Kb to 9 Kb), M for medium (5.5 Kb to 7 Kb) and S for small (0.9 Kb to 1.5 Kb) (Fig. 5-1). These segments are encapsidated and may circularize. The L segment is thought to encode at least the putative RNA polymerase L, since its coding capacity is larger than that required for the L protein. In the *Bunyavirus* genus, the M segment encodes the two glycoproteins, G_1 and G_2, and the S segment contains overlapping open reading frames and codes for the N protein and a non-structural protein, NS_s (Gentsch and Bishop, 1978, 1979; Akashi and Bishop, 1983; Bishop *et al.,* 1983). The segmented nature of the genome explains the frequent genetic reassortments observed in *Bunyaviridae* (Ushijima *et al.,* 1981). However, genomic exchange is restricted to members of the same genus and sometimes the same subgroup (Bishop *et al.,* 1981).

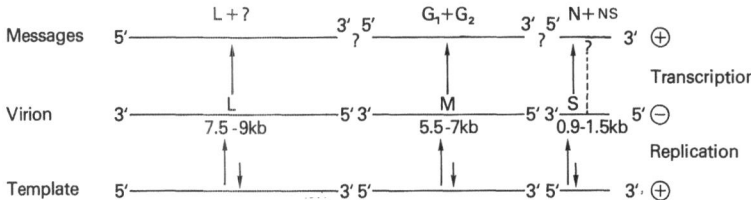

Fig. 5-1. Representation of the replication of the most studied *Bunyaviridae*. The genome consists of three negative strand segments, L, M, and S. The approximate numbers of bases are indicated below the genomic segments. The L segment codes for the putative RNA polymerase, L, and perhaps another unidentified protein. The M segment codes for the two glycoproteins, which may first be synthesized as a single precursor. The S segment codes for N protein and a second nonstructural protein, NS_s, at least in some species of *Bunyaviridae*. It is not known whether these mRNAs are capped and poly (A) tails have not been found

Naked virion RNA is not infectious. RNA replication yields full-length positive-stranded RNA copies which are encapsidated and serve as templates for the production of negative strand virion RNA segments. Virion RNA segments are also transcribed to give full-length or subgenomic complementary mRNAs (Fig. 5-1). It is not known whether the 5′ ends of these mRNAs are capped, but the 3′ ends are apparently not polyadenylated (Ulmanen *et al.,* 1981). In Uukuniemi and snowshoe hare viruses, incomplete transcripts of the S segment code for two proteins (N and NS_S), possibly by overlapping reading frames (reviewed by Strauss and Strauss, 1983; Bishop *et al.,* 1983).

Bunyaviridae replication and transcription are insensitive to α-amanitin, actino-mycin D and rifampicin (Bishop and Shope, 1979). Virus production is blocked in enucleated cells, but this may be due to damage to other cellular organelles during enucleation (Pennington and Pringle, 1978). Thus, the role of the nucleus in *Bunya-viridae* replication is still unsettled.

Virus Proteins

The structural and functional properties of *Bunyaviridae* proteins are summarized in Table 5-1.

Table 5-1. *Proteins of* Bunyaviridae[1]

Species	Gene segment	MW $\times 10^{-3}$	Molecules per virion	Location and putative function
N	S	19– 50	2100	NC protein
L	L	120–200	20–25	Polymerase, transcriptase
G_1	M	75–120	630	Surface glycoprotein
G_2	M	30– 63	630	Surface glycoprotein
NS	S	11[2]	–	?

[1] After Strauss and Strauss (1983).
[2] Fuller und Bishop (1982).

(1) *N Protein* (MW: 19 K to 50 K). N protein is involved in the early stages of
replication and is the structural protein that binds to both positive and negative
RNA strands (but not to mRNAs) to form NCs.

(2) *L Protein* (MW: 120 K to 200 K). The polymerase and transcriptase activities
are probably carried by this largest virus protein which exists in only a few copies
per virion.

(3) *Glycoproteins G_1 and G_2* (MW: 75 K to 120 K; 30 K to 63 K). Attribution of
distinct functions to each glycoprotein has not been achieved yet. These proteins
are undoubtedly responsible for attachment of the virus to cells during the adsorp-
tion-penetration phase.

(4) *Nonstructural proteins* (MW: 11 K; others?). Several polypeptides have been
described, which are specified by the virus but are present only in infected cells
(McPhee and Della-Porta, 1981; Ulmanen *et al.*, 1981; Fuller and Bishop, 1982; Smith
and Pifat, 1982). It appears that the non-structural proteins differ from one virus to
another. In addition, some bunyaviruses do not code for these proteins. Their functions are unknown, but, by analogy with other virus families, they are thought to
play roles in virus replication and transcription, or to modify host functions.

III. Intracellular Synthesis of Virus Components

N protein is produced in cells early after infection (Bishop and Shope, 1979). The
two glycoproteins, G_1 and G_2, are synthesized from the same mRNA segment on
membrane-bound ribosomes. A large 110K precursor is apparently inserted into the
membrane and is immediately cleaved, possibly by the same enzyme, the signal
peptidase, that cleaves off the insertion signal peptide (Ulmanen *et al.*, 1981).

IV. Assembly of Virus Components

Budding occurs on smooth membranes of the endoplasmic reticulum and of the Golgi apparatus. As a result, virions are shed into intracytoplasmic cisternae (Fig. 5-2) (Murphy *et al.,* 1978 a; Smith and Pifat, 1982). Thus, the *Bunyaviridae* differ

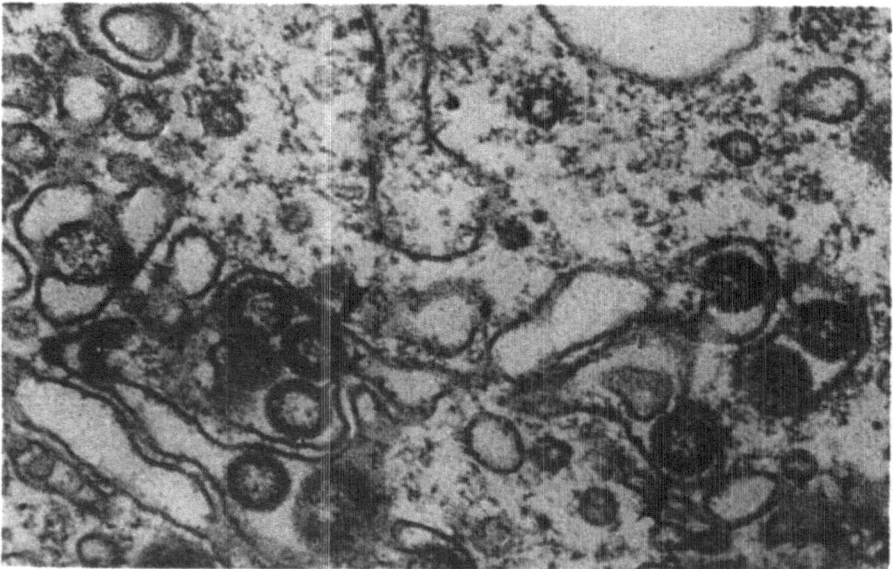

Fig. 5-2. Maturation of a phlebovirus (Punta Toro virus) in Vero cells 16 hours after infection. Viruses are budding into smooth membrane vesicles. The envelope bilayer with its spikes is very distinct (large arrowheads) and in continuity with the vesicular membrane (small arrowheads). Electron-dense material under the virion envelope probably corresponds to NC. (Courtesy of Dr. J. F. Smith, from Smith and Pifat, 1982; reproduced with permission of Academic Press, New York)

from all other negative strand enveloped RNA viruses by their site of budding and this may be correlated with the absence of a matrix (M)-like protein in the virions (Smith and Pifat, 1982). Indeed, an M protein appears to be essential for normal assembly of *Rhabdo-, Paramyxo-,* and *Orthomyxoviridae* which never bud into Golgi cisternae (Compans and Klenk, 1979) (see Chapters 2, 3, 4).

Even though NC components are synthesized earlier in infection than glycoproteins (Lazdins and Holmes, 1979), NC cannot be identified morphologically in the cytoplasm outside of the budding sites where spikes are also visible. Virus maturation is a continuous process of involution of localized areas of smooth vesicle membranes where NC and viral spikes are accumulating on opposite sides of the virus bud. NC is closely associated with the membrane during budding. No spikes can be seen in the membrane adjacent to budding sites. Since *Bunyaviridae* contain no M protein, at least one of the glycoproteins, G_1 nor G_2, must be transmembranous and establish direct contact with the NC on the cytoplasmic side of the plasma membrane (Lyons and Heyduk, 1973; Murphy *et al.,* 1973; Smith and Pifat, 1982) as in the case of coronaviruses (see Chapter 7).

V. Virus Release and Organization of the Virion

Virions are transported to the extracellular space in Golgi-derived vesicles and released by the normal process of exocytosis, *i.e.* the fusion of the vesicle with the plasma membrane (Murphy *et al.*, 1968 a, b; Holmes, 1971; Lyons and Heyduk, 1973; Murphy *et al.*, 1973; Smith and Pifat, 1982). In the *Bunyavirus* genus, absence of virus antigens on the plasma membrane, as determined by immunoferritin labeling before and after virus release, indicates that (1) virus components are not accumulating in the plasma membrane in amounts detected by this technique and (2) transport vesicles do not contain in their membranes virus antigens which could be transferred to the plasma membrane during exocytosis. Thus, virus components are clustering at the budding site in the Golgi apparatus. There is, however, one image published by Murphy *et al.* (1973) which demonstrates virus budding from the plasma membrane of a mouse neuron *in vivo*, but this is a unique event so far.

Thus, in contrast to most other enveloped viruses, some *Bunyaviruses* appear to display virus antigens only on budding and free virions and not on the infected cell

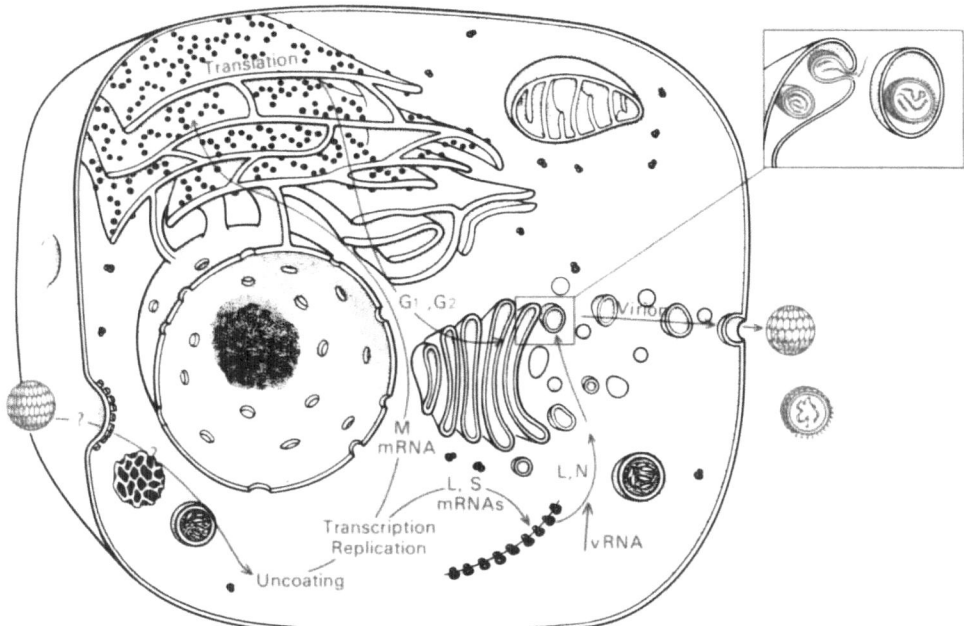

Fig. 5-3. Summary of the putative events of transcription and assembly of *Bunyaviridae*. The mechanism of entry is still unknown. L and S mRNAs are translated on polysomes and the resulting proteins move to the site of assembly in the Golgi apparatus and smooth membrane vesicles. M mRNA is translated in the rough ER. The translation product is probably a precursor protein which is immediately cleaved into the glycoproteins, G_1 and G_2. It is presumed that the three segments are contained in three NCs which are packaged into the virion as shown in the box. Virus glycoproteins are detected at the budding sites, but are apparently not seen at the cell surface. The virus travels to the cell surface in transport vesicles. The remarkable hexagonal arrays of spikes as seen in negative staining (Fig. 5-5) are shown on the envelope of the complete virus

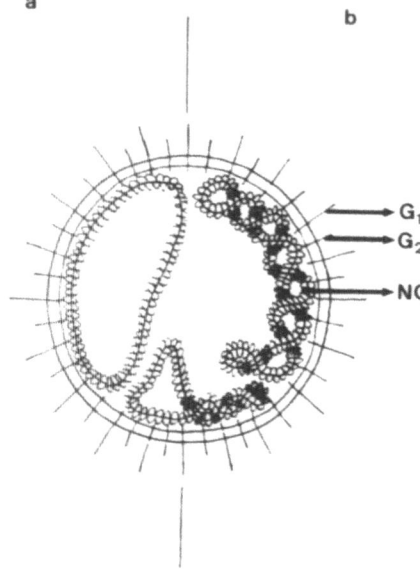

Fig. 5-4. Representation of a bunyavirion, showing its three encapsidated RNA segments which may form circles *(a)* or supercoils *(b)*. It is not known whether G_1 and G_2 reside on different spikes. G_2 has a lower molecular weight than G_1

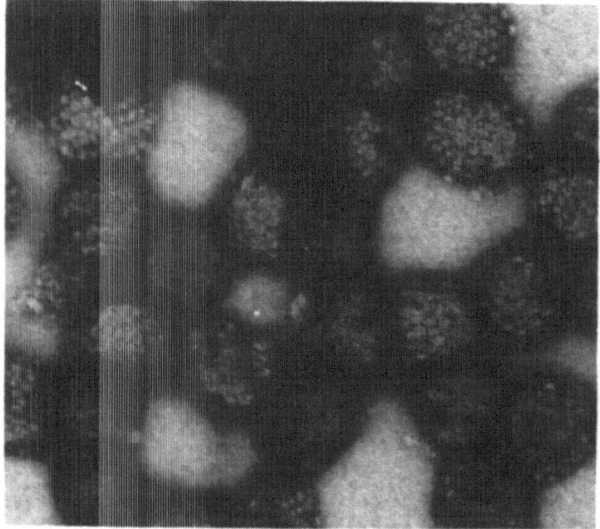

Fig. 5-5. Negative staining of virus Unkuniemi after glutaraldehyde fixation. Spikes are organized in a hexagonal array Magnification: X 120,000. (Courtesy of Dr. C. H. von Bonsdorff)

plasma membrane. Therefore, the host's immune system may not recognize infected cells and this may play a role in the pathogenicity of these viruses.

Fig. 5-3 is a summary of the events of replication and assembly of *Bunyaviridae* in the cell. The structure of the complete virus is schematized in Fig. 5-4. From studies on isolated NCs, it appears that NC may form supercoiled circles (Pettersson and von Bonsdorff, 1975; Obijeski *et al.*, 1976; Hewlett *et al.*, 1977) or superhelical filaments (Saikku *et al.*, 1971; Dahlberg *et al.*, 1977). The NC is contained in a lipid bilayer envelope which contains radial spikes 5 to 10 nm long on the outer surface (Holmes, 1971; Obijeski and Murphy, 1977). Spikes are arranged in a hexagonal array in some members of the family, such as Uukuniemi virus (von Bonsdorff and Pettersson, 1975), Punta Toro virus and Karimabad virus (Smith and Pifat, 1982) and this organization is more visible after glutaraldehyde fixation (Fig. 5-5) as represented schematically in Fig. 5-3.

6

Assembly of *Arenaviridae*

I. Introduction

Arenaviruses can infect a variety of mammals, including humans, but they usually have a principal rodent host (except for Tacaribe virus isolated from bats). The virus is maintained in nature by persistently infected animals. The prototype of arenaviruses is lymphocytic choriomeningitis virus or LCMV (originally, there were two isolates, one from a monkey and one from a human), which can infect primates, and of which the principal reservoir is the house mouse. The other pathogens are Lassa virus (in West Africa) and the viruses of the Tacaribe complex (Tamiami, Junin, Machupo, and Pichinde viruses, isolated in Florida and South America). Lassa, Junin, and Machupo viruses cause severe and often fatal hemorrhagic fevers in humans (Pedersen, 1979; Rawls and Leung, 1979; Howard and Simpson, 1980). It is well-established that LCM is an immunopathological disease (Buchmeier *et al.*, 1980), while the mechanisms of pathogenesis by *Lassavirus* and the hemorrhagic fever agents are still unclear.

Arenaviruses of these different groups all resemble each other: their sizes vary from 50 to 300 nm (viruses up to 500 nm are occasionally found); their envelopes bear spikes and surround ribosomal granules ("sandy" particles; hence, the name "arena") and filamentous NC (Murphy *et al.*, 1970; Murphy and Whitfield, 1975; Lascano and Berria, 1974).

Although the pathogenesis of LCM infection has been extensively studied in rodents (reviewed in Lehmann-Grübe, 1971; Buchmeier *et al.*, 1980), the molecular aspects of arenavirus replication have only been studied in the recent years (reviewed in Vezza *et al.*, 1978 b; Pedersen, 1979; Rawls and Leung, 1979). We will briefly summarize what is known about the molecular organization of the virus before describing its assembly.

II. Molecular Organization

Arenaviruses have a negative-stranded segmented genome (Fig. 6-1). There are two single-stranded virus RNA segments, large (L) and small (S), 6.3 to 8.4 Kb and 3.3 to

3.9 Kb, respectively (Vezza *et al.*, 1978 a, b; Strauss and Strauss, 1983). Recent studies on Pichinde virus (Vezza *et al.*, 1980) have shown that segment S codes for the two major structural proteins, the N protein and GPC, the precursor of the envelope glycoproteins GP_1 and GP_2 (Buchmeier *et al.*, 1978). On the S segment, the N protein gene is located at the 3′ end and the GPC gene at the 5′ end. These 2 genes are separated by a potential hairpin region of secondary structure (Auperin and Bishop, 1983) and may specify two separate mRNAs. It is not known whether these messages are capped. The large segment, L, codes for a protein which is probably the virus RNA-dependent RNA polymerase (Harnish *et al.*, 1981, 1983). S segments are responsible for pathogenic potential, as shown by genetic reassortment studies (Compans *et al.*, 1981). Both the L and S segments have a homologous sequence within the highly conserved 3′ region of their RNA. This sequence may be the

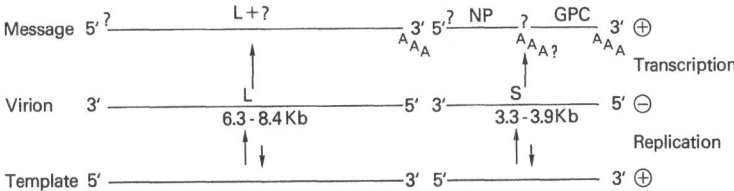

Fig. 6-1. Scheme of replication and transcription of the arenavirus prototype, Pichinde virus. The genome is composed of a large *(L)* and a small *(S)* negative strand RNA species and the approximate numbers of bases are indicated on the negative strands. The L fragment probably codes for the polymerase, L, and the S fragment for the N protein (NP) and a glycoprotein precursor, *GPC.* The latter two proteins have separate messages. It is not known whether the mRNAs are capped

recognition site for the RNA polymerase, in both transcription and replication (Auperin *et al.*, 1982). In LCMV, L and S have complementary termini with base-pairing potential (Dutko *et al.*, 1981). Indeed, circular NCs have been recovered from Tacaribe virus (Palmer *et al.*, 1977). Although the RNAs do not form covalently closed circles, circular structures with base-paired terminal stems have been seen (Vezza *et al.*, 1978 a).

Arenaviruses may depend upon nuclear functions for their replication. Alpha-amanitin, an inhibitor of DNA-dependent RNA polymerase II (Mahy *et al.*, 1972), inhibits Pichinde virus replication and α-amanitin-resistant cells, with an altered RNA polymerase II, will produce arenavirus in the presence of the drug. Arenaviruses may thus have characteristics in common with *Orthomyxoviridae*, as discussed earlier (Strauss and Strauss, 1983).

Arenaviruses also contain a large amount of host-derived ribosomal RNA species (28S, 18S, and 4 to 5S) which are released in ribosome particles by non-ionic detergent treatment of the virion. Any role of ribosomal RNA in arenavirus replication is unlikely. Variable amounts of ribosomes are packaged and the amount of 28S and 18S RNA inside the virus can vary greatly depending on the virus stock and growth conditions (Rawls and Leung, 1979). When virions contained temperature-sensitive ribosomes unable to perform protein synthesis at non-permissive temperature, no effect on the replication of Pichinde virus was observed (Leung and Rawls, 1977).

Virus Proteins

Table 6-1 summarizes the known structural proteins of five different arenaviruses (Pedersen, 1979). N protein (often called NP) is the major structural protein of NCs. P is a minor, non-phosphorylated component also associated with NCs (Compans *et al.*, 1981). N protein appears to be the major virus-specific phosphorylated product resulting from the protein kinase activity found in purified LCM virions (Howard and Buchmeier, 1983). As mentioned above, L, a large protein (MW: 200 K) has been identified in Pichinde virus (Harnish *et al.*, 1981) and is probably the RNA polymerase. Another minor component of the virion (MW: 12 K) has been identified in Pichinde virus (Ramos *et al.*, 1972).

Some arenaviruses have two glycoproteins (Pichinde-LCM) (Vezza *et al.*, 1977; Buchmeier *et al.*, 1978; Young *et al.*, 1981), others apparently have only one (Tacaribe, Tamiami, Amapari) (Gard *et al.*, 1977). In LCMV, GPC (MW: 74 K) is found in the cell and is the precursor of GP_1 (MW: 44 K) and GP_2 (MW: 35 K). Each of these three proteins has a different sugar composition (Buchmeier and Oldstone, 1979). In Tacaribe and Amapari viruses, a precursor, GPC (MW: 70 K) has also been identified and yet only one glycoprotein (MW: 42 K) was found (Saleh *et al.*, 1979). In LCMV, the exact relationship between GP_1 and GP_2 is not known, but the most peripheral protein appear to be GP_1 (Buchmeier, personal communication). The extent of penetration of the glycoprotein(s) into the virus lipid layer is still uncertain (Compans *et al.*, 1981). Additional glycoproteins of LCM have been identified recently (gp 60 and gp 85) but are not yet characterized (Bruns *et al.*, 1983). Different arenaviruses have distinct surface antigenic determinants but show cross-reactivity of their NC proteins (Buchmeier and Oldstone, 1978; Buchmeier *et al.*, 1978). In addition, there is some homology between the GP_2 glycoprotein of LCMV and the African arenavirus, Mozambique virus (Buchmeier *et al.*, 1981).

III. Intracellular Synthesis of Virus Components

The sites of synthesis of the major structural proteins are probably free ribosomes for N protein and membrane-bound ribosomes for GPC or GP. Immunofluorescence reveals both granular and diffuse staining for N protein in the cytoplasm of LCMV-infected cells (Fig. 6-2b) (Buchmeier *et al.*, 1981). GPC is found in the cytoplasm and not in the virion, and therefore may be cleaved close to the cell surface. In LCMV-infected cells, both GP_1 and GP_2 can be detected by immunofluorescence in the cytoplasm and on the surface of infected cells (Fig. 6-3a and b) (Buchmeier *et al.*, 1981).

IV. Assembly of Virus Components

NC Assembly

Available evidence suggests that NC may assemble in the cytoplasm, although no structure has been definitely identified as such in the cytoplasmic matrix. Tamiami

Table 6-1. *Proteins of Arenaviruses*[1]

Virus	Precursor	Protein	Apparent MW $\times 10^{-3}$	Location	Known or putative function
All viruses		N(P)	62–72	NC	Main structural component of NC
All viruses		P*	71–79	NC	May be related to N
All viruses?		L	200	NC	Polymerase
LCM Pichinde	GPC 74 K	GP₁, GP₂	52 to 64	} Spike protein(s)	Transmembrane? With hydrophobic domain?
Lassa			35 to 38		Glycosylated
Tamiami Others		GP	38 to 44		
Tacaribe Amapari	GPC 70 K	GP	42		

[1] From Pedersen, 1979, and from Drs. M. J. Buchmeier, C. R. Howard, and R. W. Compans, personal communication.
* P for protein, not for phosphorylated.

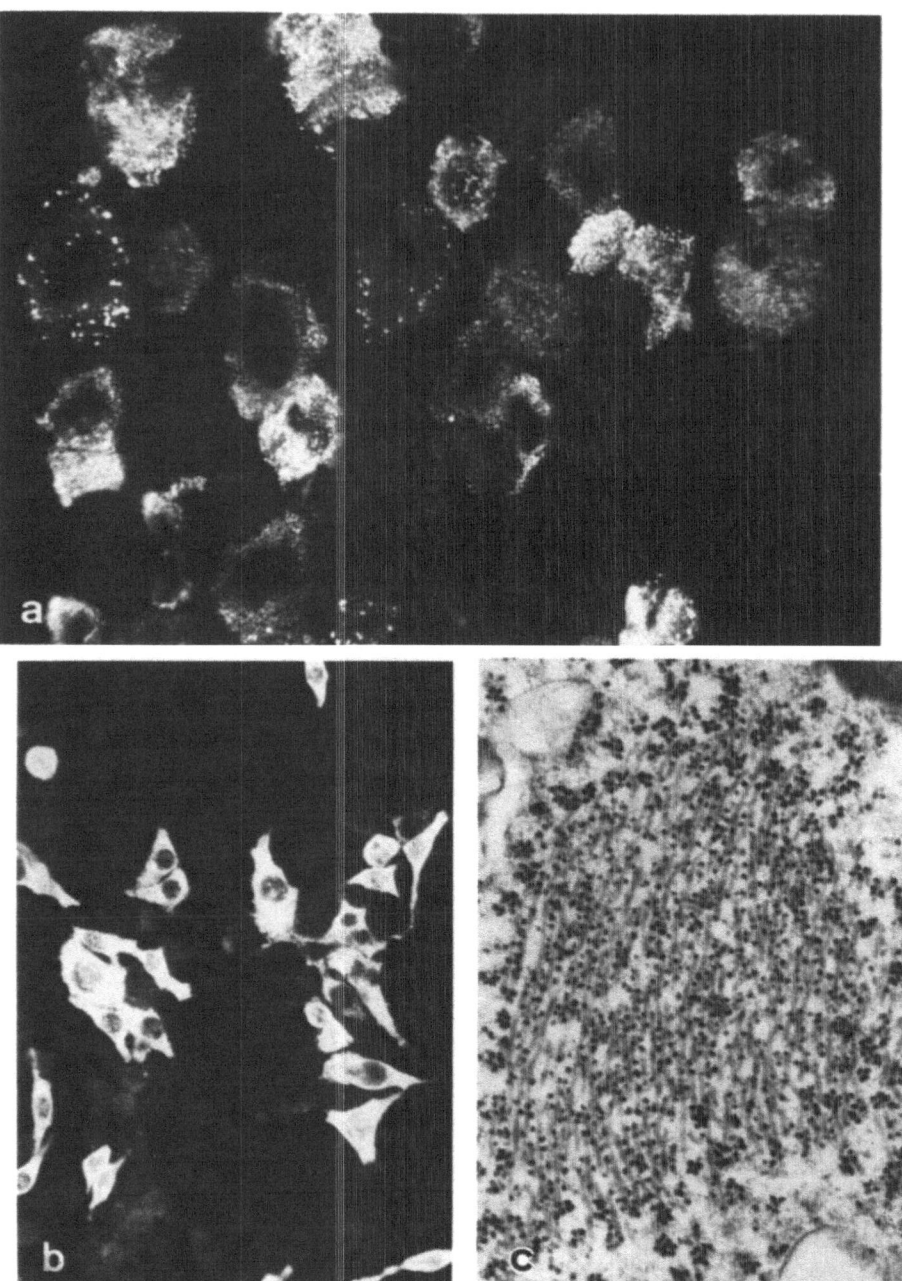

Fig. 6-2. Intracytoplasmic virus proteins and inclusions in arenaviruses. *(a)* Immunofluorescent staining of dorsal root ganglion neurons of the cotton rat infected with Tamiami virus. Internal antigens are stained in a dustlike granular aspect. *(b)* Similar granular and diffuse fluorescence is seen after staining of L929 cells infected with lymphocytic choriomeningitis virus (LCMV) with a monoclonal antibody to the N protein. *(c)* Electron micrograph showing a cytoplasmic inclusion in a nerve cell of the cotton rat infected with Tamiami virus. Fine 10 nm filaments are intermixed with ribosomal aggregates. Magnifications: (c) × 40,000. [Courtesy of (a) Dr. F. A. Murphy, from Murphy *et al.,* 1976; reproduced with permission of William and Wilkins, Baltimore. (b) Dr. M. J. Buchmeier *et al.,* 1981; reproduced with permission of Academic Press, New York. (c) Dr. F. A. Murphy, from Murphy and Whitfield, 1975; reproduced with permission of World Health Organization, Geneva]

virus-infected cells often show a dustlike granular aspect of internal fluorescence when stained with antiviral antibodies (Fig. 6-2a) (Murphy *et al.,* 1976) and N protein of LCMV is also present in abundance in the cytoplasm of infected cells (Buchmeier and Oldstone, 1981; Buchmeier *et al.,* 1981) (Fig. 6-2b). By electron microscopy, intracytoplasmic inclusions are frequently detected in infected cells *in vitro* and *in vivo* and consist of accumulations of single ribosomes within a moderately electron-dense matrix which reacts with antiviral antibodies (Abelson *et al.,* 1969). Recent studies have shown N protein staining exclusively on ribosomes of infected cells in LCMV infected mice (Rodriguez *et al.,* 1983). In LCM and Tamiami virus infections, early inclusions contain fine filaments intermixed with ribosomes (Murphy *et al.,* 1970, 1976; Murphy and Whitfield, 1975) (Fig. 6-2c). These filaments may represent NCs.

Virus Budding

Early in infection, the plasma membrane at budding sites is modified by the presence of surface spikes and of electron-dense material on the inner side of the virus envelope (Fig. 6-3c). Dense granules are occasionally lined up in this dense layer (Murphy *et al.,* 1975). Late in infection, very large areas of the plasma membrane become modified: a thick layer of amorphous material containing dense granules lies under the membrane (Murphy *et al.,* 1970). These granules may be ribosomes or may correspond to the beads seen on isolated NCs (see below).

The nature of early interactions between NCs and glycoprotein(s) at budding sites are not known. In the arenaviruses which have only one glycoprotein, the carboxyl-terminal portion of this putative transmembrane protein may interact with NCs unless a yet undefined protein triggers the assembly process. In viruses with two glycoproteins, GP_2 might be localized more on the inner portion of the plasma membrane than GP_1, as suggested by failure of GP_2 on the cell surface to be labeled after surface radiolabeling of intact viruses and infected cells (Buchmeier *et al.,* 1981; Bruns *et al.,* 1983).

Defective Budding

In the nervous system of mice persistently infected with LCMV, N protein is found in abundance in the cytoplasm of infected neurons, which do not express surface glycoproteins and do not produce complete virions (Rodriguez *et al.,* 1983). This suggests that N protein is synthesized preferentially and that assembly is defective in these chronically infected nerve cells.

V. Organization of the Virion

The structure of a released virion is shown in Fig. 6-3d. Virion size may vary considerably, the largest virions being shed late in infection. Many virions incorporate ribosomal granules (1 to 8). These granules disappear after RNAse treat-

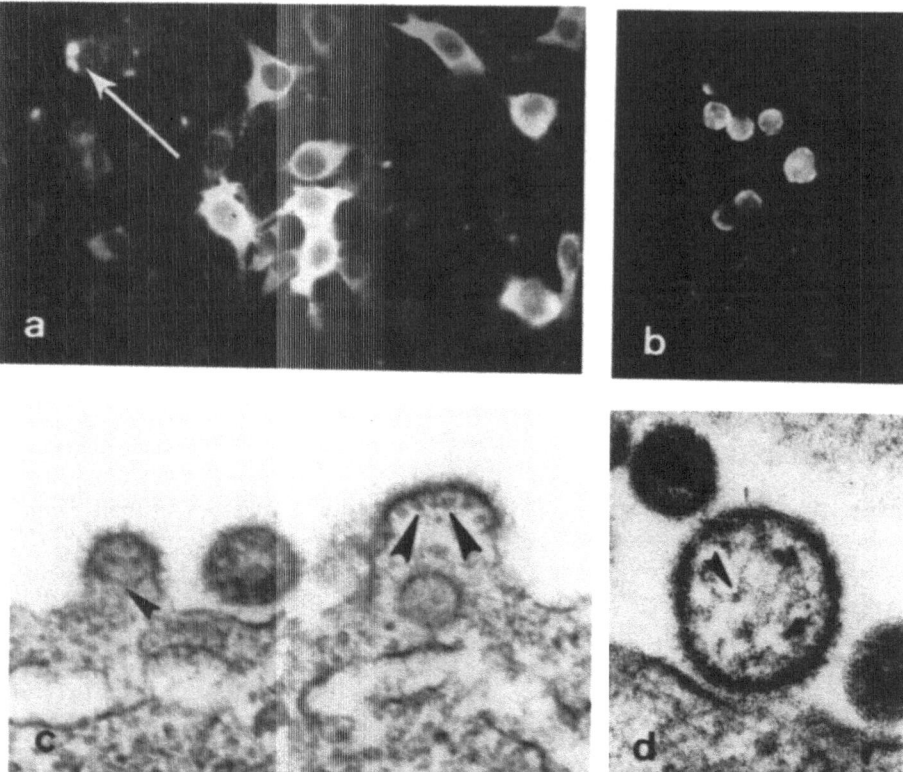

Fig. 6-3. Envelope glycoproteins and maturation of arenaviruses: In a and b, immunofluorescent staining of LCMV glycoproteins is detected in the cytoplasm (a) and on the surface (b) of infected L929 cells. In *(a)*, GP$_1$ staining of fixed cells is seen in the perinuclear area of some cells (arrow) and the entire cytoplasm in others. *(b)* GP$_1$ is detected on the surface of living cells. (c) and (d) show thin sections of Vero cells infected, respectively, with Pichinde (c) and Tamiami (d) viruses. In *(c)*, three different stages of budding at the plasma membrane are observed. A fine strand, possibly corresponding to NC, appears folded in one viral bud (small arrowhead). Dense granules are aligned under the envelope in the larger bud (large arrowheads). In *(d)*, a released virus is shown covered with spikes (short arrow) and containing rare granules and a fine filamentous structure 10 nm in diameter thought to be the NC (arrowhead). Magnifications: (c) 84,000; (d) 137,000. [Courtesy (a) and (b) of Dr. M. J. Buchmeier from Buchmeier *et al.*, 1981; reproduced with permission of Academic Press, New York; (c) of Drs. P. R. Young and C. R. Howard; (d) of Dr. F. A. Murphy, from Murphy and Whitfield, 1975; reproduced with permission of World Health Organization, Geneva]

ment of virions in thin sections (Dalton *et al.*, 1968). NC structure can rarely be identified in thin sections, probably because of the thin diameter of the NC strand. In some particles, however, one can distinguish a thin, loose strand (less than 20 nm in diameter), apparently connecting the granules (Fig. 6-3d). There is a large and unexplained discrepancy between the amount of filamentous NC seen in thin sections and by negative staining (see below). NCs with attached ribosomes have been isolated only rarely from Pichinde virus (Young and Howard, 1983), and there is probably no functional significance to this finding.

When virions disrupt spontaneously or are broken by osmotic shock or by detergent treatment, NCs up to 450 nm long appear either tightly coiled (helix

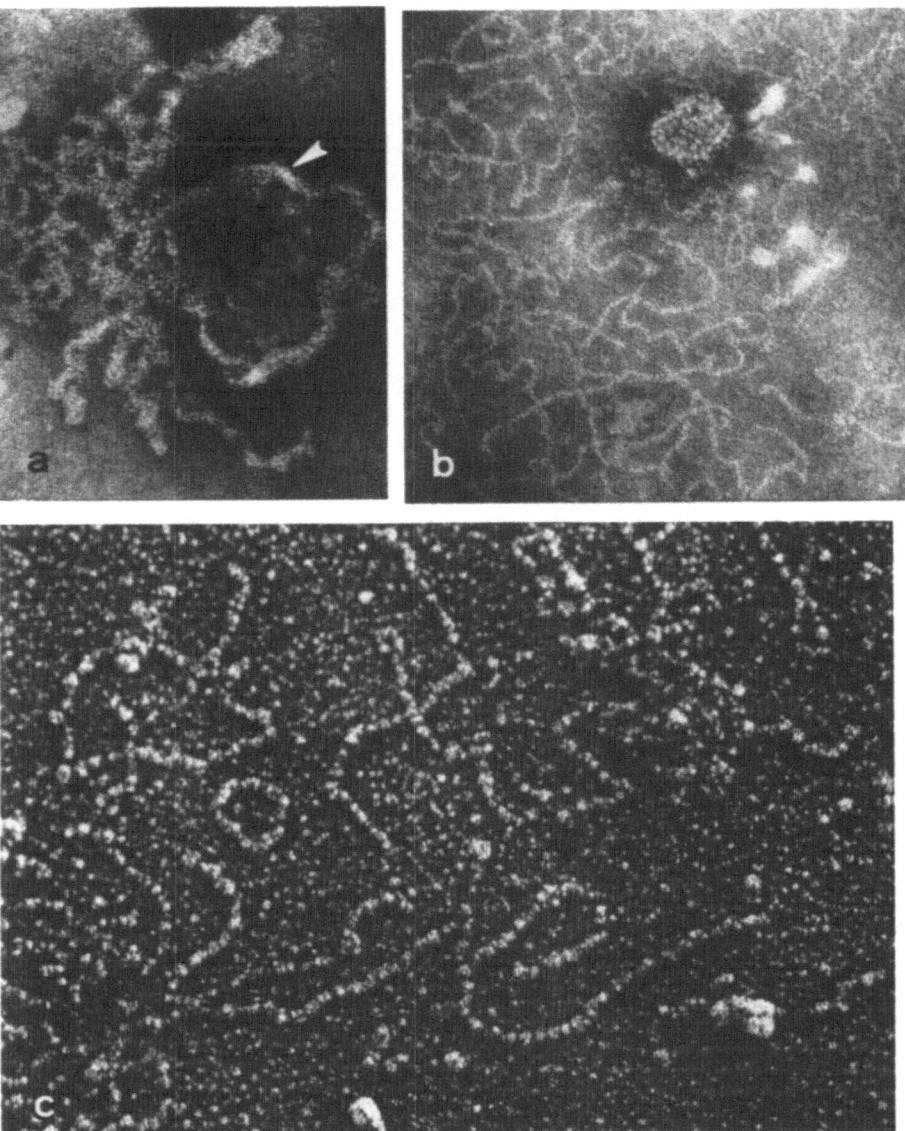

Fig. 6-4. Structure of the virion and NC of Pichinde virus, a prototype arenavirus. *(a)* Purified Pichinde virus lysed by osmotic shock and stained with phosphotungstic acid (pH 5). NC filaments, 12 to 15 nm in diameter, have been released. Envelope spikes appear club-shaped (arrowhead). *(b)* Extensive unfolding of these filaments reveal that NC is formed of a linear array of beads 4 to 5 nm in diameter and approximately 6 nm apart. This organization is reminiscent of the nucleosomes. *(c)* Purified NCs have been rotary-shadowed with gold-palladium at an angle of 10 degrees. They have the appearance of a string of beads. Linear arrays of 15 nm beads as well as isolated circles with various degrees of super-coiling are present. Magnification: (a), (b), (c) ×132,000. (Courtesy of Dr. P. R. Young and C. R. Howard, from Young and Howard, 1983; reproduced with permission of Cambridge University Press, England)

diameter, 20 nm) or as strands 9–15 nm in diameter with a beaded string appearance (Palmer *et al.,* 1977; Pedersen, 1979; Compans *et al.,* 1981; Young and Howard, 1983) (Fig. 6-4a, c). *Bunyaviruses* also have NCs with a beaded appearance (see Chapter 5). The beads may represent supercoiling or association between adjacent turns of the helix. Beads vary in size (5 to 15 nm), probably proportionally to the degree of supercoiling (Young and Howard, 1983). In unfolded NC, beads 5 nm in size form linear arrays and are reminiscent of the chromatin nucleosomes (Fig. 6-4b). The beads are larger and accentuated when NCs are reacted with monoclonal anti-N protein antibodies. Since circular strands were also found (Palmer *et al.,* 1977), it was proposed that the twisting of these circular NCs results in the formation of the 20 nm tightly coiled helix described above (Young and Howard, 1983).

The envelope spikes are club-shaped and 10 nm long (Fig. 6-4a). A hollow center in the spike may be seen when projections are viewed on end in negatively-stained preparations (Murphy and Whitfield, 1975). Spikes can be arranged in 5- and 6-fold symmetry, creating a regular lattice array (Young and Howard, 1983). The width of the envelope lipid bilayer appears increased compared to membranes of uninfected cells and the two leaflets are denser than usual (Murphy and Whitfield, 1975).

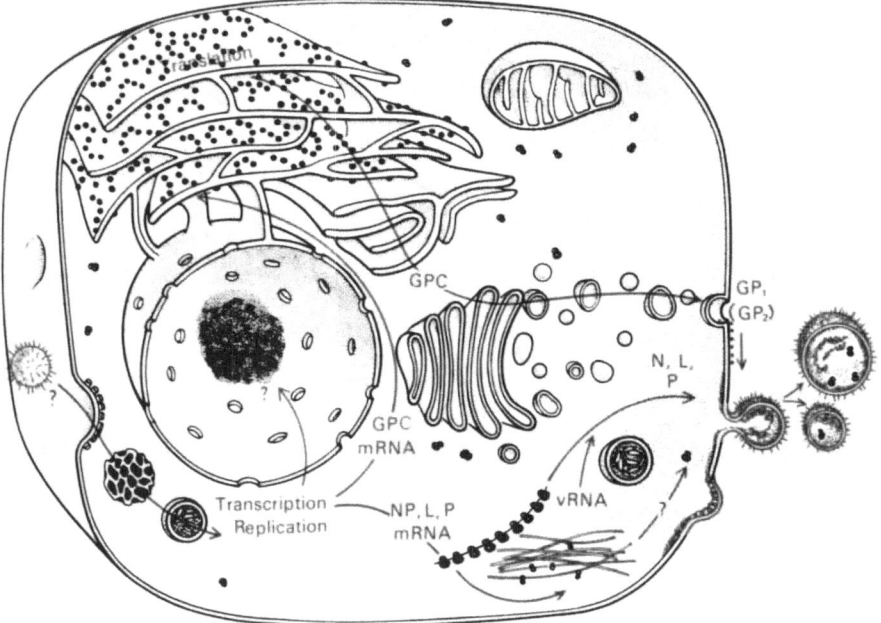

Fig. 6-5. Putative events of replication and assembly of arenaviruses. There are several question marks in this scheme. Arenaviruses may enter the cell in vacuoles or coated pits (Chanas *et al.,* 1980). Transcription of the two negative strand segments results in the synthesis of at least three mRNAs. Genome-associated proteins are made on free polysomes while glycoprotein mRNA is undoubtedly translated in the rough ER. The resulting precursor of the glycoprotein is probably transported to the Golgi and the cell surface in transport vesicles as is the case for glycoproteins of other viruses. How NCs are assembled and whether they are present in cytoplasmic inclusions is not known. NCs and ribosomes appear to align under the membrane during budding. The incorporation of ribosomes into the virion is a hallmark of arenaviruses

Understanding of the molecular interactions occurring during arenavirus assembly awaits further molecular analysis of the structure, function, and binding sites of all proteins involved in the process. The putative events of replication and assembly of *Arenaviridae* in the cell are summarized in Fig. 6-5. A model of arenaviruses is presented in Fig. 6-6 (Compans *et al.*, 1981).

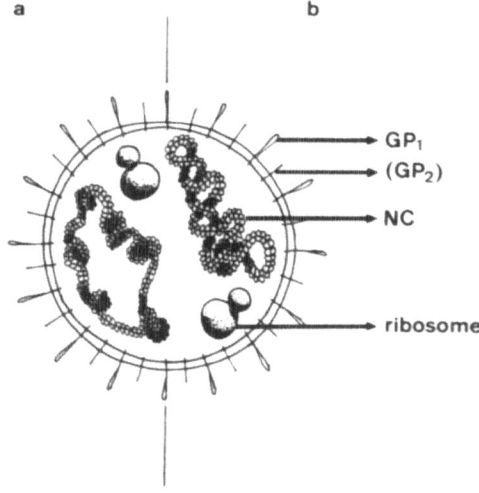

Fig. 6-6. A model of the organization of arenaviruses. Virions probably contain two NC species, one for the L segment and one for the S segment. NC may be circular and supercoiled in some places. They are clearly detected only when the virion is disrupted. The string of beads appearance is shown in *(a)* and the supercoiled NC in *(b)*. Only some virus strains have a second glycoprotein (GP₂) and, in this case, it is not known whether the two glycoproteins reside in one spike or two different spikes (modified from Compans *et al.*, 1981)

7
Assembly of *Coronaviridae*

I. Introduction

Coronaviridae are enveloped RNA viruses which mature by budding into intracytoplasmic membranes. Coronaviruses cause respiratory and/or enteric infection in humans and many domestic animals (reviewed by Wege *et al.*, 1982). The prototype coronavirus is avian infectious bronchitis virus (IBV, Tyrrell *et al.*, 1978). *Coronaviridae* exhibit rather fastidious requirements for the species and tissue types which they will infect. Because of the difficulty of isolating coronaviruses, most of

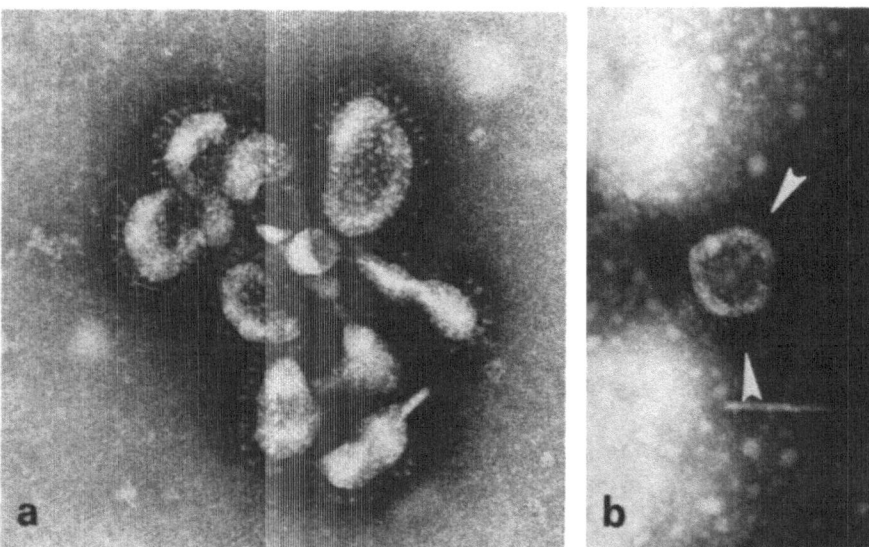

Fig. 7-1. Negatively stained coronavirions. *(a)* shows a cluster of human respiratory coronaviruses with the large, widely-spaced peplomers or spikes that form the "corona" around the virion. *(b)* illustrates a human enteric coronavirus with delicate peplomers (arrowheads). Such viruses have frequently been visualized in preparations of human fecal material, but they are very difficult to propagate *in vitro*. Magnifications: (a) ×110,000, (b) ×90,000. [Courtesy of Dr. L. Oshiro, (a), and Dr. O. Caul, (b)]

them were first classified by their virion morphology in negatively stained preparations (Tyrrell *et al.*, 1975). The virions are characterized by large, club-shaped peplomers or spikes about 20 nm long and 7 nm wide at their tips (Almeida and Tyrrell, 1967; Tyrrell *et al.*, 1968; Fig. 7-1). Avian coronaviruses vary in diameter from 70 to 120 nm, but murine coronaviruses have more uniform diameters, approximately 90 nm. More recently, common features of virus replication and biochemistry have confirmed the assignment of coronaviruses into a single virus family (Tyrrell *et al.*, 1978; ter Meulen *et al.*, 1981). There are at least 3 subgroups of *Coronaviridae*, based on their mutual lack of antigenic cross-reactivity (Pedersen *et al.*, 1978; Sturman and Holmes, 1983). Avian coronaviruses, such as infectious bronchitis virus (IBV), do not share antigenic determinants with mammalian coronaviruses. Mammalian coronaviruses fall into two distinct groups: one includes mouse hepatitis virus (MHV), bovine coronavirus (BCV), hemagglutinating encephalomyelitis virus of swine (HEV) and human respiratory coronavirus OC43; and a second includes human respiratory coronavirus 229E, transmissible gastroenteritis virus of swine (TGEV), canine coronavirus (CCV), and feline infectious peritonitis virus (FIP). There are also coronaviruses antigenically unrelated to these three major subgroups, such as porcine enteropathic coronavirus, CV777 (Pensaert *et al.*, 1981).

II. Molecular Organization

We will summarize the molecular composition of coronavirions and events in coronavirus replication and then consider the morphology of the virions and their budding mechanisms in detail.

Transcription

The coronavirus genome is a single-stranded molecule of RNA of M.W. 5.4 to 6.9×10^6 which is capped at its 5′ end and polyadenylated at its 3′ end (Lai and Stohlman, 1978; Stern and Kennedy, 1980 a; Siddell *et al.*, 1982; Fig. 7-2). The isolated genomic RNA is infectious and serves as an mRNA within the infected cell. Thus, coronaviruses are positive-stranded RNA viruses. The genomes of IBV and MHV have been mapped by comparing oligonucleotides of genomic RNA with oligonucleotides of 3′ co-terminal, polyadenylated RNAs of different lengths generated by RNAse digestion of virion RNA. These maps have then been compared with the oligonucleotide patterns of the subgenomic mRNAs found in infected cells (Stern and Kennedy, 1980 b, Lai *et al.*, 1981).

There are 5 subgenomic mRNAs for IBV, and 6 for MHV. The mRNAs of coronaviruses are unusual in that they form a nested set of molecules of varying length which share a common, polyadenylated 3′ end (Stern and Kennedy, 1980 a and b; Weiss and Leibowitz, 1981; Cheley *et al.*, 1981; Fig. 7-2). Except for a short sequence of nucleotides at the 5′ end of each mRNA, the oligonucleotides of each mRNA are completely identical to the 3′ end of the next larger mRNA species (Lai *et al.*, 1982 a).

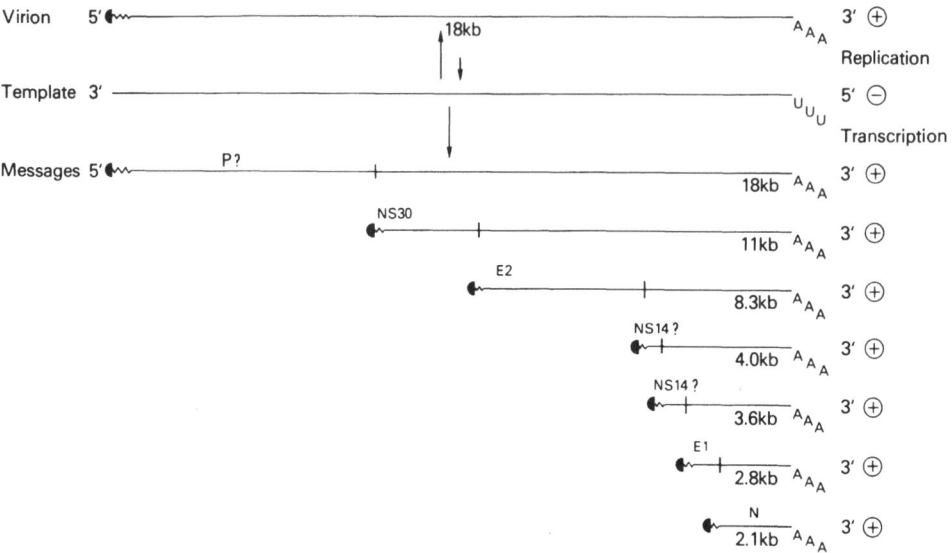

Fig. 7-2. Scheme of the transcription and replication of coronaviruses, using mouse hepatitis virus as a model. The 18 kb positive-stranded genomic (virion) RNA serves as a messenger RNA in the cell to direct the synthesis of RNA-dependent RNA polymerase which uses the genomic RNA as a template to make a full-length negative-stranded RNA. From this template, new 18 kb genomic RNA and 6 sub-genomic mRNAs are made; all species form a nested set with common 3' ends. All of these positive-stranded RNA species are capped (◆———) and polyadenylated (AAA). A short leader sequence, indicated by the wavy line, is found on the 5' end of each mRNA but occurs only once on the genomic strand. For each mRNA, only the gene at the 5' end (to the short vertical line) is translated; the protein product of each mRNA is indicated by the letters above the 5' gene. The two smallest mRNAs have been cloned and sequenced. (Adapted from Sturman and Holmes, 1983)

The 5' end of each mRNA consists of a cap plus a short common nucleo-tide sequence referred to as the leader sequence (Lai *et al.*, 1983).

In vitro translation of the 6 subgenomic mRNAs of MHV has demonstrated that each species codes for only one polypeptide and that the translated gene is at the 5' end of the mRNA (Siddell *et al.*, 1980, 1981 c; Leibowitz and Weiss, 1981; Leibowitz *et al.*, 1982; Rottier *et al.*, 1981 a). A tentative map based upon these studies is shown in Fig. 7-2.

The structural proteins of different coronaviruses appear to follow a common pattern, and the terminology for MHV proteins will be used here. There are gener-ally three structural polypeptides: a nucleocapsid protein N, and two envelope gly-coproteins, E1 and E2 (Garwes, 1980; Siddell *et al.*, 1982; Sturman and Holmes, 1983 a).

Replication

Coronavirus replication has recently been the subject of several comprehensive re-views (ter Meulen *et al.*, 1980; Siddell *et al.*, 1982 and 1983; Sturman and Holmes, 1983 a; Rottier et al., 1984), and the reader is referred to these for details which are beyond the scope of this chapter. The major features of coronavirus replication and transcription will be summarized here briefly.

Because of the positive-stranded nature of its genome, the coronavirion contains no RNA-dependent RNA polymerase. This enzyme is synthesized in cells soon after infection and first directs the synthesis of full-length negative strands (Brayton *et al.*, 1982; Dennis and Brian, 1981, 1982; Lai *et al.*, 1982 b). Synthesis of negative strands is complete within 5–6 hours after infection; it is not clear how continued synthesis of negative strands is inhibited. From this template, the RNA-dependent RNA polymerase directs the synthesis of genomic RNA and 5 subgenomic mRNAs (IBV; Stern and Kennedy, 1980a and b) or 6 subgenomic mRNAs (MHV; Spaan *et al.*, 1981, 1982; Wege *et al.*, 1981). A cap and the same short leader nucleotide sequence are added to the 5′ end of each mRNA, either by RNA splicing or by allowing the common leader sequence to serve as a primer for synthesis of each mRNA (Baric *et al.*, 1983). The primer model is supported by the observation that caps and leaders are found on nascent mRNA strands (Lai *et al.*, 1983 b), but no free leader RNA has yet been isolated from infected cells.

Two distinct peaks of RNA polymerase activity occur during virus replication (Brayton *et al.*, 1982). The first involves primarily synthesis of the full-length negative-stranded template, and the second coincides with synthesis of new genomic RNA and mRNAs. The virus protein species responsible for polymerase activity have not yet been identified.

Throughout the replicative cycle, the ratios of the different subgenomic RNAs remain constant (Stern and Kennedy, 1980a; Wege *et al.*, 1981; Spaan *et al.*, 1981; Leibowitz *et al.*, 1981). Each of the mRNAs appears to direct the synthesis of a single viral polypeptide (Fig. 7-2).

Virus Proteins

Table 7-1 summarizes the properties of the 3 structural proteins of MHV. The functions of the non-structural proteins of corona-viruses are not known.

1. *N protein* (MW: 50 K). The NC protein, N, is phosphorylated (Stohlman and Lai, 1979). With the positive-stranded viral genomic RNA, the N protein forms a helical NC which is RNAse-sensitive. Regulatory or enzymatic functions of N have not been identified.

2. *E1 glycoprotein* (MW: 23 K). E1 is an unusual virus glycoprotein in several ways. It has 3 domains: external, intramembranous, and cytoplasmic (Sturman and Holmes, 1977, 1983). The short amino-terminal external domain (about 5 K) contains all of the E1 oligosaccharides (Sturman and Holmes, 1977). In MHV, short oligosaccharides are O-glycosidically linked to serine or threonine residues of E1 (Niemann *et al.*, 1982), unlike the complex oligosaccharides N-glycosidically linked to asparagine residues which are found on nearly all other virus glycoproteins (Sharon and Lis, 1981). The E1 glycoprotein of IBV contains only N-linked oligosaccharides (Stern *et al.*, 1982; Stern and Sefton, 1983). The significance of this diversity of glycosylation patterns among coronaviruses is not known.

The E1 molecule is very hydrophobic, particularly when its external domain has been removed. Like a small number of integral membrane proteins, it tends to

Table 7-1. *Proteins of coronavirus MHV*[1]

Name	Coded by mRNA Number	Apparent MW $\times 10^{-3}$	Known or putative function
N	7	50–60	NC protein
E1	6	23	Membrane glycoprotein Determines budding site? Binds to NC
E2	3	180	Peplomeric glycoprotein Binds to cell-surface receptors Causes cell fusion (protease activated) Elicits neutralizing antibody Elicits natural cell-mediated cytotoxicity
NS14	4 or 5	14	?
NS30	2	30–35	?
NS200	1	200	Polymerase?

[1] After Sturman and Holmes (1983).

self-aggregate spontaneously and forms dimers, trimers, and tetramers when boiled with SDS and β-mercaptoethanol (Sturman and Holmes, 1977). Recent evidence from cloning of the E1 genes of MHV and IBV indicates that there are two long stretches of hydrophobic amino acids in E1, suggesting that E1 may traverse the lipid bilayer several times (Armstrong *et al.* and Cavanagh *et al.*, personal communication). Within infected cells, E1 accumulates within the Golgi apparatus (Holmes *et al.*, 1981 a, and 1984). We have suggested that E1 functions as a matrix glycoprotein for coronaviruses (Holmes *et al.*, 1981 a, and 1984; Sturman and Holmes, 1983 a). Thus it interacts with NCs by its cytoplasmic domain(s), determines the intracellular localization of virus budding, and, with the lipid bilayer, forms the virus envelope.

3. *E2 glycoprotein* (MW: 180 K). E2 forms the large peplomers or spikes characteristic of coronavirions. E2-molecules can be removed from virions by protease digestion (Garwes and Pocock, 1975; Sturman and Holmes, 1977), although a small hydrophobic domain is believed to remain embedded in the virus envelope. It is not certain whether this anchor region penetrates the membrane completely to form a cytoplasmic domain. E2 probably interacts with E1 and/or N during virus budding.

E2 contains a large amount of N-linked oligosaccharide (Sturman, 1981; Niemann and Klenk, 1981). E2 is acylated, containing covalently bound palmitic acid (Niemann *et al.*, 1982; Schmidt, 1982 a), presumably located at or near the lipid bilayer (Schmidt, 1982 a and b). E2 is proteolytically cleaved at a late stage in virus maturation (Sturman and Holmes, 1983, and 1984; Holmes *et al.*, 1984). In MHV-A59, cleavage appears to be dependent upon the host cell and yields two molecules, 90 A and 90 B, which comigrate in electrophoresis in SDS-containing polyacrylamide gels with an apparent molecular weight of 90 K. These have different amino acid compositions and only 90A contains covalently bound palmitic acid. Cleavage of the large peplomeric glycoprotein of IBV in chicken cells may be very efficient, since only the two cleavage products of E2, gp 93 and gp 84, are found on the virion (Cavanagh, 1981; Wadey and Westaway, 1981; Stern et al., 1982).

E2 is a large, multifunctional molecule. It is responsible for virus-induced cell fusion (Holmes et al., 1981 b; Collins et al., 1982) and its cleavage is required for cell fusion (Sturman and Holmes, 1983, and 1984). E2 on the plasma membrane renders cells subject to cell-mediated cytotoxicity (Welsh et al., 1983). It induces neutralizing antibody (Garwes et al., 1976, 1978/79; Hasony and Macnaughton, 1982; Schmidt and Kenny, 1981, 1982) and binds the virion to receptors on the plasma membranes of susceptible cells (Holmes et al., 1981 b). E2 also plays an important role in the pH-dependent thermolability of coronaviruses (Sturman, 1981). The anchoring region of E2 probably interacts with E1 in the virus envelope.

III. Synthesis, Transport and Processing of Virus Proteins

The synthesis of N protein is directed by the smallest mRNA, which is derived from the 3' end of the viral genome (Siddell et al., 1980; Rottier et al., 1981 a). Synthesis of N occurs on free polysomes, although even in very short pulse-labeling experiments some N is found on RER and smooth membranes isolated from homogenized cells (Frana and Holmes, unpublished observations). Thus, some mRNAs on membrane-bound polysomes may direct the synthesis of N protein. Immunofluorescence studies with anti-N antibody reveal delicate flecks of fluorescence in the cytoplasm by 2 to 3 hours after virus inoculation. These foci increase in size and number during infection.

The N protein is phosphorylated at serine and threonine residues (Stohlman and Lai, 1979). Although protein kinase activity has been demonstrated in the virion (Siddell et al., 1981 a), it is not yet clear whether this activity represents a virus-specific or cellular enzyme. The role (if any) of phosphorylation and dephosphorylation in modulation of the biological functions of N is not known. Pulse-chase studies show that although large amounts of N are synthesized in the infected cell, only a very small fraction of N is chased out of the cell into mature virions (Holmes et al., 1984; Rottier et al., 1981 b; Siddell et al., 1981 b). This suggests that N protein may have several functions during virus replication in addition to formation of NCs. Only one species of N appears in the virion of each coronavirus strain. However, even between closely related strains of MHV the apparent molecular weight of the virion-associated N varies, ranging from 50 K to 60 K (Siddell et al., 1982). Whether this difference is due to phosphorylation has not been determined. Several species of N which migrate more rapidly than the virion N have been detected within infected cells late in the infectious cycle or following immunoprecipitation (Cheley and Anderson, 1981). Their functions are unknown.

The synthesis of the peplomeric glycoprotein E2 (Rottier et al., 1981 a; Siddell et al., 1981 b; Leibowitz et al., 1981), encoded in mRNA 3 occurs on RER membranes (Niemann et al., 1982). By immunofluorescence, E2 can first be detected as fine cytoplasmic granules at 4 hours after infection. The glycoprotein then disperses rapidly throughout the cytoplasm, presumably upon intracytoplasmic membranes (Holmes *et al.,* 1981 b). After 7 hours, delicate flecks of fluorescent staining are observed on the plasma membrane. These gradually spread to cover the cell surface.

The first species of E2 to be detected in pulse-label experiments is a 150 K form (Siddell *et al.,* 1981 b) which is cotranslationally glycosylated at asparagine residues by transfer of core oligosaccharides from dolichol-phosphate intermediates. This form of E2 is sensitive to endoglycosidase H. It is transported to the Golgi apparatus where further trimming and glycosylation of the N-linked oligosaccharides and acylation occur under the direction of cellular enzymes (Niemann *et al.,* 1982). Pulse-chase studies show that in the 17 Cl 1 line of BALB/c mouse fibroblasts, synthesis of E2 is well-balanced with its release into virions (Holmes *et al.,*1981 b); labeled E2 is quantitatively chased into released virions within 2 hours. At least early in infection, there is no large excess of E2 in the cell. At a late stage in the processing of E2, when or just before it reaches the plasma membrane, proteolytic cleavage by a trypsin-like cellular enzyme occurs (Holmes *et al.,* 1984). In tunicamycin-treated cells, E2 is synthesized as a 120 K non-glycosylated polypeptide which is not incorporated into virions (Niemann and Klenk, 1981).

The matrix glycoprotein, E1, of MHV-A 59 is synthesized on membrane bound ribosomes as a non-glycosylated 20 K polypeptide (Niemann *et al.,* 1982). It is transported to the Golgi, where cell fractionation studies show that it is post-translationally glycosylated by the addition of several sugars to serine or threonine residues. This process of O-linked glycosylation, which is probably done by cellular enzymes, is not well understood. Each molecule of E1 receives from 1 to 3 oligosaccharide chains.

Immunofluorescence studies with monospecific or monoclonal antibodies to E1 and E2 have shown that the intracellular transport of the two coronavirus glycoproteins is different. E2, like the peplomeric glycoproteins of most other enveloped viruses, is transported from the RER through the Golgi to the plasma membrane. In contrast, the E1 glycoprotein is transported only as far as the Golgi apparatus, where it accumulates during virus infection (Fig. 7-3A and B). This has been demonstrated by simultaneously staining E1 with immunofluorescent antibody and marking the *trans* cisternae of the Golgi by a histochemical reaction for thiamine pyrophosphatase (Sturman and Holmes, 1983; Doller *et al.,* in preparation). It appears likely that the restricted intracellular transport of the E1 glycoprotein may account for the intracellular budding site of coronaviruses (Holmes *et al.,* 1981 b; Sturman and Holmes, 1983). It is interesting to speculate whether other virus groups which mature by budding from intracellular membranes, such as bunya- and flaviviruses may show a similarly restricted pattern of intracellular transport of a NC-binding protein.

Coronaviruses cause cell fusion *in vitro* and *in vivo.* This cell fusion appears to be mediated by E2 glycoprotein on the plasma membrane of infected cells, since incubation in the presence of anti-E2 antibody prevents virus-induced cell fusion (Holmes *et al.,* 1981 b; Collins *et al.,* 1982). Fluorescent antibody staining and immunoelectron microscopy (Figs. 7-3 C-E) show that the E2 glycoprotein is on the plasma membrane in large amounts late in the infectious cycle, whereas, as noted above, relatively little E1 migrates to the plasma membrane (Holmes *et al.,* 1981 b). E2 cleavage appears to be dependent upon the host cell, since virions from different cell types show different degrees of cleavage (Holmes *et al.,* 1983).

Proteolytic cleavage of E2 glycoprotein is required for coronavirus-induced cell fusion (Sturman and Holmes, 1983 and in press; Holmes *et al.,* in press). This was shown by experiments in which concentrated, purified MHV did not cause rapid

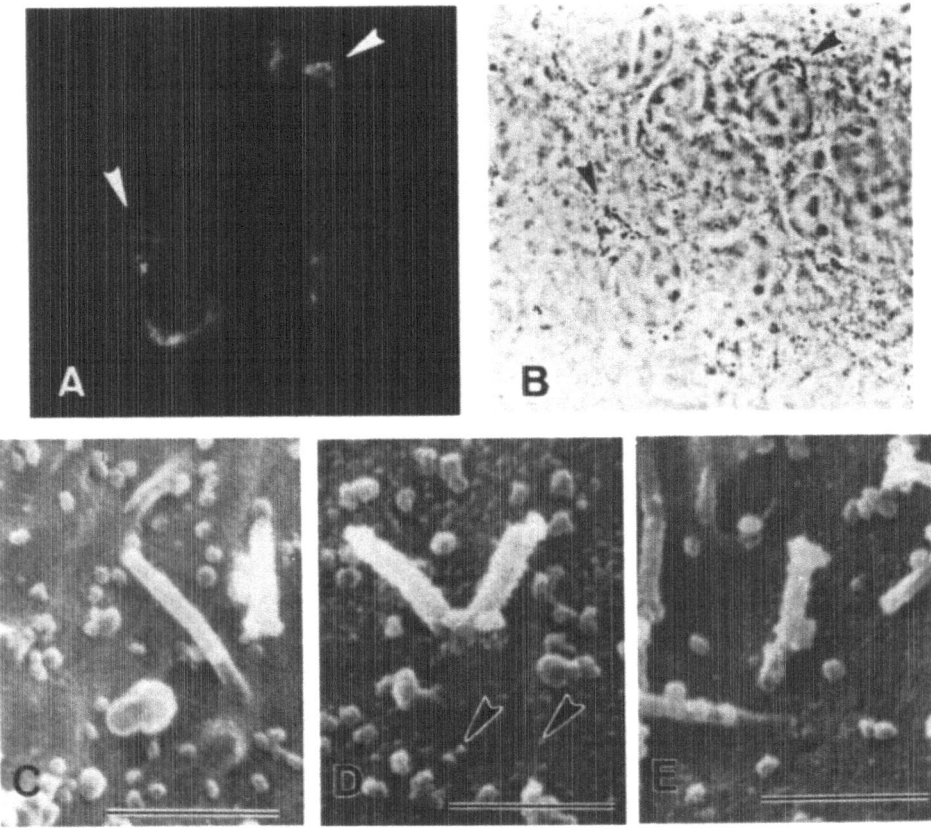

Fig. 7-3. Differential intracellular localization of the two coronavirus envelope glycoproteins. *(A)* and *(B)* show that E1 accumulates within the Golgi apparatus. Cells infected at a low multiplicity with MHV were fixed in formaldehyde at the end of the virus latent period, permeabilized with detergent and labeled with monospecific anti-E1 antibody and fluorescent anti-rabbit IgG (A). The same cells were reacted with a cytochemical marker for the Golgi apparatus, thiamine pyrophosphatase (B). The E1 antigen in the cells of (A) (arrowheads) is located in the Golgi region shown in (B) (arrowheads). E2, the peplomeric glycoprotein, does not accumulate in the Golgi, but is transported to the plasma membrane, as shown in *(C)* to *(E)*. MHV-infected cells 24 hours after infection were fixed and labelled with normal rabbit serum (C), or monospecific anti-E2 antibody (D), or anti-E1 antibody (E) followed by *Staphylococcus aureus* protein A conjugated to peroxidase. The presence of antigen is shown by development of a granular reaction product following reaction with hydrogen peroxide and 3,3′ diaminobenzidine. The cells were coated with metal and examined by scanning electron microscopy. E2 is transported to the plasma membrane as shown by the granular labeling in (D) (arrowheads), but E1 is not transported to the plasma membrane, because panel (E), like the normal rabbit serum control, (C), shows no granular reaction product on the plasma membrane. Magnifications: (A) and (B) × 960; (C) to (E) × 23,000. (Courtesy of Dr. E. Doller)

fusion of susceptible cells in the presence of cycloheximide, whereas trypsin-treated, purified and concentrated MHV caused rapid cell fusion. (The only effect of trypsin on the virions was the quantitative cleavage of E2 180 to E2 90). Activation of plaque formation, hemagglutination and/or virus cytopathic effect by trypsin have been shown for several coronaviruses (reviewed in Sturman and Holmes, 1983). Whether cleavage of the E2 glycoprotein is required for virus infectivity is not yet clear, however, since at present there is no method available to obtain homogeneous populations of virions with uncleaved E2. Some coronaviruses which have been difficult to propagate *in vitro,* such as human enteric coronaviruses (Fig. 7-1b), may undergo a single cycle of infection yielding large numbers of non-infectious virions, as shown in negatively stained preparations (Caul and Egglestone, 1977). It may be necessary to identify a cell type in which cleavage of E2 occurs or to identify a different protease which may activate infectivity in order to propagate these fastidious viruses *in vitro.*

IV. Assembly of Virus Components

In this section, we will review the assembly of the A59 strain of MHV, since it has been studied most extensively. Where other coronaviruses differ from this model, this will be noted in the text.

NC Assembly

The specific interaction of N with genomic RNA has not yet been studied. It is likely that only genomic RNA is encapsidated, since encapsidation of mRNAs would interfere with their translation. If this is true, we postulate that an encapsidation signal is present near the 5' end of the genomic RNA, since all other regions are shared by subgenomic mRNAs.

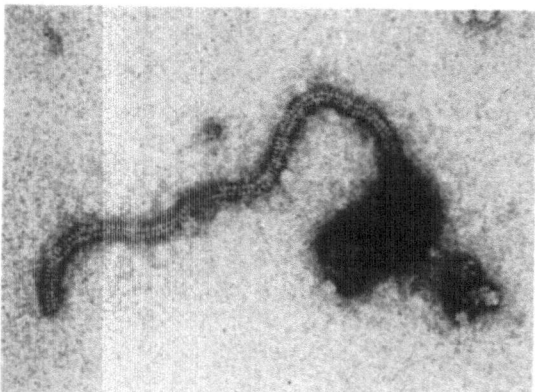

Fig. 7-4. Helical NC from the coronavirion. A long strand of helical NC purified from a human coronavirus by density gradient ultracentrifugation is shown in a negatively stained preparation. Such images have been difficult to obtain with many coronaviruses. Magnification: × 205,000. (Courtesy of Dr. O. Caul, from Sturman and Holmes, 1983, with permission of Academic Press)

The NC of coronaviruses is helical, but the turn-to-turn bonding of the NC strands appears to be very weak. The tubular appearance of the NC is readily observed in thin sections of budding virions (David-Ferriera and Manaker, 1965; Becker *et al.*, 1967; Oshiro *et al.*, 1971), but is not often apparent in released virions, suggesting that it may undergo a conformational change after virus budding and release (Holmes and Behnke, 1981). Intracytoplasmic inclusions of coronavirus NCs were not recognized for many years, perhaps in part due to the great flexibility of the NC strands. Large inclusions of NCs have been observed in coronavirus-infected cells late in the infectious cycle (Caul and Egglestone, 1977). NCs with a density of 1.25 to 1.30 g/cc have been isolated from virions or from infected cells and shown to consist of genomic RNA and N protein with or without E1 (Kennedy and Johnson-Lussenberg, 1975/76; Sturman *et al.*, 1980). Coronavirus NCs are apparently not tightly coiled, so that isolated NCs often appear as thin, kinky strands rather than helical coils. Only when virions are gently or spontaneously disrupted, is it possible to observe or to purify long helices of NC with a diameter of 9 to 11 nm (Macnaughton *et al.*, 1978; Davies *et al.*, 1981; Caul *et al.*, 1979; Fig. 7-4).

Interaction of NC and Envelope Proteins and Virus Budding

Coronaviruses mature by budding into the lumina of RER or Golgi cisternae (Figs. 7-5 and 7-6). Virions form when helical NCs align under regions of intra-

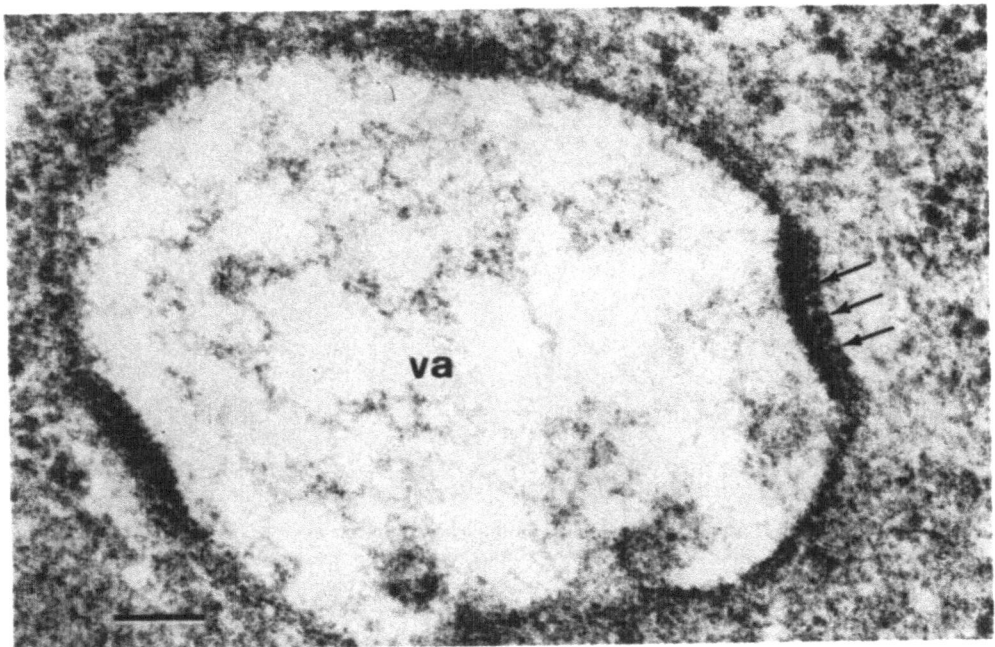

Fig. 7-5. Association of the virus NC with peplomers at smooth-walled intracellular membranes. Regions where virus buds are forming are visible at each side. Arrows indicate cross-sections of the helical NC which has formed a regularly spaced array under the membrane. The fuzzy coat on the lumenal surface probably represents virus peplomers. Magnification: ×130,000; Bar=100 nm. (Courtesy of Dr. S. Dales, from Massalski *et al.*, 1982, with permission from S. Karger AG, Basel)

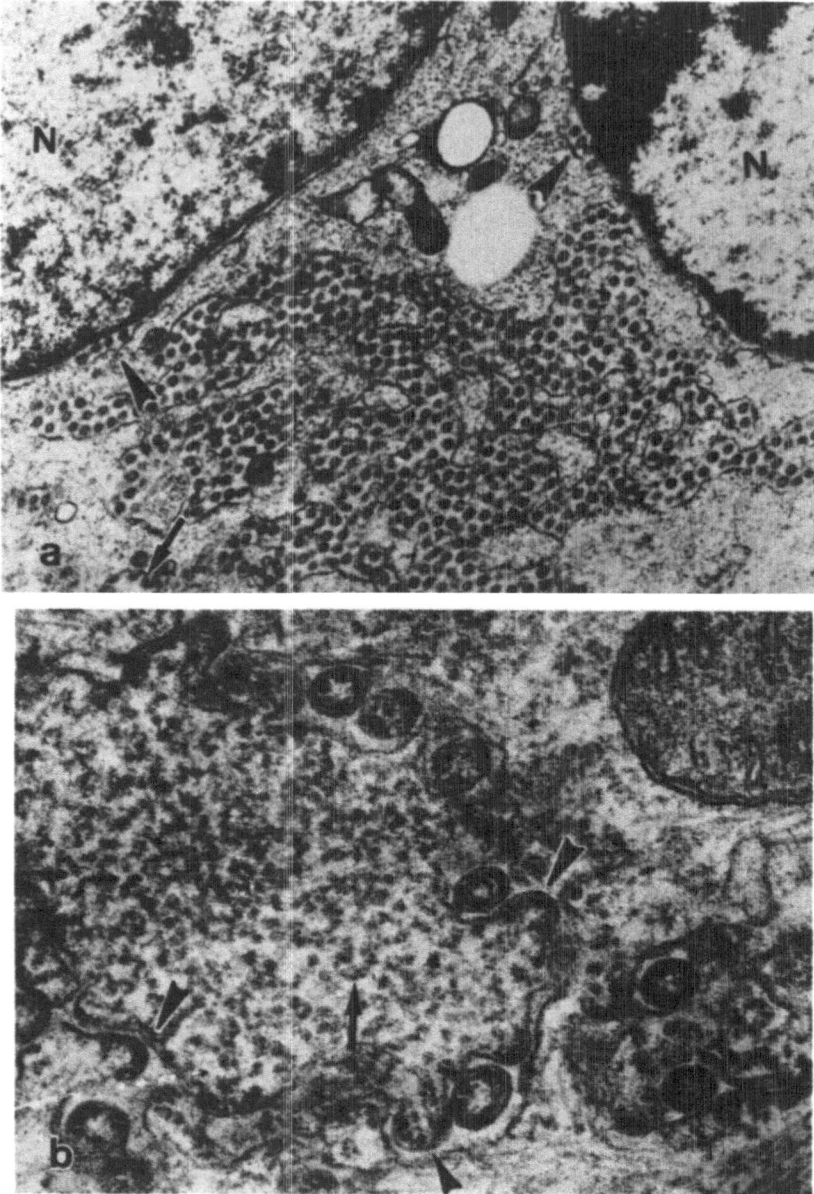

Fig. 7-6. A binucleate cell infected with MHV in which large numbers of virions have accumulated in the lumen of the RER, where a budding virus is seen (arrow). The virions are also commonly seen in the lumen of the nuclear envelope (arrowheads). In *(b)*, MHV virions bud (arrowheads) from an area of cytoplasm containing numerous strands of NC (arrow). The centers of the virions are electron-lucent, since the NC is attached to the virus membrane. Magnifications: (a) ×15,000; (b) ×80,000. [(a) Courtesy of J. N. Behnke; (b) from Dubois-Dalcq *et al.*, 1982, with permission from Academic Press]

cellular membranes containing virus glycoproteins (David-Ferriera and Manaker, 1965; Oshiro, 1973; Massalski *et al.*, 1981, 1982; Ducatelle, 1981; Dubois-Dalcq *et al.*, 1982). NCs may associate with the forming envelope by interacting with the cytoplasmic domain of the E1 glycoprotein (Holmes *et al.*, 1981 b). Virions form and are released from cells treated with tunicamycin, although these cells contain little E2 and such virions are completely devoid of E2 (Holmes *et al.*, 1981 a). Thus E2 is evidently not required for the formation of coronavirions. Virions synthesized in the presence of tunicamycin contain normal amounts of N and fully glycosylated E1, but are not infectious, probably because, lacking E2 peplomers, they cannot attach to virus receptors.

Virions in the RER appear to have large peplomers on their envelopes (Chasey and Alexander, 1976). An important but still unanswered question is whether the glycoproteins E1 and E2 are processed on intact virions as these migrate through the Golgi apparatus, or whether the glycoproteins are fully processed first and then assembled into virions. For the latter to be true, budding of virions from the RER would require migration of glycolysated E2 from the Golgi back to the RER, but such retrograde transport has never been demonstrated. This question can be resolved by isolation of virions from intracellular compartments and characterization of their structural proteins.

Virions are released from membranes by pinching off into the lumina of the RER or Golgi apparatus. Large numbers of virions may accumulate within the lumen of the RER (Fig. 7-6a). In the RER, the virions are spherical particles with electron-lucent centers and NC strands beneath the envelope (Fig. 7-6a and b). Release of virions into the lumen apparently does not involve cellular actin, since actin has not been detected routinely in coronavirions, with the possible exception of IBV (Lomniczi and Morser, 1981).

Release of Virions from the Cell and Post-Release Maturation

Following their release into the RER and Golgi, coronavirions appear to escape from the cell within smooth-walled vesicles that migrate to the cell membrane and fuse with it (Doughri *et al.*, 1976; Oshiro *et al.*, 1971). Time lapse cinematography of MHV-infected cells shows that during the period of maximum virus release, cells are not lysed, although they may undergo fusion (Holmes, unpublished observation). This process of release is similar to secretion of zymogens from pancreatic acinar cells (Holmes *et al.*, 1981 a). Thus, coronaviruses apparently depend on several specialized host cell functions for virus maturation and release, including two mechanisms of protein glycosylation, two pathways of intracellular transport of glycoproteins, and a cellular secretion pathway.

After coronaviruses are released from the cell, many virions remain adsorbed to the plasma membrane (Oshiro *et al.*, 1971; Oshiro, 1973; Doughri and Storz, 1977; Sugiyama and Amano, 1981; Fig. 7-7a–c). It is not clear why such large numbers of virions bind to infected cells. Are they simply virions adherent to the smooth-walled vesicles which have not detached from the membrane following fusion of the vesicle with the plasma membrane? Or were these virions first released into the

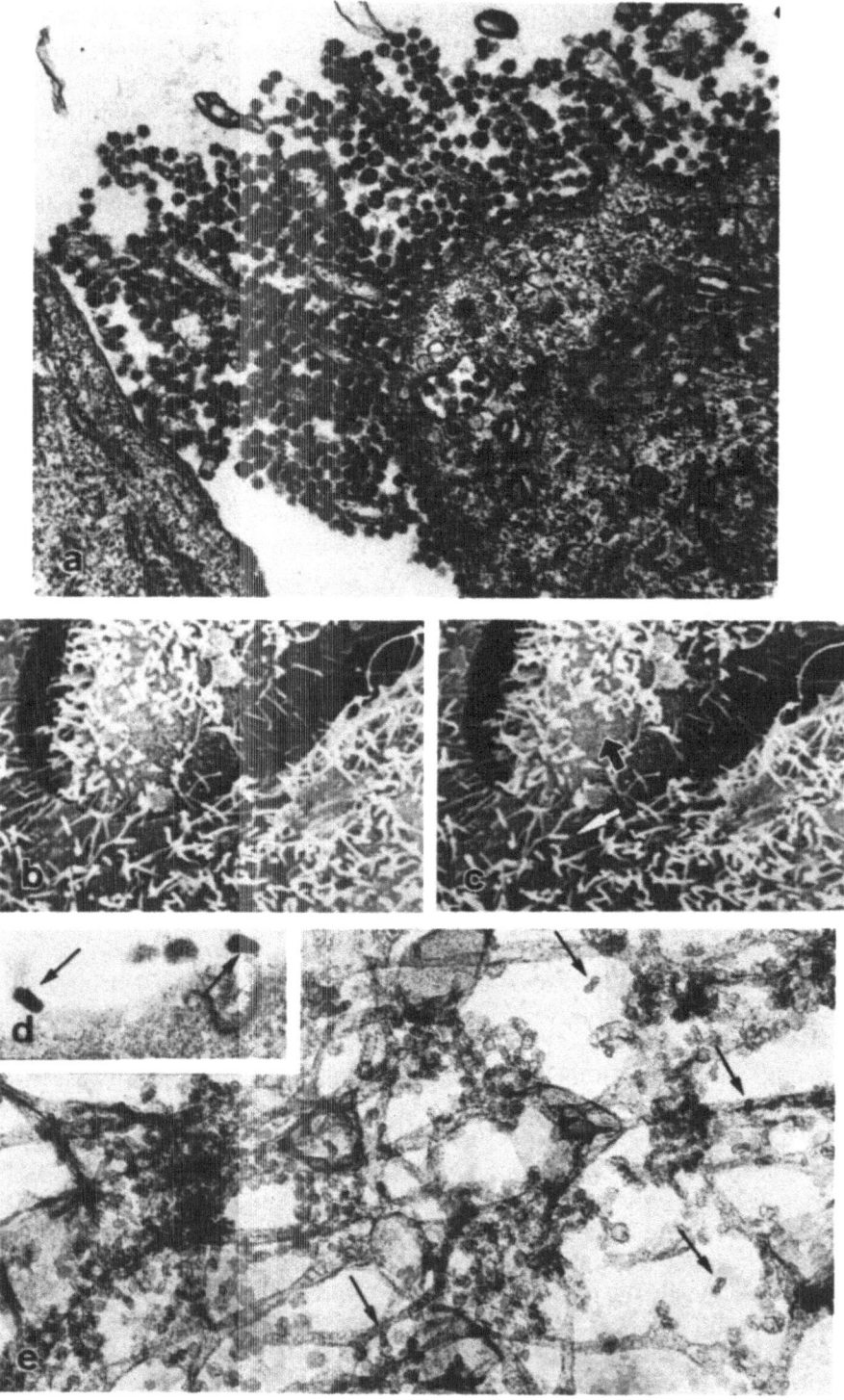

medium and then bound to receptors on the plasma membrane? If the latter is true, then why are there very different numbers of virions bound to cells in the same culture (Fig. 7-7b and c)? Do the virions bind to cellular receptors which are differently expressed in different cells, possibly due to cell cycle differences, or do they bind to a virus glycoprotein such as E2, which is present on the plasma membrane?

As virions of the A59 strain of MHV are transported through the Golgi-associated vesicles in some cell types, they become flattened, disc-shaped and electron-dense (Holmes *et al.*, 1981a; Fig. 7-7d and e). Possibly, these changes could be associated with release of the NC from its attachment to the envelope. They might be triggered by changes in pH or ionic concentrations in different intracellular membrane compartments. Alternatively, they may reflect processing of virus components which takes place as virions migrate through different intracellular compartments. For instance, since cleavage of the 180 K E2 to 90 K E2 appears to be an event which occurs immediately prior to release of the virions (Holmes *et al.*, 1984), it is possible that the condensation of virions may be triggered by that event. Since proteolytic cleavage of E2 appears to be dependent upon the host cell, this might explain why post-release condensation of virions is not seen for all coronaviruses or even for MHV-A59 in all cell types.

Many enveloped viruses show polarized budding from epithelial cells (Rodriguez-Boulan and Sabatini, 1978; Chapter 1). Since coronaviruses do not bud from the plasma membrane, variations in the transport of the E2 glycoprotein to apical or basolateral domains of the plasma membrane would be expected to have no effect on virus budding. However, many polarized cells show marked polarity in secretion of cellular proteins and this polarized secretion might affect release of coronaviruses from post-Golgi vacuoles. Although extensive studies on the release of coronaviruses from cells in epithelial tissues have not been done, release of porcine coronavirus from both apical and basal regions of intestinal cells has been demonstrated (Doughri and Storz, 1977). It is clear that MHV virions may be released from the lower surface of mouse fibroblasts, since membrane fragments of cells which remain attached to the substrate after virus-infected cells peel away show large numbers of adherent, flattened, disc-shaped virions (Fig. 7-7e, Holmes and Turner, unpublished observations).

Fig. 7-7. Large numbers of coronavirions adsorbed to the plasma membrane of infected cells. *(a)* shows human respiratory coronaviruses attached to cells *in vitro*. *(b)* and *(c)* are a stereo pair of scanning electronmicrographs of MHV-infected cells demonstrating that adjacent cells may show great differences in the number of adherent virions. On the upper cell, virions (large arrow, c) are tightly packed on the membrane, whereas on the lower cell, virions (small white arrow, c) are widely scattered over the plasma membrane and microvilli. *(d)* and *(e)* show that MHV A59 virions on the plasma membrane often appear flattened and disc-shaped. In sections of infected cells observed by transmission electron microscopy (d), virions (arrows) at the plasma membrane are flattened and electron-dense. In (e), flattened virions are adsorbed to fragments of cell membrane left adhering to the substrate after an MHV-infected cell has detached. The preparation was fixed, dehydrated, dried by the critical point method, and observed with a high voltage electron microscope. Large numbers of virions (arrows) adhere to the membranes beneath the cells. Magnifications: (a) ×35,000; (b) and (c) ×9400; (d) ×52,000; (e) ×33,000. [(a) Courtesy of Dr. L. Oshiro, from Oshiro *et al.*, 1971, with permission of Cambridge University]

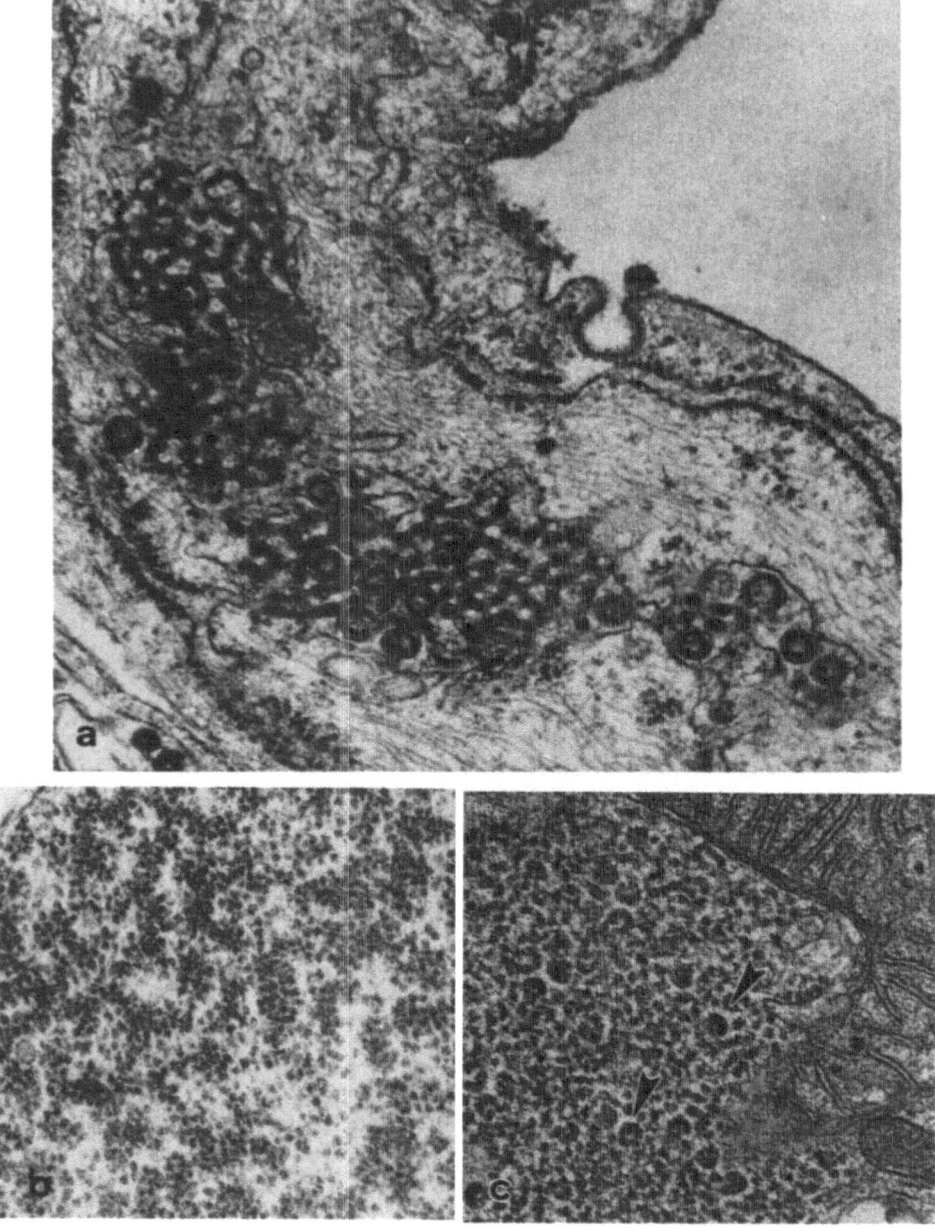

Fig. 7-8. Inclusions characteristic of coronavirus-infected cells. In *(a)*, reticular inclusions formed of specialized infoldings of RER membranes are shown in a cell infected with a human coronavirus. *(b)* and *(c)* show cytoplasmic inclusions of MHV NC seen in differentiated nerve cells *in vitro*. Inclusions formed by the wild Type JHM strain of MHV contain core-like structures composed of twisted NC strands (arrowheads) in *(c)*. Inclusions formed by a ts mutant of the JHM strain are mostly made of granules (b). Magnifications: (a) × 50,000; (b) × 51,000; (c) × 68,200. [(a) from Oshiro *et al.*, 1971, with permission of Cambridge University; (b) and (c) from Dubois-Dalcq *et al.*, 1982; with permission of Academic Press]

Defective Assembly

In this section, we will consider circumstances in which the synthesis of virus components is not synchronized with release of mature virions. Under these conditions, morphological events can be observed which may provide insight into virus assembly. Coronaviruses do not normally bud from the plasma membrane (Oshiro, 1973), although such budding is occasionally seen late in the infectious cycle (Dubois-Dalcq *et al.,* 1982) when some E1 may have been transported to the plasma membrane. After release of virions has ceased, cytoplasmic inclusions of viral NCs have been observed (Caul and Eggleston, 1977; Watanabe, 1969; Massalski *et al.,* 1982; Dubois-Dalcq *et al.,* 1982; Figs. 7-6b and 7-8b and c). These inclusions vary considerably in appearance and their protein composition has not yet been determined. Reticular cytoplasmic inclusions, which consist of a convoluted network of tightly apposed membranes continuous with the RER, are another characteristic of coronavirus infection (David-Ferriera and Manaker, 1965; Oshiro *et al.,* 1971; Fig. 7-8a).

Coronavirus-infected cells often contain spherical inclusions about 100 to 400 nm in diameter, which contain fine fibrils and are bounded by double membranes apparently derived from RER (David-Ferriera and Manaker, 1965; Sturman and Holmes, 1983; Fig. 7-9). These closely resemble structures called type 1 cytopathic vesicles (CPV-I) seen in cells infected with alpha- or flaviviruses (Murphy, 1980; Grimley *et al.,* 1972; see Chapter 8). They may represent aberrant virion formation or be associated with virus RNA synthesis, as has been suggested for alphaviruses (Grimley *et al.,* 1972).

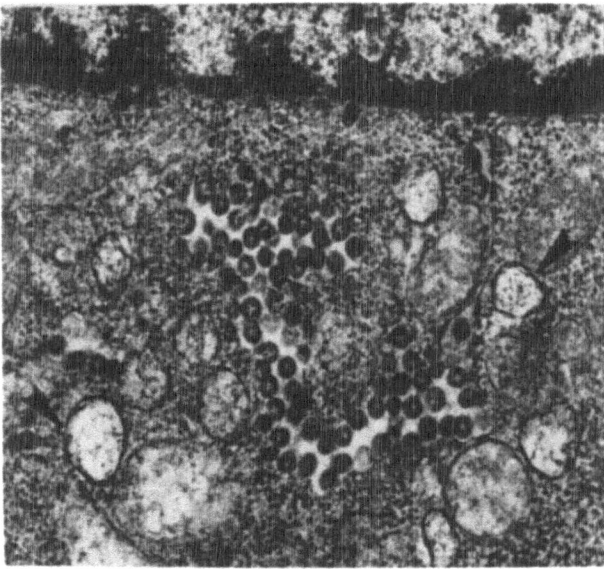

Fig. 7-9. Membrane-bound inclusions (arrowheads) containing thin electron-dense filaments. These inclusions in coronavirus-infected cells resemble the type 1 cytopathic vesicles seen in cells infected with alpha- and flaviviruses (see Chapter 8, Figs. 8-5a and 8-13). They may be involved in virus RNA synthesis. Magnification: ×75,000

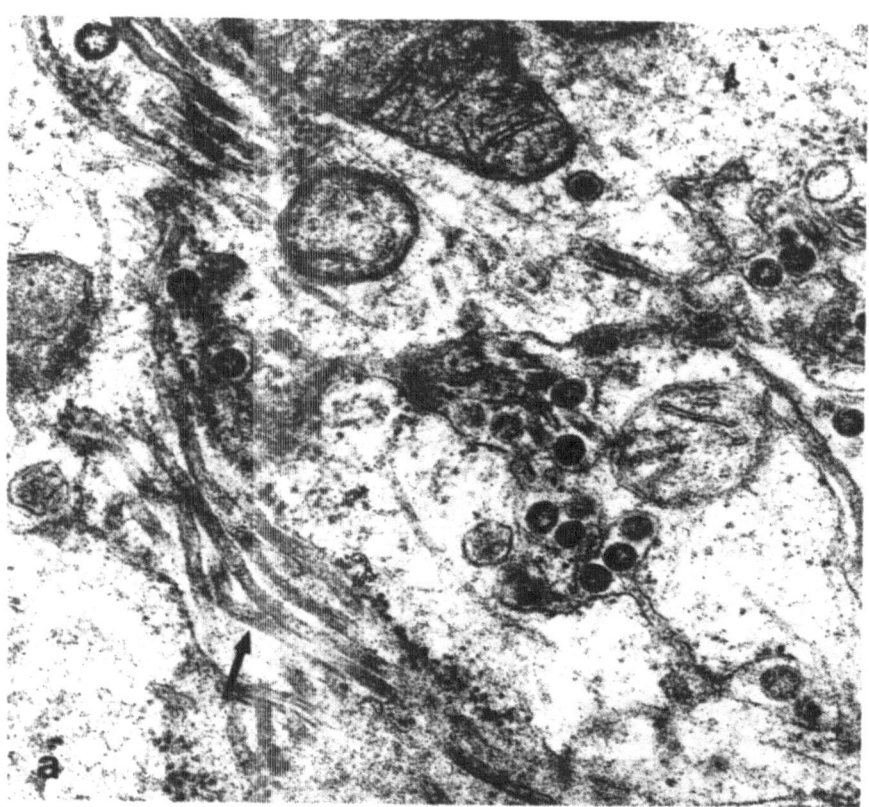

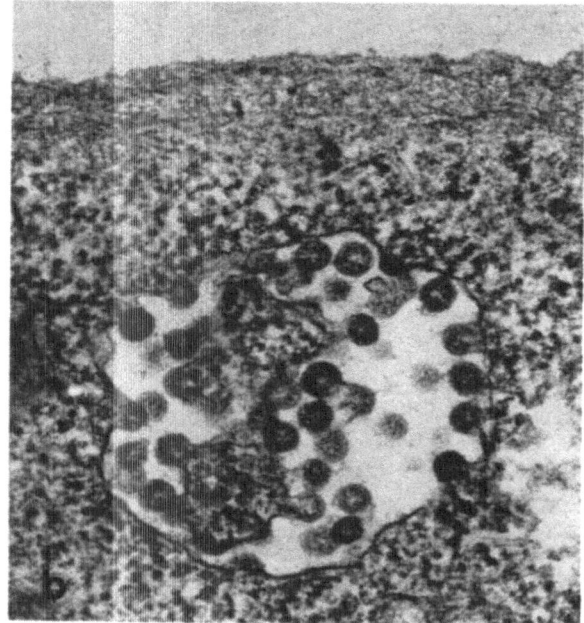

Late in infection with MHV, and in tunicamycin-treated MHV-infected cells, long rigid tubules about 50 nm in diameter are seen in the lumen of the RER and in smooth-walled vesicles (Dubois-Dalcq *et al.,* 1982; Holmes *et al.,* 1981 a; Fig. 7-10a). Similar tubules have been assembled *in vitro* from the matrix protein of a para-myxovirus (Heggeness *et al.,* 1982 a). Since E1 functions as a M-like protein which associates with membranes, it has been suggested that the coronavirus tubules are formed of excess E1 (Sturman and Holmes, 1983).

Empty virus particles have been isolated from coronavirus-infected cells by density gradient ultracentrifugation (Macnaughton and Davies, 1980), but the mechanism of their formation and release is not understood.

To analyze interactions of NCs with virus-specific membrane proteins, investigators of other virus systems have studied the formation of pseudotypes with different viruses. Attempts to isolate pseudotypes of coronaviruses with other virus types have been limited to pseudotypes of MHV with murine leukemia virus (Yoshikura and Taguchi, 1978). Further studies of coronavirus pseudotypes with other viruses that bud into the RER and Golgi may elucidate the assembly of coronaviruses.

Treatment of MHV-infected cells with monensin arrests virus budding (Niemann *et al.,* 1982; Fig. 7-10b), and removal of monensin results in rapid budding of virions into the Golgi apparatus. These effects are probably caused by monensin-induced inhibition of intracellular transport of MHV glycoproteins.

Ts mutants have been used to study the morphogenesis of several groups of en-veloped viruses. Although several independent collections of *ts* mutants of corona-viruses have been generated (Robb *et al.,* 1979; Haspel *et al.,* 1978; Koolen *et al.,* 1983), most of these mutants are of the RNA negative (RNA–)phenotype. The few RNA+ mutants which make structural proteins at the non-permissive temperature have not been examined for defects in virus maturation.

V. Organization of the Virion

A summary of the events in coronavirus assembly is shown in Fig. 7-11.

A model of the organization of the structural components of MHV is shown in Fig. 7-12. This general model seems to be valid for most coronaviruses (Sturman and Holmes, 1983), with the possible exception of BCV, which may have an additional glycoprotein (King and Brian, 1982). In negatively stained preparations, corona-virions have only one type of peplomer, with the possible exception of BCV and

---

Fig. 7-10. Late in MHV infection, long tubules, 25 to 30 nm in diameter, may accumulate in the lumen of the RER (arrow *a*). These may represent excess E1 glycoprotein in association with membranes. *(b)* shows that monensin may inhibit the maturation of coronaviruses. In monensin-treated cells, partially budded virions are seen in large numbers in smooth-walled vesicles associated with the Golgi. Magnifi-cations: (a) × 53,000; (b) × 60,000. [(a) from Dubois-Dalcq *et al.,* 1982; with permission of Academic Press]

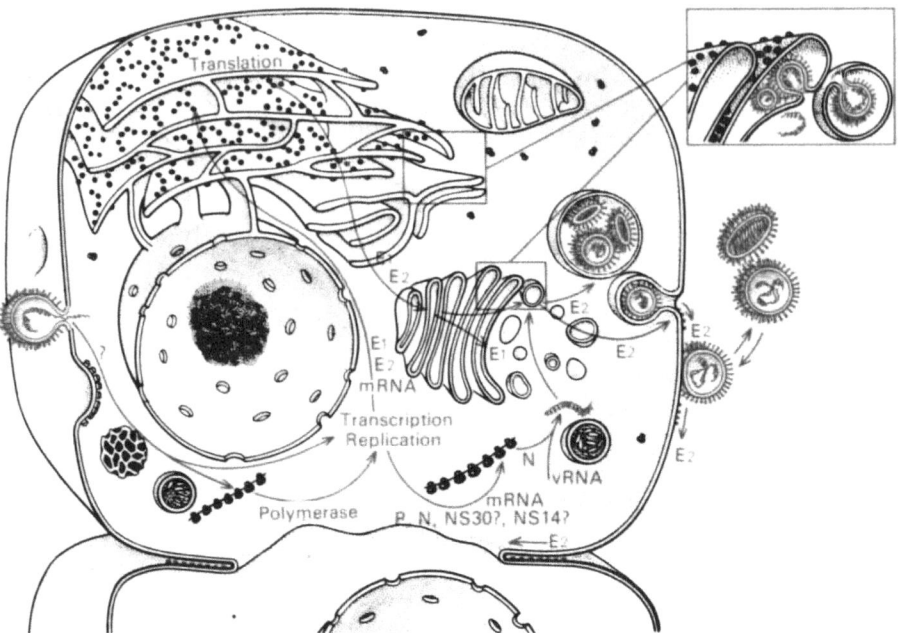

Fig. 7-11. A model summarizing the events in replication, transcription and assembly of mouse hepatitis virus. It is likely that the virus RNA enters the cell by fusion of the virus envelope with the plasma membrane. Primary translation yields the RNA polymerase which copies the genomic RNA to make a full length negative-stranded template. Then, synthesis of the 6 subgenomic mRNAs occurs and the 5' terminal end of each of these is translated to make a single protein species. The mRNAs for E1 and E2 envelope glycoproteins are translated on membrane-bound ribosomes. E2 is glycosylated with N-linked oligosaccharides and is acylated. In contrast, E1 is glycosylated with O-linked oligosaccharides and is not acylated. E2 is transported through the Golgi to the plasma membrane, but intracellular transport of E1 is terminated at the Golgi. Meanwhile, the other mRNAs appear to be translated on free polysomes to yield the non-structural proteins and the NC protein N. N assembles with new virus genomic RNA to form a helical NC. The NC probably interacts with the cytoplasmic domain of E1 in the membranes of the RER or Golgi at sites where E2 is also present. Budding of virions into the RER or Golgi is followed by their release from the intact cell by a process analogous to protein secretion (see text). The window shows virions budding from the RER and Golgi membranes. Cleavage of the E2 glycoprotein by a host cell protease appears to occur just prior to virus release. E2 is transported to the plasma membrane where it may induce fusion with adjacent cells

one strain of MHV (Bridger *et al.,* 1978; Greig *et al.,* 1971; Sugiyama and Amano, 1981). The smaller type of peplomer may contain the additional glycoprotein. Some glycosaminoglycan, a host cell component, has been detected even in highly purified coronaviruses (Garwes *et al.,* 1976; Sturman, 1981). No function for this has been identified.

The molar ratio of structural proteins within the virion is N : E1 : E2 = 8 : 16 : 1 (Sturman, 1981). This model indicates that the E1 glycoprotein interacts with both the NC and the E2 glycoprotein. The organization of the NC within the released virion is uncertain. Although the NC in the budding virus appears to be a well-formed helix attached to the inside of the virus envelope, in the released virion the NC may be less firmly bound to the membrane and may be partially uncoiled.

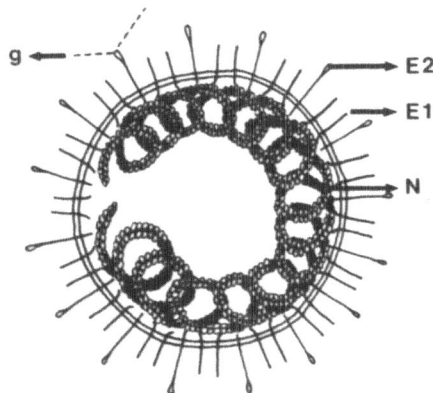

Fig. 7-12. A model of the organization of the MHV virion. The N protein is associated with the virus RNA genome in the helical NC. During budding, the NC probably associates with the cytoplasmic domain of the E1 glycoprotein which acts as a matrix protein. The peplomers are formed of the large glycoprotein E2, which probably interacts with the E1 in the virus envelope. A small amount of host glycosaminoglycan *(g)* is associated with purified coronavirions. (After Sturman and Holmes, 1983)

8
Assembly of *Togaviridae*

Introduction

The *Togaviridae* are a family of 4 genera of enveloped RNA viruses with icosahedral nucleocapsids: *Alphaviruses, Flaviviruses, Rubiviruses* and *Pestiviruses* (Melnick, 1980). The *Alpha-* and *Flaviviruses* are arthropod-borne viruses which can replicate in both vertebrate and arthropod cells. Both *Alpha-* and *Flaviviruses* are endemic in bird and animal populations and normally infect humans only as an incidental host (Shope, 1980). In humans, *Alphaviruses* cause fever, encephalitis, arthritis and rashes as well as inapparent infections. The *Flaviviruses* also cause fever and encephalitis as well as some hemorrhagic fevers. *Rubi-* and *Pestiviruses* are not arthropod-borne and apparently replicate only in vertebrate cells (Shope, 1980). The *Rubiviruses* are represented by rubella virus, which causes a mild febrile exanthem with teratologic effects in humans. *Pestiviruses* are responsible for hog cholera, bovine diarrhea and border disease of sheep, important diseases of domestic animals. Among the *Togaviridae*, the replication and maturation of *Alpha-* and *Flaviviruses* have been studied most extensively and will be considered separately in this chapter.

Alphaviruses

I. Introduction

The *Alphaviruses* are usually mosquito-borne. Each virus is maintained in a single vector species and in a single or several closely related species of vertebrate host, although other animals may be infected incidentally. Infection of the gut of a mosquito by virus-laden blood results in virus replication with minimal cytopathic effects. When virus infection spreads to the salivary glands, the insect becomes capable of transmitting the virus. Following infection of a vertebrate by the bite of an infected mosquito, virus replicates in neural and extraneural tissues. Infection may be associated with high levels of viremia and usually elicits protective immunity. Only a small proportion of humans infected with *Alphaviruses* develops clinically apparent disease.

Six subgroups of *Alphaviruses* have been defined by serologic studies (Schlesinger, 1980). These subgroups are those which contain Venezuelan equine encephalitis virus (VEE), Eastern equine encephalitis virus (EEE), Western equine encephalitis virus (WEE), Middleburg virus, Ndumu virus, and Semliki Forest virus (SFV). At the molecular biological level, the most extensively studied of the *Alphaviruses* are Sindbis virus, a member of the WEE subgroup, and Semliki Forest virus.

The virions of *Alphaviruses* are spherical particles with a diameter of 60—65 nm (Murphy, 1980). The peplomers form an indistinct, fuzzy surface coat around the virion. Detailed studies of purified virions show that the peplomers are 3—3.5 nm wide, 6—10 nm long and have a spacing of 4—6 nm from center to center. The virions have a NC with cubic (icosahedral) symmetry (Horzinek and Mussgay, 1969; Horzinek, 1973 a and b; Enzmann and Weiland, 1979) and the arrangement of peplomers on the virus envelope also appears to show cubic symmetry (Von Bonsdorff and Harrison, 1975; Enzmann and Weiland, 1979; Murphy, 1980).

II. Molecular Organization

A comprehensive book (Schlesinger, 1980) and several recent reviews (Kääriäinen and Söderlund, 1978; Garoff *et al.,* 1982; Strauss and Strauss, 1983) provide detailed information on *Alphavirus* replication. Therefore, the general features of the genomic organization and replicative cycle will be only briefly summarized here. The single stranded 4.4×10^6 MW genomic RNA is capped and polyadenylated (Kennedy, 1980). Naked virion RNA is infectious, although isolated nucleocapsids are non-infectious.

Adsorption of *Alphaviruses* to the plasma membrane is followed by endocytosis, uptake of virions into clathrin-coated pits which fuse with lysosomes. Triggered by the low pH in the endosomes, which may cause a conformational change in the virus glycoproteins (Edwards *et al.,* 1983), the virus envelope fuses with the vesicle membrane to release the virus NC into the cytoplasm (Helenius *et al.,* 1980, 1982; Helenius and Marsh, 1982; Simons *et al.,* 1982).

The transcription and replication of *Alphaviruses* are summarized in Fig. 8-1 (Kennedy, 1980). Upon entering the cell, virion RNA acts as mRNA to direct the synthesis of a 200 K to 230 K dalton precursor to the non-structural proteins. This large precursor is proteolytically cleaved to yield several non-structural proteins which comprise the viral RNA-dependent RNA polymerase complex. The polymerase copies the positive-stranded genomic RNA to make a full length negative-stranded template RNA (Sawicki and Sawicki, 1980). The recognition sites for the minus strand replicase may be a highly conserved sequence of 19 nucleotides at the 3′ end of the genome in concert with a 51 nucleotide sequence located 150 nucleo-

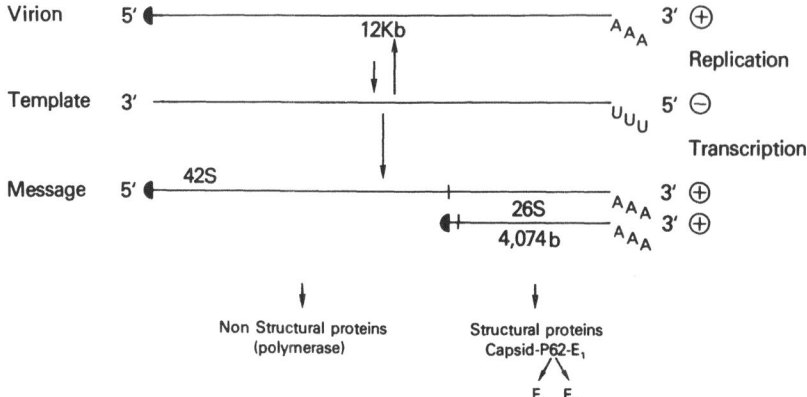

Fig. 8-1. A scheme of the transcription and replication of *Alphaviruses*. The positive-stranded 12 Kb genomic RNA in the virion serves as mRNA in the infected cell and directs the synthesis of non-structural proteins with RNA polymerase activity. This enzyme complex copies the positive-stranded genomic RNA to make a full-length negative-stranded RNA with poly(U) (UUU) at the 5' end. This is a template for synthesis of new genomic RNA and 26S RNA which serves as a polycistronic mRNA for synthesis of all structural proteins of alphaviruses. Both the genomic RNA and 26S RNA are capped (◀—) at the 5' end and polyadenylated (AAA) at the 3' end. The vertical line on the full length mRNA indicates the end of the non-structural coding region and the beginning of the genes for structural proteins. The vertical line on the 26S mRNA indicates the beginning of the open reading frame. (Adapted from Strauss and Strauss, 1983)

tides from the 5' end of the genome (Ou *et al.*, 1983; Strauss and Strauss, 1983). Synthesis of the minus strand template is completed by 4 hours after virus inoculation (Sawicki and Sawicki, 1980). Synthesis of positive-stranded 42S genomic RNA and 26S mRNA then occurs. The 42S RNA is transcribed by a positive-strand replicase while the 26S mRNA is transcribed by the transcriptase (Gomatos *et al.*, 1980). The transcriptase may initiate mRNA synthesis at a highly conserved 21 nucleotide sequence near the beginning of the 26S coding region on the negative-stranded template (Strauss and Strauss, 1983). The 26S mRNA is polycistronic and directs the synthesis of a polyprotein which is cleaved to yield all of the structural proteins. Although the nucleus is not required for replication of *Alphaviruses* in vertebrate cells, in cells of *Aedes albopictus,* replication of Sindbis virus is reported to require the host cell nucleus (Erwin and Brown, 1983).

Virus Proteins

The proteins of the *Alphaviruses* have been studied in great detail as models for the translation, membrane translocation, intracellular transport and processing of cellular proteins as well as to elucidate important processes in *Alphavirus* assembly (Bonatti *et al.*, 1979; Simons *et al.*, 1980; Kondor-Koch *et al.*, 1982). Important properties of the non-structural and structural proteins are summarized in Table 8-1. The virion is formed of equimolar ratios of each of the structural proteins.

(1) *The capsid protein, C* (MW: 30 K) interacts with the virus genomic RNA and also with the carboxyl-terminal domains of the E_2 glycoprotein. C also appears to

Table 8-1. *Proteins of* Alphaviruses

Usual name	Apparent MW $\times 10^{-3}$	Number of amino acids	Copies per virion	Known or putative function
C	30	267	250	NC protein Protease activity?
E1	49–59	438	250	Surface glycoprotein Hemagglutinin Membrane fusion?
E2	47–52	422	250	Surface glycoprotein Interaction with core
E3	10	63	250	Surface glycoprotein Signal peptide for p62
6K	4.2–6	55–60	–	Signal peptide for E_1
NS70	70		–	Replicase/transcriptase complex (protease?)
NS86	86		–	Replicase/transcriptase complex (protease?)
NS72	72		–	Replicase/transcriptase complex (protease?)
NS60	60		–	Replicase/transcriptase complex (protease?)

have protease activity which cleaves C protein from the large protein precursor. Complete amino acid sequences of C proteins from several *Alphaviruses* have been deduced from nucleotide sequences (Boege *et al.,* 1980; Garoff *et al.,* 1980 a; Rice and Strauss, 1981).

(2) *E1* (MW: 49–59 K) is a transmembrane envelope glycoprotein component of the virus peplomer or spike. The carboxyl end is anchored in the lipid bilayer (Garoff and Soderlund, 1978). E1 has hemagglutinating activity and appears to be responsible for virus-induced cell fusion. It induces cross-reactive, non-neutralizing antibodies (Dalrymple *et al.,* 1976). The amino acid sequences of several *Alphavirus* glycoproteins have been determined (Garoff *et al.,* 1980 b; Rice and Strauss, 1981).

(3) *E2* (MW: 47–52 K) is a transmembranous glycoprotein which induces type-specific neutralizing antibody (Dalrymple *et al.,* 1976). E2 interacts with E1 and E3 to form the viral peplomers. The carboxyl-terminal domain of E2 interacts with the C protein of the viral NC.

(4) *E3* (MW: 10 K) is a peripheral glycoprotein associated with the spikes of SFV. E3 is released from Sindbis virions (Welch and Sefton, 1979) and WEE (Simizu *et al.,* 1983) and does not appear to have an essential function in the virion. However, during synthesis of the viral envelope glycoproteins, the E3 region of p62, the precursor for E3 and E2 , contains the signal sequence for p62 (Mayne *et al.,* in press).

(5) Several non-structural proteins are found in cells infected with *Alphaviruses.* They are produced by proteolytic cleavage of the polyprotein encoded in the 5′ region of the 42S RNA and they appear to be involved in virus RNA dependent RNA polymerase activity.

III. Protein Synthesis, Transport and Post-Translational Modifications

The synthesis and transport of the virus structural proteins are somewhat different in the cells of vertebrates and invertebrates. Infection of invertebrate cells results in intracellular budding of virions and is more likely to lead to persistent infection, whereas infection of cells from vertebrates results in budding of virions from the plasma membrane and leads to cell death (Stollar, 1980 b; Lehane and Leake, 1982). This chapter will focus on *Alphavirus* replication in cells from vertebrates, and compare specific differences observed in invertebrate cells.

The 26S mRNA which directs the synthesis of the viral structural proteins is made in great excess over the 42S genomic RNA. Translation of the 26S mRNA has been studied extensively in cell-free translation systems, in microinjection studies, in pulse-labeling experiments, and in cell fractionation studies (Clegg, 1975; Bonatti *et al.*, 1979; Schlesinger and Kääriäinen, 1980; Kondor-Koch *et al.*, 1982). Translation of the 26S RNA begins on free ribosomes with the synthesis of the capsid protein. The nascent capsid protein is cleaved from the growing polypeptide chain, apparently by proteolytic activity of the capsid protein itself. This reveals a signal sequence at the amino terminus near the cleavage site. This signal sequence allows the ribosome-26S RNA complex to attach to the RER via signal recognition particles. Following insertion of the signal sequence into the RER membrane, a large glycoprotein is synthesized on membrane-bound ribosomes and translocated across the RER membrane. During its synthesis, this precursor is cleaved in two places, releasing a 6k protein and yielding the p62 and E1 glycoproteins. Mannose-rich core oligosaccharides are added cotranslationally to asparagine residues on p62 and E1 in the RER.

The glycoproteins are then transported to the Golgi apparatus where fatty acids are added (Schmidt and Schlesinger, 1980) in the *cis* or medial cisternae (Griffiths *et al.*, 1983). Trimming of the high-mannose oligosaccharides and addition of terminal sugars occur in the *trans* cisternae (Pesonen *et al.*, 1981; Griffiths *et al.*, 1982; Pesonen and Kääriäinen, 1982). A final step in the maturation of the virus glycoproteins occurs in the Golgi or at the plasma membrane where p62 is cleaved to yield E2 and E3 (Brown, 1980; Ziemiecki *et al.*, 1980; Mayne *et al.*, in press). Details of virus glycoprotein processing differ in arthropod cells, in that the virus glycoproteins obtain different oligosaccharides using host cell glycosyl transferases and trimming enzymes (Stollar, 1980 b).

Fluorescent antibody labeling, immunoelectron microscopy and cell fractionation experiments have been used to study the intracellular transport of *Alphavirus* glycoproteins (Erwin and Brown, 1980; Scheefers *et al.*, 1980; Saraste *et al.*, 1980 a, b; Green *et al.*, 1981 a; Griffiths *et al.*, 1982, 1983; Arias *et al.*, 1983; Fig. 8-2). Using gold- or ferritin-labeled antibody directed against the glycoproteins in cells treated with cycloheximide to inhibit further virus protein synthesis, these studies have shown that these molecules pass from the RER through the Golgi and then to the plasma membrane (Green *et al.*, 1981 a; Griffiths *et al.*, 1983). On the plasma membrane of cells from vertebrates, the virus glycoproteins are found in patches over areas where NC has bound to the cytoplasmic side of the membrane, as well as in budding

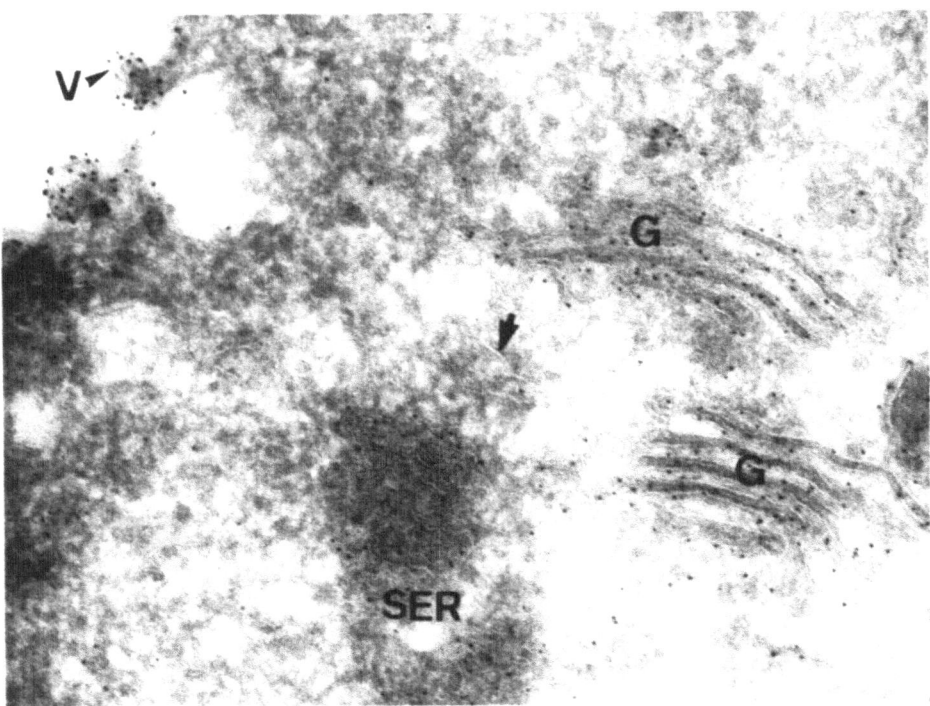

Fig. 8-2. Localization of virus spike glycoproteins of Semliki Forest virus (SFV) in the smooth endo-plasmic reticulum *(SER)* and Golgi apparatus. A frozen thin section of an SFV-infected cell was immu-nolabeled with an affinity-purified antibody against the virus spike proteins followed by *Staphylococcus aureus* protein A-gold complex. Extensive labeling is seen over Golgi stacks *(G)* as well as over SER. The labeling of the rough ER (arrow) is always considerably less than that of the SER. Labeled budding virions *(V)* are seen at the plasma membrane. Magnification: × 80,000. (Courtesy of Dr. G. Warren from Griffiths *et al.,* 1983, with permission from the Rockefeller University Press)

virions (Griffiths *et al.,* 1983; Fig. 8-2). In contrast, in cells from invertebrates, rela-tively little virus glycoprotein appears on the plasma membrane (Lehane and Leake, 1982).

Apparently, p62 and E1 form a complex immediately after synthesis (Ziemiecki *et al.,* 1980). Complexes of p62 and E1 can be isolated from the RER and these pro-teins can be chemically cross-linked to RER membranes. However, as will be shown below, mixed infection with different *Alphaviruses* or mutants of *Alphaviruses* can result in formation of pseudotype virions which contain E2 and E1 from different parental viruses. This suggests either that there is reassortment between E1 and E2 on the membranes, or that intact spikes from different viruses may be mixed in one virus envelope. Fluorescence photobleaching recovery experiments suggest that the *Alphavirus* glycoproteins are initially mobile in the plasma membrane, but become increasingly immobilized as virus budding proceeds (Johnson *et al.,* 1982). Complexes of E1, E2, and E3 are quite stable. They can be solubilized from virions (Helenius and von Bonsdorff, 1976; Ziemiecki and Garoff, 1978), formed into micelles (Morein *et al.,* 1978; Simons *et al.,* 1978), reinserted into membranes (Helenius *et al.,* 1981) and associated *in vitro* with NCs (Helenius and Kartenbeck, 1980).

IV. Assembly of Virus Components

Morphogenesis of *Alphaviruses* has been studied extensively and is the subject of many detailed reviews (Kääriäinen and Söderlund, 1978; Garoff, 1979; Brown, 1980; Murphy, 1980; Simons and Garoff, 1980; Garoff *et al.,* 1982; Simons *et al.,* 1982). Elucidation of the molecular interactions associated with *Alphavirus* budding has provided important information about cellular processes required for maturation of enveloped viruses.

NC Assembly

Newly synthesized C protein quickly and efficiently assembles with the 42S genomic RNA in the cytoplasm to form NCs (Söderlund, 1973). This interaction is believed to occur between a cluster of basic amino acids and proline near the amino terminus of the C protein and a region near the 5' end of the 42S RNA (Rice and Strauss, 1981). No free C protein can be isolated from *Alphavirus*-infected cells. A transient 90S precursor of Semliki Forest virus NC isolated from infected cells contains 42S RNA with a small amount of capsid protein, but it is difficult to detect after capsid protein is being made in large amounts (Ulmanen, 1978).

NCs of *Alphaviruses* have been purified following disruption of the virus envelope by enzymes or detergents. In negatively stained preparations, the NCs appear spherical and rather deformable and have diameters of 35–39 nm (Horzinek and Mussgay, 1969; Murphy, 1980). Fixation with formaldehyde yields NCs with symmetrically arranged subunits in an icosahedral lattice with a 3-fold axis of symmetry. At acid pH, the NC of SFV undergoes an irreversible contraction (Söderlund *et al.,* 1971). This may occur in the acid environment of the endocytic vesicles following virus uptake, and might facilitate separation of the NC from the virus envelope which fuses with the vesicular membrane.

In vitro assembly of NCs has been achieved (Wengler *et al.,* 1982). Purified C protein of Sindbis can react with isolated viral genomic RNA, with other single-stranded RNAs or even with single-stranded DNA of a variety of sizes to form core-like NCs with the size, density, and sedimentation velocity of the NCs isolated from virions by detergent disruption. These observations suggest that C is a nucleic acid-binding protein which can self-aggregate to form a capsid with cubic symmetry only when it is bound to nucleic acid. This would explain why no empty capsids are formed during *Alphavirus* replication. In the cell, assembly of NCs is very specific and only the 42S genomic RNA or deletions of the 42S RNA which yield defective interfering virions are encapsidated (Stollar, 1980 a).

The formation of an icosahedral NC from 42S RNA and C protein depends upon electrostatic interactions between RNA and basic amino acids, and upon protein-protein interactions which have not yet been defined. In the icosahedral capsid, C protein is probably oriented with the amino-terminal end toward the interior and the carboxyl-terminal end on the outside of the capsid where it can interact with virus glycoproteins (Rice and Strauss, 1981; Strauss and Strauss, 1983).

It is not clear how the NCs of *Alphaviruses* move to the membrane which is the virus budding site. Is the NC assembled *in situ* on the membrane or does the newly formed NC containing the genomic RNA and C protein quickly migrate to the membrane after assembly?

Interactions Between NC and Envelope Proteins

In cells from vertebrates, *Alphaviruses* mature by budding from the plasma membrane (Figs. 8-2 to 8-5). NCs associate with regions of the plasma membrane which contain virus glycoproteins (Fig. 8-2). This association appears to occur via the cytoplasmic carboxyl-terminal domain of the p62 glycoprotein. Binding of the NC to the membrane appears to require a critical concentration of this glyco-protein on the membrane, since mutants which fail to transport the glycoproteins to the plasma membrane do not bind NCs to the plasma membrane (Erwin and Brown, 1980).

Detailed models for *Alphavirus* budding have been proposed (Simons and Garoff, 1980; Brown, 1980; Garoff *et al.,* 1982; Strauss and Strauss, in press). Following the first interaction of the NC with the cytoplasmic domain of p62, the precursor is proteolytically cleaved to yield E2 and E3. This results in a conforma-tional change of E2 (Edwards *et al.,* 1982) and may strengthen the association of E2 with E1. Additional dimers of E1 and p62 glycoproteins which randomly move in the plane of the membrane into this region are then co-operatively bound to the NC-glycoprotein complex. This process forms the budding virion and results in the exclusion of cellular membrane proteins from the virus envelope. Because the virion contains an equimolar ratio of all of the virus structural proteins, it appears likely that one spike may bind to each C protein of the NC. Indeed, the lipid bilayer may merely provide a matrix for forming a planar array of glycoproteins, and lipid *per se* may not participate in the interaction of the NC with the glyco-proteins. For example, *in vitro,* all virus lipids can be removed from virions with the detergent octyl glucoside and the virus glycoproteins remain attached to the NC. Also, the spikes can be dissociated and reassociated with the NC in the absence of lipid (Helenius and von Bonsdorff, 1976; Helenius and Kartenbeck, 1980).

Negative staining and freeze etching studies have clearly demonstrated that, unlike most of the other enveloped RNA viruses, the envelope of *Alphaviruses* exhibits icosahedral symmetry (T=4) (von Bonsdorff and Harrison, 1975). This may be due to the strong transmembrane interactions between the peplomers and the virus nucleocapsid.

Why do *Alphaviruses* normally bud from the plasma membrane rather than from intracytoplasmic membranes? Probably NCs can bind to any membrane where sufficient virus glycoproteins are present. During the period of exponential virus release, the virus glycoproteins are on the plasma membrane, so binding of NCs occurs there. However, when large amounts of virus glycoproteins are present in intracytoplasmic membranes, such as late in the infectious cycle or in cells infected with virus mutants defective in glycoprotein transport, binding of NCs may occur at these intracellular membranes. Budding of *Alphaviruses* into intracyto-plasmic membranes has been observed under these conditions (Saraste *et al.,* 1980 a and b). Brown (1980) has suggested that budding from internal membranes might

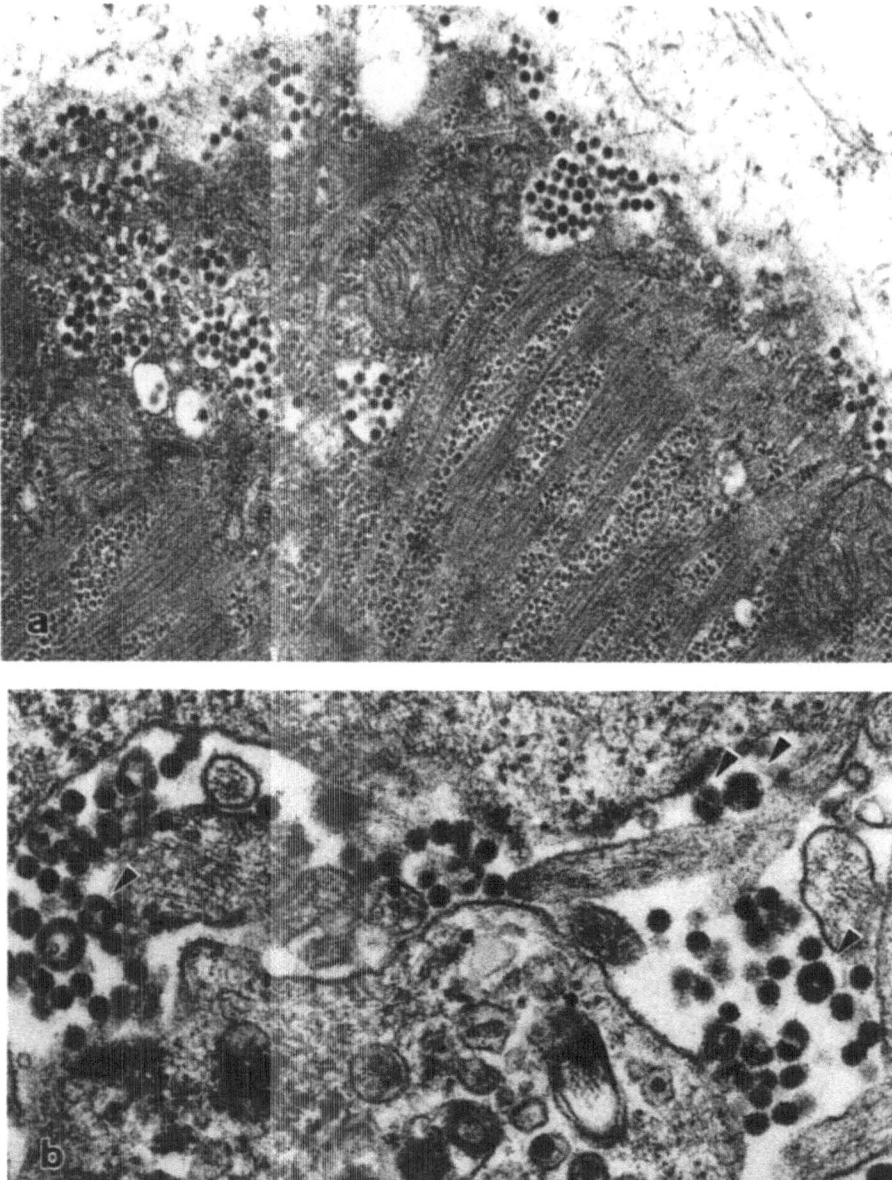

Fig. 8-3. *Alphaviruses* at the plasma membrane of infected cells. In *panel (a)*, SFV virions released from a striated muscle cell are trapped near the cell membrane. *(b)* shows normal Western equine encephalitis virus (WEE) virions and virions containing multiple NCs (arrowheads). Such multiploid virions occur occasionally in normal infections but are very common in cells infected with certain virus mutants.
Magnifications: *(a)* ×32,800; *(b)* ×47,500. (Figs. 8-3 to 8-5 courtesy of Dr. P. Grimley)

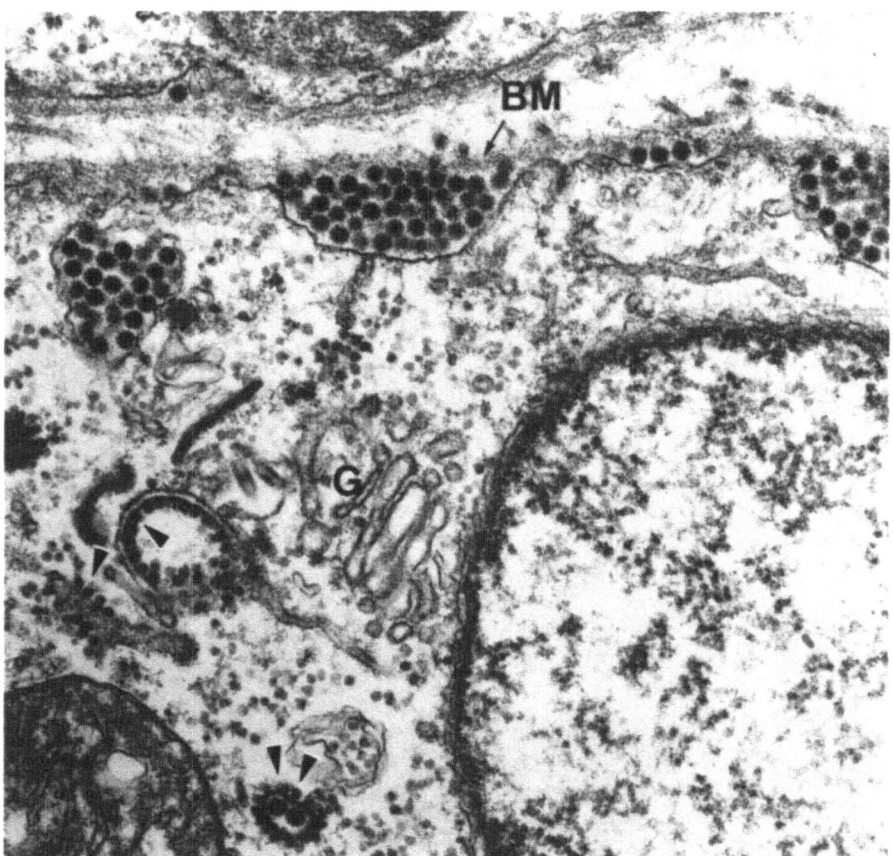

Fig. 8-4. SFV infection of a cell in the tongue of a mouse. Numerous virions are seen at the plasma membrane beneath the basement membrane *(BM)*. Near the Golgi apparatus *(G)*, NCs (arrowheads) are attached to smooth membranes in the cytoplasm in large numbers and mature virions lie within some of these vacuoles, which are called type 2 cytopathic vesicles (CPV-2). Magnification: × 40,000.
(From Grimley and Friedman, 1970 a, with permission from University of Chicago Press)

be inhibited by host cell membrane proteins present in that location. Another hypothesis to explain the preferential binding of NCs to, and budding of virions from, the plasma membrane suggests that, because cleavage of p62 is required for virus budding, the location of the protease activity on the plasma membrane determines the budding site (Brown, 1980). However, recent evidence suggests that cleavage of p62 may occur in the Golgi like that of secretory proproteins and myxo- and paramyxovirus glycoproteins (Rice and Strauss, 1981; Mayne *et al.*, in press; Strauss and Strauss, in press). Cleavage can occur under conditions in which budding is inhibited, such as in medium of low ionic strength (Mayne *et al.*, in press). Taking all these observations together, the site of *Alphavirus* budding probably depends upon both cleavage of p62 and presence in the membrane of adequate amounts of E1 and E2.

In permissive cell types, early after infection, synthesis of virus components appears to be balanced with the release of virions from the plasma membrane, so

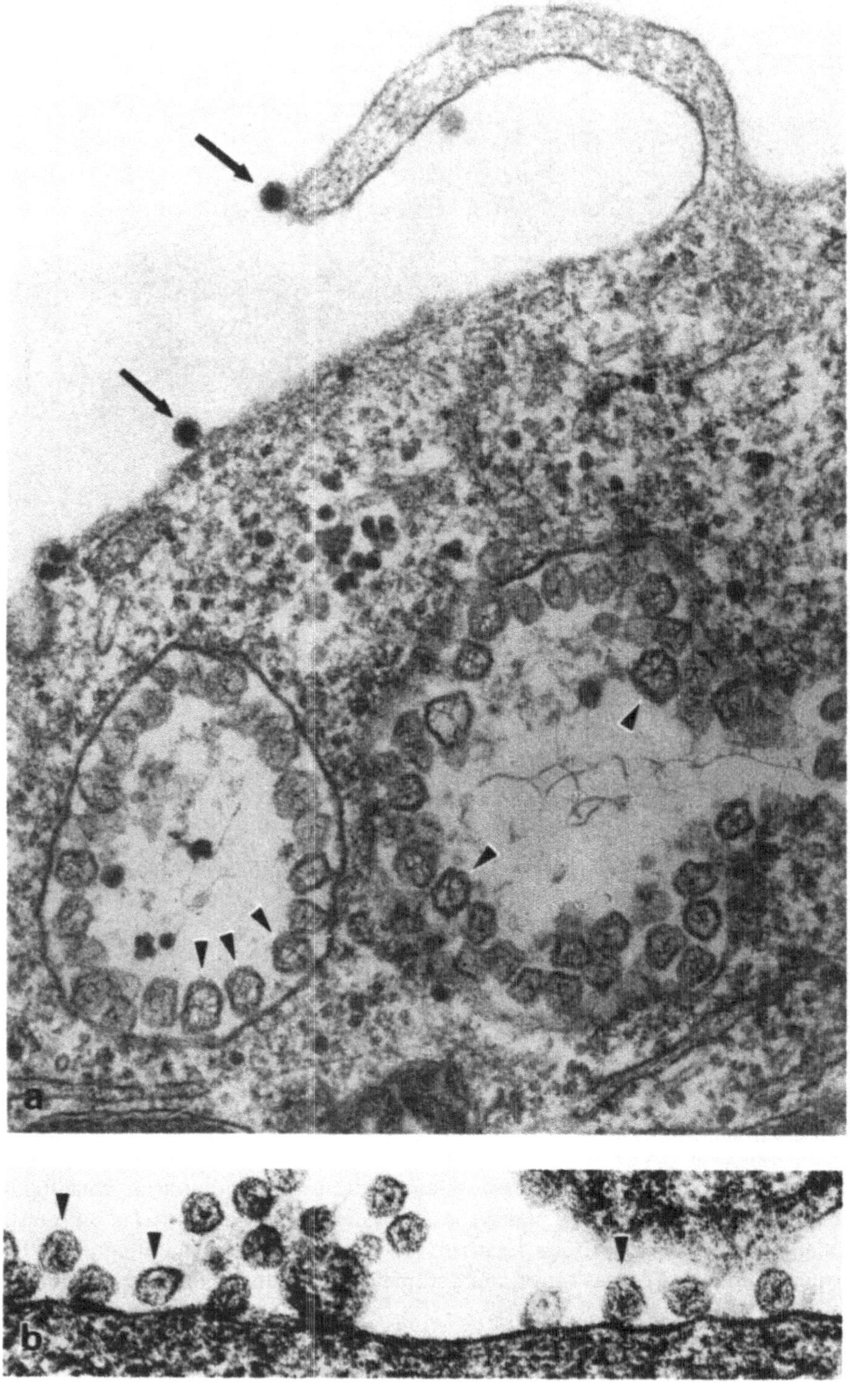

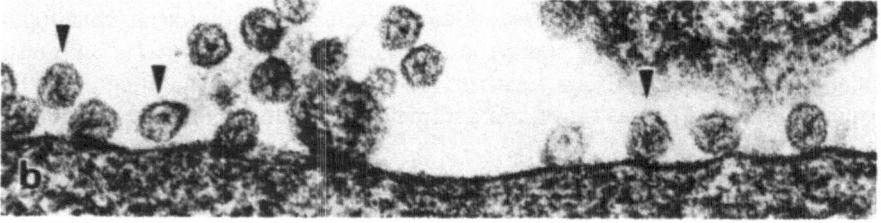

that little accumulation of virus components within the cells is observed. NCs are seen attached to the cytoplasmic side of the plasma membrane and virus buds form at the plasma membrane (Acheson and Tamm, 1967; Brown *et al.*, 1972). Later, excess virus nucleocapsids accumulate in the cytoplasm attached to membrane vesicles.

Alphaviruses are released from the plasma membrane after budding and may be found in large numbers at the cell surface (Figs. 8-3, 8-4, and 8-5). Release of virions which bud into intracellular membranes, as in insect cells, is believed to be *via* fusion of virus-filled vesicles with the plasma membrane. Following release of virions, the binding of the NC to the viral envelope appears to be decreased, in comparison to the strong interactions which occur during the formation of the virus bud (Brown, personal communication). This might facilitate release of infecting NCs into cells following their entry through endocytic vesicles.

V. Defective Assembly

During the early stages of *Alphavirus* infection, often preceding formation of NCs, small (60—70 nm) membranous spheres containing fibrillar material have been observed attached to the lumenal wall of the RER and to the plasma membrane (Grimley *et al.*, 1968, 1972, 1973; Fig. 8-5). These complexes are called type 1 cytopathic vesicles (CPV-1). Virus RNA synthesis is required for their formation and virus RNA has been localized in these vesicles by autoradiography (Grimley *et al.*, 1968). It has been postulated that they are sites of virus RNA synthesis. Similar vesicles have been observed in cells infected with flaviviruses and coronaviruses (see Figs. 7-9 and 8-13).

Several different types of virus mutations affect the formation of virions (reviewed by Strauss and Strauss, 1980). Mutants in the C protein have been described which give rise to multiploid virions containing multiple NCs, possibly by faulty interaction with the glycoprotein. Such multiploid virions are also seen occasionally in infections with wild-type *Alphaviruses* (Fig. 8-3b). Ts mutants in the envelope proteins have been described which inhibit intracellular transport (Braha and Schlesinger, 1978; Erwin and Brown, 1980; Saraste *et al.*, 1980 a, b). Sequencing of their cloned genes has suggested that intracellular transport depends upon a proper 3-dimensional configuration of the E1 glycoprotein (Arias *et al.*, 1983). Mutants in which cleavage of p62 to E2 and E3 is prevented at the non-permissive temperature may permit the transport of E1 to the membrane and binding of the NC to the membrane, but no budding is observed until the cells are shifted to the permissive temperature (Brown, 1980). Similarly, attachment of NCs to the plasma membrane in the absence of virus bud formation was seen when a host-dependent mutant of

---

Fig. 8-5. *Alphavirus* budding and formation of type 1 cytopathic vesicles. *(a)* shows two virions budding at the plasma membrane (arrows) and demonstrates the type 1 cytopathic vesicles (CPV-1) seen in *Alphavirus*-infected cells shortly after infection. Spherical buds of membranes which contain electron-dense fibrillar material (arrowheads) are seen in smooth-walled vesicles in a cell infected with WEE. In *panel (b)*, similar small membranous buds form at the plasma membrane of a cell infected with SFV. Magnifications: *(a)* ×70,000; *(b)* ×107,500

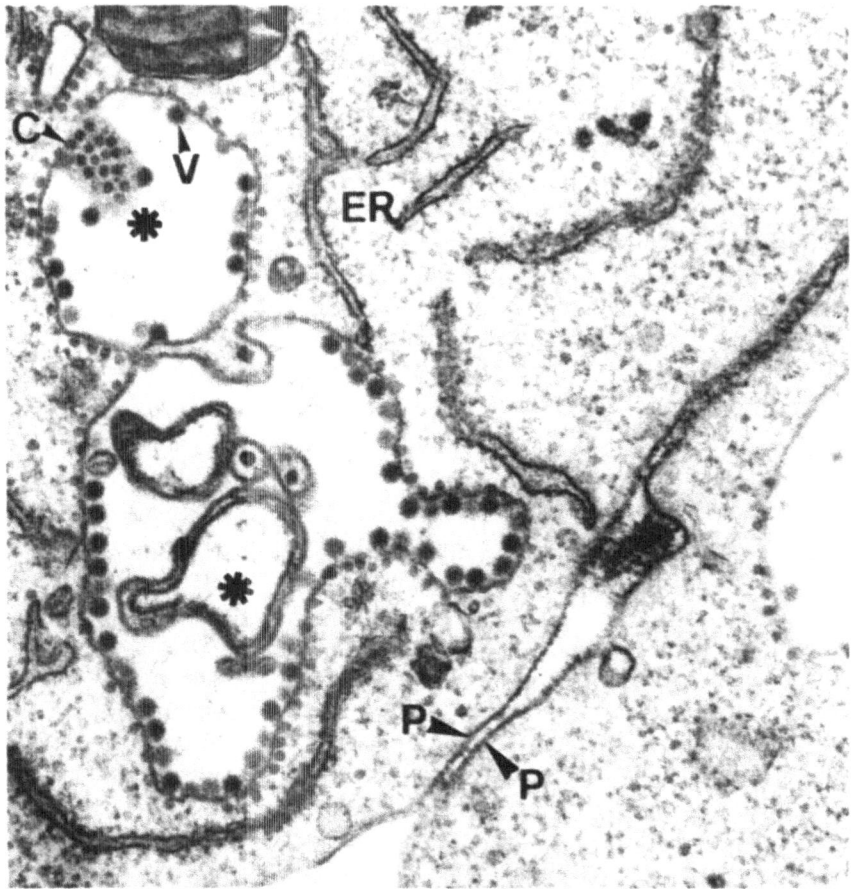

Fig. 8-6. An SFV-infected BHK cell treated with monensin, which arrests the intracellular transport of the viral glycoproteins and prevents the addition of terminal sugars to E1 and the cleavage of p62 to E2 and E3. In this cell, dilated Golgi cisternae (asterisks) contain many virions *(V)*, and many NCs *(C)* are bound to the cytoplasmic side of the membranes. *ER* is the endoplasmic reticulum and *P* indicates the plasma membrane. Magnification: × 50,000. (Figs. 8-6 and 8-7 courtesy of Dr. G. Warren from Griffiths *et al.,* 1983, with permission from the Rockefeller University Press)

Fig. 8-7. Cytochemical and immunoelectron microscopic identification of the region of the Golgi which shows capsid binding activity and virus budding in monensin-treated cells. *(a)* shows electron-dense thiamine pyrophosphatase (TPPase) reaction product in the *trans* Golgi cisternae (large arrowhead) but not in the cisternae (asterisk) containing the budding virions (small dark arrowheads). *(b)* a frozen thin section shows that virions bud into the medial compartment of the Golgi into cisternae which can be labeled with gold-labeled antibodies to *Ricinus communis* agglutinin 1 *(RCA)*. The smaller gold marker *(SFV,* thin arrows) indicates antibodies to virus spike glycoproteins on budding virions and on the lumenal side of the membrane (ME). NCs *(C)* can be seen bound to the cytoplasmic side of the membrane. Magnifications: *(a)* × 39,000; *(b)* × 110,000

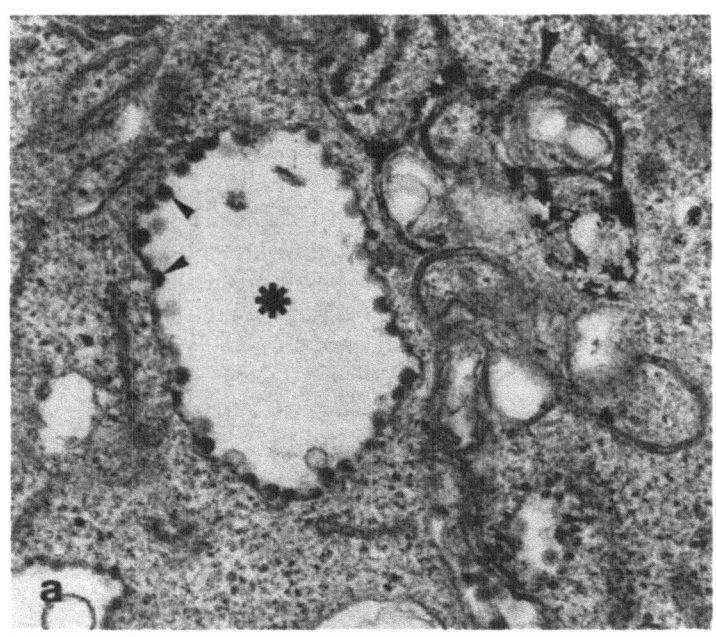

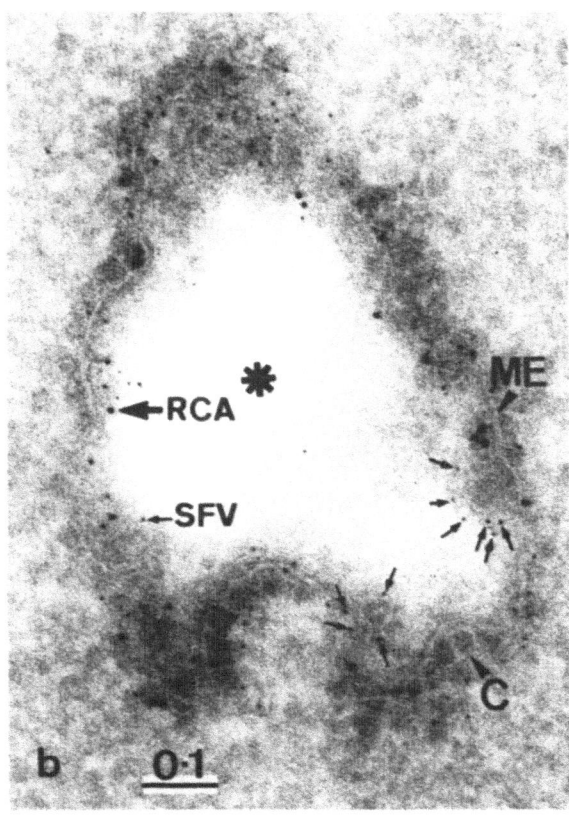

Sindbis virus was grown in cells from vertebrates. Although this mutant replicates normally in insect cells, an alteration in the E2 glycoprotein appears to be responsible for the inability of the mutant to bud from cells of vertebrates (Durbin and Stollar, personal communication).

Budding of *Alphaviruses* can be prevented by treatment of infected cells with high salt (Bell *et al.*, 1978). Under these conditions, NCs are observed aligned under the plasma membrane and p62 is cleaved to yield E2 and E3, but no budding occurs. When the normal medium is restored, a brief period of highly synchronized virus budding and release occurs. This modulation of virus budding appears to be due to a charge-dependent alteration in the conformation of the p62 glycoprotein (Strauss *et al.*, 1980). Brief treatment of cells with trypsin also inhibits *Alphavirus* budding (Adams and Brown, 1982).

Monensin has been used to study the intracellular transport of *Alphavirus* envelope proteins (Kääriäinen *et al.*, 1980; Griffiths *et al.*, 1983; Quinn *et al.*, 1983). In BHK cells, monensin arrests the intracellular transport of p62 and E1 in the medial cisternae of the Golgi apparatus as shown by the elegant immunoelectron microscopic and cytochemical studies of Griffiths *et al.* (1983; Figs. 8-6 and 8-7). In these cells, large dilated vacuoles are formed which bind large numbers of NCs on

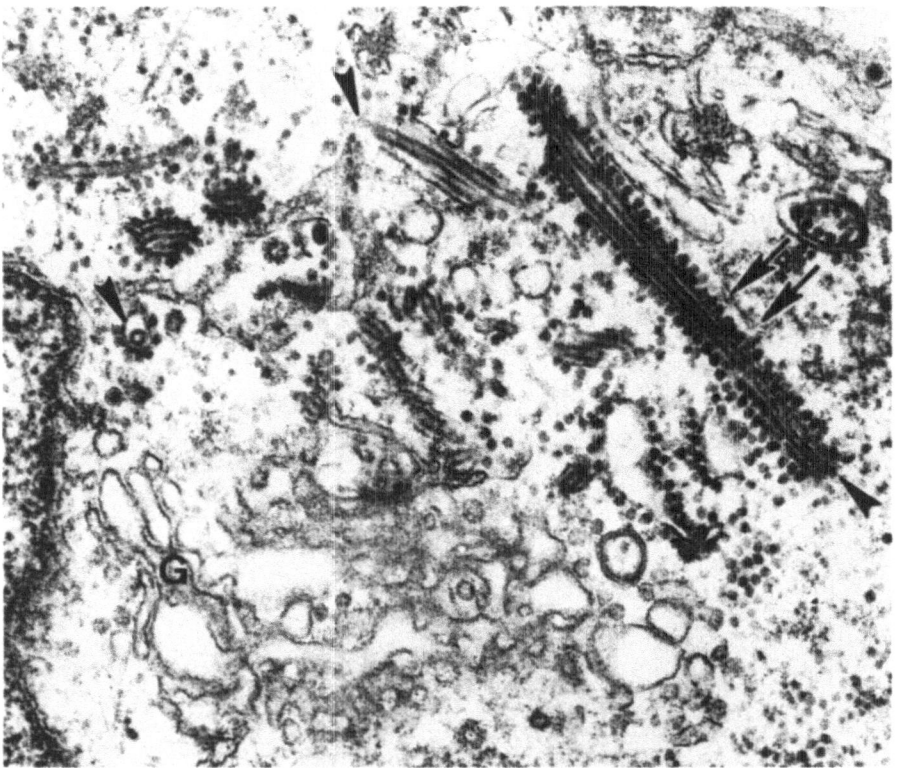

Fig. 8-8. A cell 24 hours after inoculation with SFV which contains long rigid tubules (arrowheads) within membranes to which numerous NCs (arrows) have bound. *G* indicates the Golgi. Magnification: ×40,000. (Courtesy of Dr. P. Grimley)

their cytoplasmic surface. These vacuoles resemble the type 2 cytopathic vesicles (CPV-2) seen late in normal *Alphavirus* infections (Grimley and Friedman, 1970 a and b; Grimley *et al.*, 1973; Figs. 8-4 and 8-8). Possibly, accumulation of virus glycoproteins in the medial Golgi cisternae occurs late in infection and leads to the formation of the CPV-2. The monensin-induced membranes with their bound NCs can be isolated from other membranes of homogenized cells by density gradient ultracentrifugation (Quinn *et al.*, 1983). Formation of virions can occur in these vacuoles (Figs. 8-6 and 8-7). It will be of great interest to determine whether virions formed in these vacuoles have p62 cleaved to E2 and whether they show normal infectivity. Long rigid tubules of unknown composition sometimes develop in CPV-2 late in the infectious cycle (Grimley and Friedman, 1970 b; Fig. 8-8).

In cells from invertebrates, in which persistent infection is much more common, virions frequently form in the RER or nuclear envelope, although early in infection some budding from the plasma membrane is also seen. Release of these intracellular virions from the cell has been postulated to result from fusion of vesicles filled with virions with the plasma membrane (Stollar, 1980 b).

Several *Alphaviruses* have been reported to cause intranuclear accumulation of NCs (Murphy, 1980), but these have not yet been shown to contain *Alphavirus* antigens. The nucleus is not required for the maturation of *Alphaviruses* in vertebrate cells. However, in insect cells, some function of the cell nucleus is required for *Alphavirus* replication since enucleation prevented *Alphavirus* replication but not rhabdovirus replication (Erwin and Brown, 1983).

Although the binding of *Alphavirus* envelope proteins with NCs appears to be a very specific process, pseudotypes can be formed between different *Alphaviruses* (Strauss and Strauss, 1980; Strauss *et al.*, 1983). For example, the glycoproteins of WEE and Sindbis viruses can substitute for each other in virions containing either genome. Thus, the glycoproteins of these two viruses are able to interact effectively with NC protein during virus budding. Further studies of the E1 and E2 glycoproteins of the parental viruses in such pseudotypes may clarify how soon after synthesis these two glycoproteins become associated and at what point they form spikes as they are transported through the cell. Pseudotypes of *Alphaviruses* with rhabdoviruses, defective Rous sarcoma virus and LDH virus have been formed, but these appear to contain only the NCs of the other virus with *Alphavirus* glycoproteins. No *Alphavirus* NCs with glycoproteins of other virus families have been detected. These observations suggest that the NCs of *Alphaviruses* exhibit stringent binding specificity for their homologous glycoproteins, whereas *Alphavirus* glycoproteins can be incorporated into virions of other viruses.

VI. Organization of the Virion

A model describing the replication, transcription and assembly of *Alphaviruses* is shown in Fig. 8-9. Postulated structural relationships between virion components are shown in Fig. 8-10. Further studies on the morphology of *Alphaviruses* will probably focus upon the formation and symmetry of the NC, specific domains in the structural proteins which interact with RNA, lipid, and other proteins or which

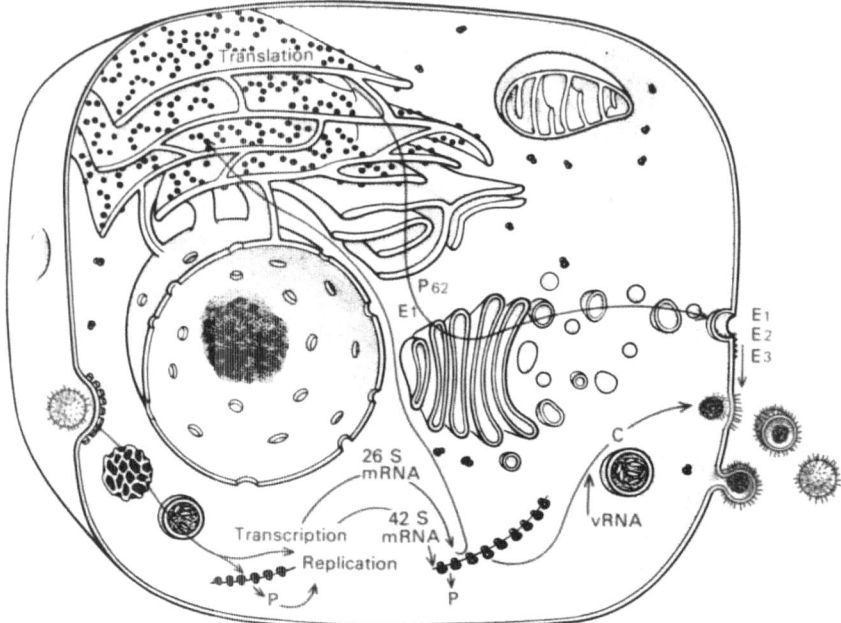

Fig. 8-9. A model summarizing the events in the replication, transcription and assembly of *Alpha-viruses*. Virions enter the cell through clathrin-coated vesicles and virus RNA is released into the cyto-plasm by fusion of the virus envelope with the endosomal membrane. RNA-dependent RNA poly-merase is synthesized using the genomic RNA as mRNA. Primary transcription yields a full length negative-stranded RNA which is copied to make genomic RNA and 26S RNA. The structural proteins are translated from the polycistronic 26S RNA. The NC protein, *C*, is made first on free polysomes and the growing polypeptide chain is cleaved to reveal a signal sequence which inserts into the RER. Then, p62 and E1 are made on membrane-bound polysomes, glycosylated, acylated, and transported through the Golgi to the plasma membrane. Cleavage of p62 to yield E2 and E3 is a late step which occurs in association with virus budding.

direct intracellular transport of these proteins (Arias *et al.*, 1983). Molecular genetic techniques including formation of chimeric proteins or proteins with site-specific mutations are expected to provide greater insight into *Alphavirus* morphogenesis.

Flaviviruses

I. Introduction

The *Flaviviruses* are a group of more than 50 viruses which are similar biochemically and antigenically (Russell *et al.*, 1980). The group includes mosquito-borne viruses such as yellow fever and dengue viruses, tick-borne viruses including tick-borne encephalitis, and non-arthropod-borne viruses such as Dakar bat virus. Where they have been studied, these viruses are all small and enveloped with positive-stranded RNA genomes, capsids with cubic symmetry, and three structural protein species,

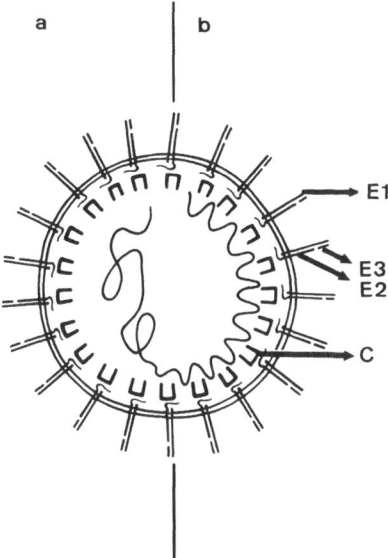

Fig. 8-10. A model of the organization of the *Alphavirus* SFV. The virus envelope contains three glycoproteins. E3 is peripherally located and E1 and E2 are transmembrane glycoproteins. The cytoplasmic
domain of E2 interacts with the C protein of the icosahedral NC, causing an icosahedral arrangement
of peplomers on the viral envelope. The arrangement of the genomic RNA within the virion is not
known. *(a)* is a model in which the RNA is free within the NC and *(b)* is a model in which the RNA
interacts in a regular manner with C protein

one of which is a glycoprotein. The virions are spherical particles 36–44 nm in
diameter (Murphy, 1980; Fig. 8-11a and b). The peplomers (spikes) are clearly
resolved in virions isolated from sucrose gradients and they form a thin layer
around the virion. On some flaviviruses, ring-like structures 7 nm in diameter are
observed (Fig. 8-11c and d). The virus core is an electron-dense spherical structure
27 nm in diameter.

II. Molecular Organization

The *Flaviviruses* have not been as extensively studied by molecular biological
techniques as *Alphaviruses* for several reasons. They have a latent period of at least 12
hours and they do not shut off host cell macromolecular synthesis (Westaway,
1980). Thus pulse-labeling experiments early in the infectious cycle do not yield
clear results. Finally, the virions are somewhat fragile and difficult to characterize.
Nevertheless, significant information about the structural proteins is now
available, and cloning of *Flavivirus* genes is in progress.

The replication of *Flaviviruses* has been reviewed by Schlesinger (1977) and
Westaway (1980). The *Flavivirus* genome is a single stranded 4×10^6 dalton RNA
molecule of positive or message sense. It is capped at the 5' end, but is not
polyadenylated (Wengler and Wengler, 1981). Recent evidence (J. Dalrymple,

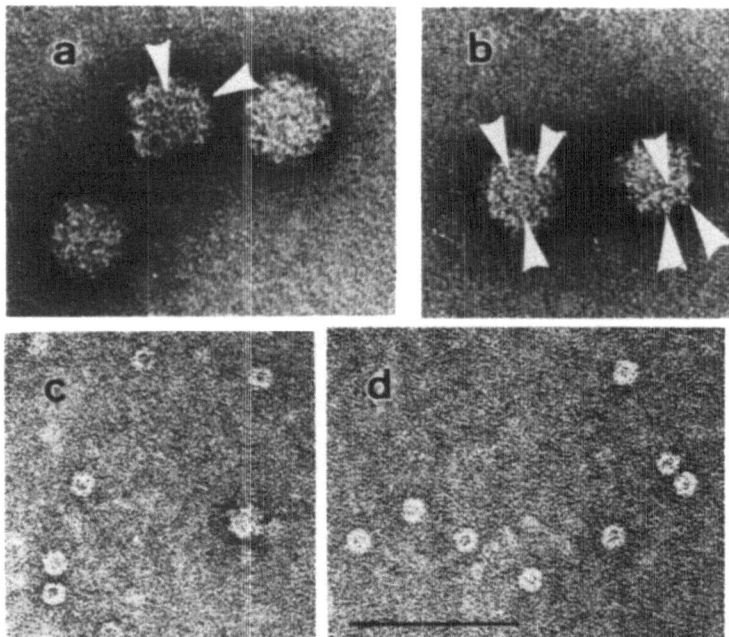

Fig. 8-11. Structure of *flavivirion* and subviral particles. *(a)* and *(b)* show that dengue virions are spherical particles which have circular subunits (arrowheads) 7 nm in diameter on the surface of the virus envelope. Slowly sedimenting hemagglutinin (SHA) particles, which are released during viral infections, are shown in *(c)* and *(d)*. These contain V3, V1 and NV2 proteins in a 14 nm ring-shaped structure. Magnification: ×140,000. (Figs. 8-11a–d; courtesy of Dr. W. Brandt, from Murphy, 1980; with permission from Academic Press)

personal communication) shows that the 3′ ends of at least three different *Flaviviruses* share a sequence of 9 to 22 nucleotides. The genes for the structural proteins are located at the 5′ end of the virus genome, and the NC protein is the 5′ terminal protein (J. H. Strauss, personal communication).

Replication and transcription of *Flaviviruses* differ markedly from *Alphaviruses* (Westaway, 1980; Strauss and Strauss, 1983). Within flavivirus-infected cells, the only virus-specific RNAs to be identified are 42-44S single stranded RNA and a 20S double stranded replicative intermediate. No subgenomic mRNAs have been detected in infected cells.

No virion-associated RNA-dependent RNA polymerase activity has been characterized, although it appears likely, by analogy with *Alphaviruses,* that several non-structural proteins may participate in this function in the infected cells. The virus replicase/transcriptase activity appears to be associated with membranes (Westaway and Ng, 1980). Autoradiography shows that virus RNA synthesis occurs in a sharply circumscribed area of the cytoplasm near the nucleus (Matsumura *et al.,* 1971).

There are three structural polypeptides in *Flaviviruses* whose properties are summarized in Table 8-2 (Russell *et al.,* 1980). The exact locations and interactions of these proteins within the virion have not been determined as precisely as for the *Alphaviruses.*

Table 8-2. *Structural Proteins of* Flaviviruses

Usual name	Apparent MW X10^{-3}	Known or putative function
V2	14	NC protein
V1	8	Non-glycosylated membrane protein May function as M protein
V3	51—59	Peplomeric glycoprotein Hemagglutinin Binds to cell surface receptors

Adapted from Westaway, 1980.

(1) *V2* (MW: 14 K) is the NC protein. It binds to virus RNA and forms the NC which lies within the viral envelope.

(2) *V1* (MW: 8 K) is a non-glycosylated membrane protein which appears to lie beneath the membrane in intact virions, since it cannot be labeled with [125]I. V1 may function like an M protein. Amino acid sequences deduced from cloned flavivirus genes suggest that V1 is probably an integral membrane protein derived from a glycosylated precursor by proteolytic cleavage (J. H. Strauss, personal communication).

(3) *V3* (MW: 51 K—59 K) is a glycoprotein which comprises the peplomers. It is the virus hemagglutinin and probably binds the virus to receptors on host cells. How V3 interacts with V1 and V2 is not known.

(4) Non-structural proteins. Westaway (1980) summarizes the data on many non-structural virus-specific proteins in *Flavivirus*-infected cells. Work on the identification of structural relationships and functions of these proteins is in progress.

III. Protein Synthesis, Transport and Post-Translational Modifications

Fluorescent antibody labeling (Fig. 8-12) shows that the non-structural proteins are made in the perinuclear area. Then the structural proteins become dispersed through the cytoplasmic membranes, while the non-structural proteins remain associated with the perinuclear membrane (Cardiff *et al.*, 1973). The first morphological change seen in cells infected with *Flaviviruses* is proliferation of smooth and granular membranes in the perinuclear area (Baruch, 1963; Leary and Blair, 1980; Murphy, 1980; Figs. 8-13 and 8-14). Possibly these intracellular membranes are required for *Flavivirus* RNA polymerase activity.

Cell fractionation studies suggest that all of the virus proteins are synthesized in association with cellular membranes (Westaway, 1980) and that translation of virus structural and non-structural proteins occurs on heavier membranes than translation of cellular proteins (Westaway and Ng, 1980). This is markedly different from the other enveloped RNA viruses in which only the glycoproteins are synthesized on membranes. Preliminary *in vitro* translation studies have shown that

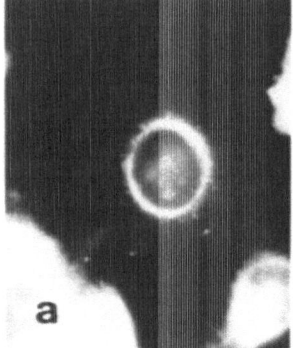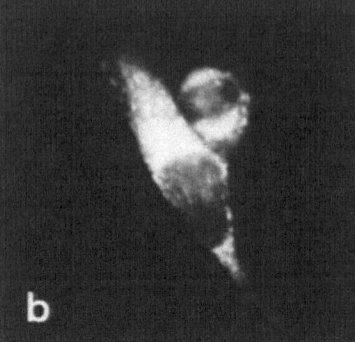

Fig. 8-12. Distribution of the antigens of the *Flavivirus* dengue-2 in infected cells. In *(a)*, fluorescent labeling with antibody to non-structural proteins (soluble complement-fixing antigen, SCF) yields sharply demarcated perinuclear staining, whereas labeling with antibody to virion structural proteins (rapidly sedimenting hemagglutinin) *(b)* gives diffuse, granular cytoplasmic staining. Magnification: ×1,470. (Courtesy of Dr. W. Brandt, from Cardiff *et al.,* 1973, with permission from American Society for Microbiology)

the 42S RNA from infected cells can direct the synthesis of a large polyprotein which contains sequences of some of the structural proteins (Wengler *et al.,* 1979).

Pulse labeling studies have not yet yielded a clear picture of the precursor-product relationships of the virus-specific proteins. These studies and pactamycin mapping studies suggest that for *Flaviviruses,* unlike other RNA viruses, there may be multiple internal initiation sites for translation (Westaway, 1980). Alternatively, proteolytic cleavage of a polyprotein may be required to generate the structural proteins. Cloning of a *Flavivirus* genome and determination of its nucleotide sequence is likely to resolve this central question in *Flavivirus* replication.

Only the V3 protein is glycosylated. It has been detected on the plasma membrane of infected cells by immunofluorescence and immunoelectron microscopy.

IV. Assembly of Virus Components

The assembly of *Flaviviruses* has not been extensively studied at the molecular level. The NCs from disrupted virions have been visualized in negatively stained or shadowed preparations. Because of the small size, lability and indistinct capsomers of *Flavivirus* NCs, it has been difficult to determine the precise geometry of the cubical symmetry of the NCs.

In contrast to *Alphaviruses, Flaviviruses* have no free NCs in infected cells. Isolated *Flavivirus* NCs are 27 nm in diameter, significantly smaller than *Alphavirus* NCs and similar to the size of ribosomes. Possibly, some of the rough membranes seen in *Flavivirus*-infected cells are not RER, but membranes with bound *Flavivirus* NCs. This question may be resolved by immunoelectron microscopy.

Virtually nothing is yet known about the way in which the NC protein, V2, binds to the genomic RNA or to other virus structural proteins. Indeed, as Murphy points out in his comprehensive review of *Flavivirus* morphogenesis (1980), it has

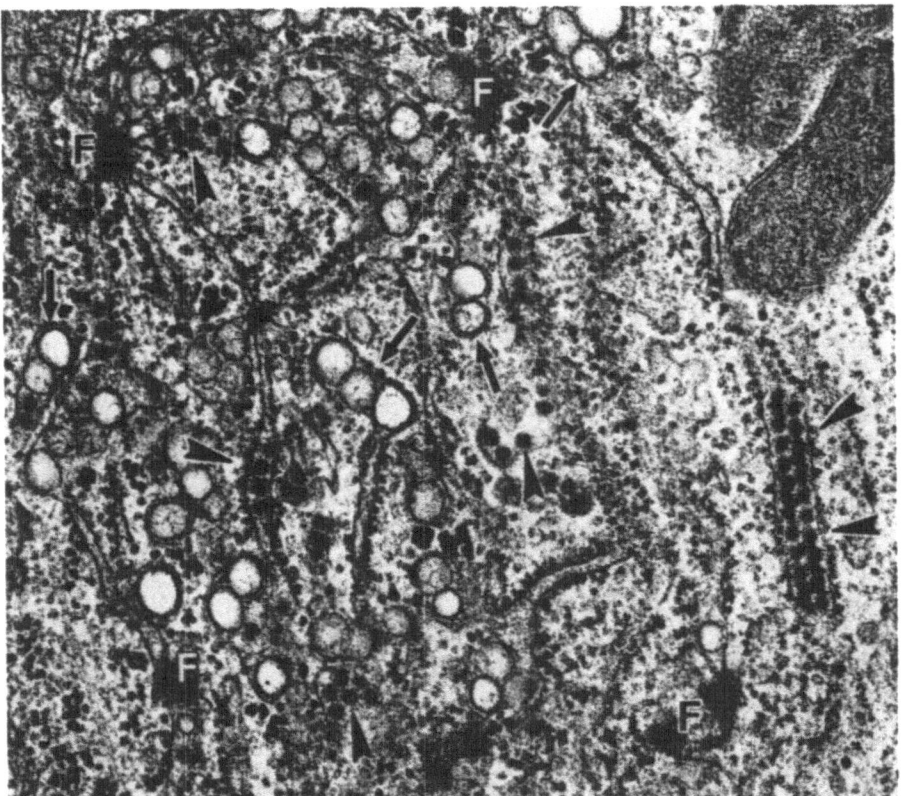

Fig. 8-13. Cells infected with the *Flavivirus,* yellow fever virus. Enveloped virions are seen in the lumen of the RER (arrowheads) both singly and in tightly packed arrays. The RER also contains membranous spheres (dark arrows) called type 1 cytopathic vesicles (CPV-1) which contain fine electron-dense filaments. *Flavivirus*-infected cells frequently contain dense filamentous material *(F)* in association with extensive rough and smooth intracytoplasmic membranes. Magnification: × 23,000. (Figs. 8-13 to 8-16, are courtesy of Dr. P. Grimley)

not yet been definitely shown that *Flaviviruses* mature by budding like other enveloped RNA viruses. Budding certainly is the most probable mechanism for the formation of a virion with this structure, but thin sections of *Flavivirus*-infected cells do not clearly show partially formed virus buds either on the intracellular membranes or the plasma membrane (Deubel *et al.,* 1981; Figs. 8-13 to 8-15), although some observers have claimed to see rare budding virions (Matsumura *et al.,* 1977; Murphy, 1980; Bhatt *et al.,* 1981). As an alternative to budding from membranes, assembly of virions from components within cytoplasmic vesicles has been proposed (Leary and Blair, 1980). This seems unlikely, in light of information about the assembly of other enveloped RNA viruses. Failure to visualize virus budding could be explained if budding were a very rapid event. The very small size of the *Flavivirion* may also be a factor which limits ability to visualize budding virions. Since fixatives such as glutaraldehyde and osmium tetroxide, do not fix lipids rapidly, limited flow of membrane components is possible even after fixation. This could allow virus buds to be flattened or pinched off after fixation. Indeed, images of fusing pinocytic vesicles at the plasma membrane were rarely observed until the development of fast-freezing (see Chapter 1), a new technique which freezes cells in milliseconds. This technique might be useful for visualizing budding of *Flaviviruses.*

Regardless of their mechanism of formation, it is clear that *Flaviviruses* are released into RER, Golgi (Baruch, 1963; Blinzinger, 1972; Figs. 8-13 to 8-15) and probably also from the plasma membrane (Murphy, 1980). In cultures of vertebrate cells infected with *Flaviviruses,* large numbers of virions in tightly packed hexagonal arrays or in crystals have frequently been observed at plasma membranes, although

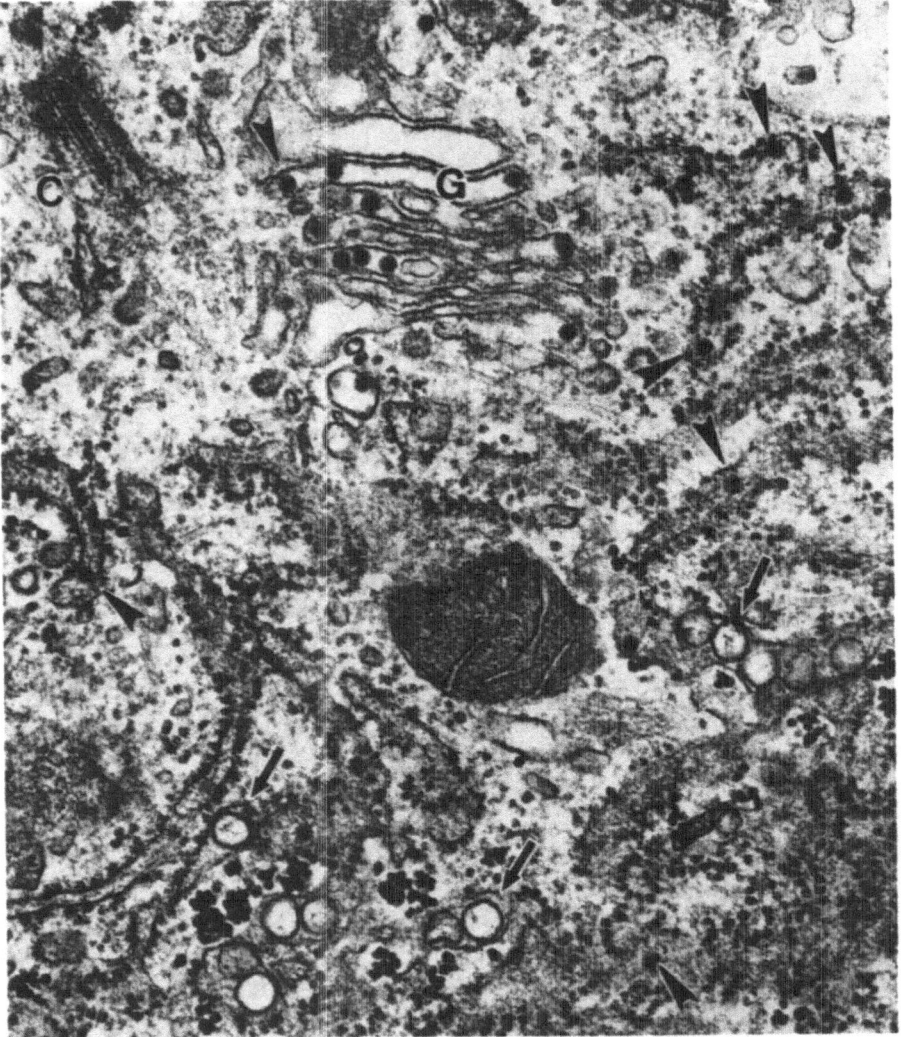

Fig. 8-14. A cell infected with St. Louis encephalitis virus (SLE). Enveloped virions are seen in the Golgi cisternae *(G)* and in the lumen of the RER (small arrowheads) and CPV-1 are seen in the RER (arrows). In *Flavivirus*-infected cells, budding virions have almost never been visualized. *C* indicates a centriole. Magnification: ×47,500

no budding virions were observed (Fig. 8-15a). In insect cells, large numbers of *Flavivirions* may accumulate within RER lamellae (Sinarachatanant and Olson, 1973; Deubel *et al.*, 1981; Fig. 8-15b). What factor(s) determines the location of virus development in vertebrate and arthropod cells infected with *Flaviviruses* is not known.

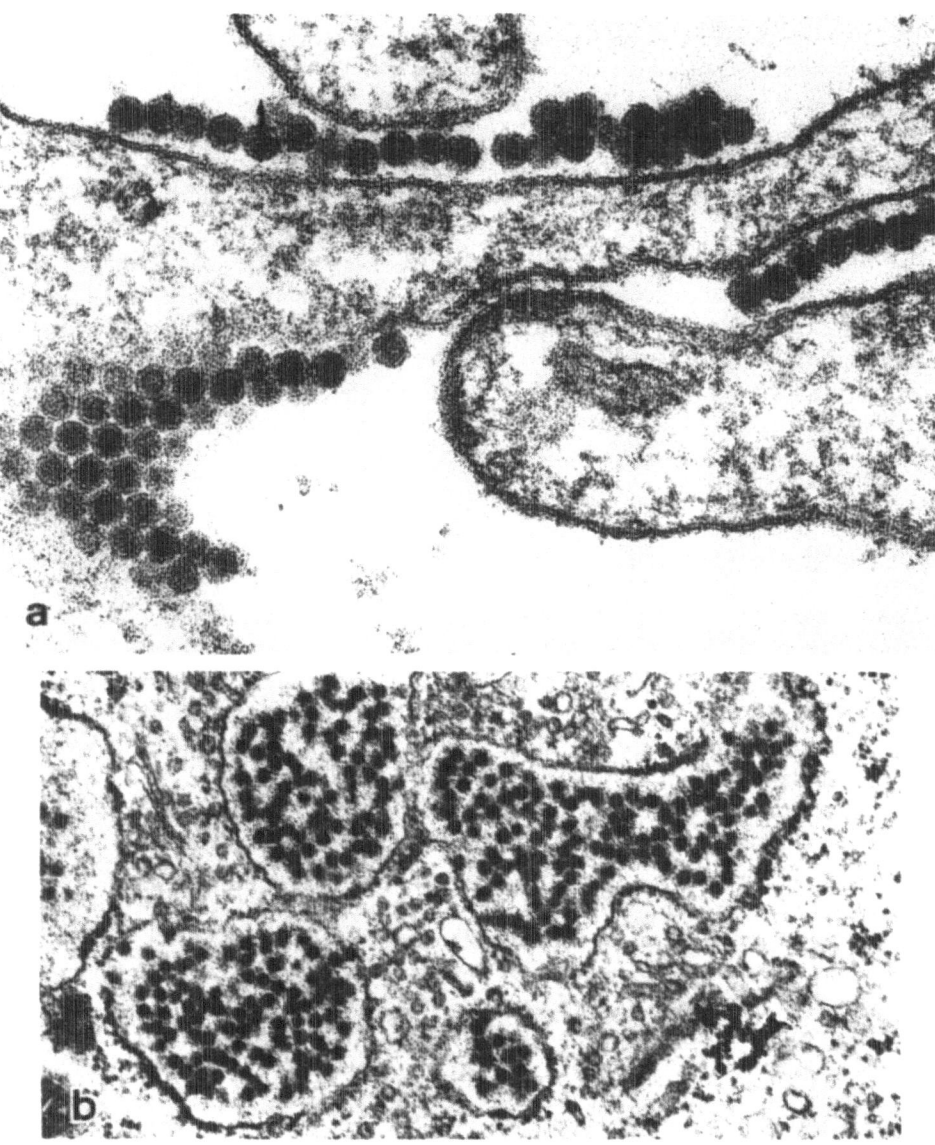

Fig. 8-15. Comparison of *Flavivirus* infection in cells from vertebrates and invertebrates. In *(a)*, yellow fever virions are seen in tightly packed hexagonal array on the plasma membrane of a cell from a vertebrate and some virions are also seen in intracytoplasmic vacuoles as observed with other viruses (Figs. 8-13d, 8-14). In contrast, in a mosquito cell, large numbers of yellow fever virions accumulate in the lumen of the RER *(b)* and fewer virions are seen at the plasma membrane. Magnifications: *(a)* ×115,000; *(b)* ×40,000

The intracellular virions of dengue type-2 virus and Japanese B encephalitis virus have been isolated from disrupted cells and compared with released virions (Shapiro *et al.*, 1972 and 1973). The intracellular virions contained V3, V2, and NV2, a non-structural protein. These virions, of much lower specific infectivity than the released virions, may represent precursors to the infectious released virions (Westaway, 1980).

Two mechanisms for release of intracellular virions from infected cells have been proposed (Schlesinger, 1977; Murphy, 1980; Westaway, 1980). Virus-filled vacuoles may migrate through the cytoplasm to fuse with the plasma membrane or virions may be released by lysis of infected cells (Demsey *et al.*, 1974). Release of virions through membranous channels connected to the external medium has been demonstrated by serial thin sectioning, but cell lysis may also release some intracellular virions.

Viruses which bud from the plasma membrane often show preferential budding from the apical or basolateral domains of the membrane of epithelial cells (Rodriguez-Boulan, 1983). Although *Flaviviruses* appear to form at intracellular membranes, some polarity in virus release from epithelial cells has been detected. In insect salivary gland, SLE virus release occurred preferentially from the apical region of the cell rather than from the basolateral region (Whitfield *et al.*, 1973).

V. Defective Budding

Some membrane vesicles, like the CPV-1 vesicles seen in cells infected with *Alphaviruses,* have been observed in *Flavivirus*-infected cells (Murphy, 1980; Fig. 8-13). The function of these small vesicles within cytoplasmic membranes is not known and the delicate strands of material in the vesicles have not been characterized. By analogy to *Alphaviruses* (Grimley *et al.*, 1972), these structures might play a role in RNA synthesis.

A unique striated filamentous inclusion has been observed in *Flavivirus*-infected cells by many investigators (McGavran and Easterday, 1963; Blinzinger *et al.*, 1971; Matsumura *et al.*, 1977; Oshiro *et al.*, 1978; Bhatt *et al.*, 1981; Grimley and Henson, 1983; Fig. 8-16a and b). The inclusion consists of long fibers or sheets of electron-dense strands in a crystalline pattern with fine, short filaments which are perpendicular to the long strands and regularly spaced along them. The composition of these filaments is not known.

The proliferation of membranes in *Flavivirus*-infected cells sometimes leads to the formation of complex tubular cytoplasmic inclusions. Fig. 8-16c shows such tubular inclusions in a tightly packed honeycomb pattern. The composition of these tubules has not been determined.

In preparations of *Flaviviruses,* a subviral component has been identified and designated the slowly sedimenting hemagglutinin (SHA). When isolated by density gradient ultracentrifugation, the 70S SHA was found to consist of doughnut shaped structures 14 nm in diameter (Smith *et al.*, 1970; Fig. 8-11c and d) and to contain V3, V1, and NV2 proteins. The doughnut-shaped structures on the surface of *Flavi-*

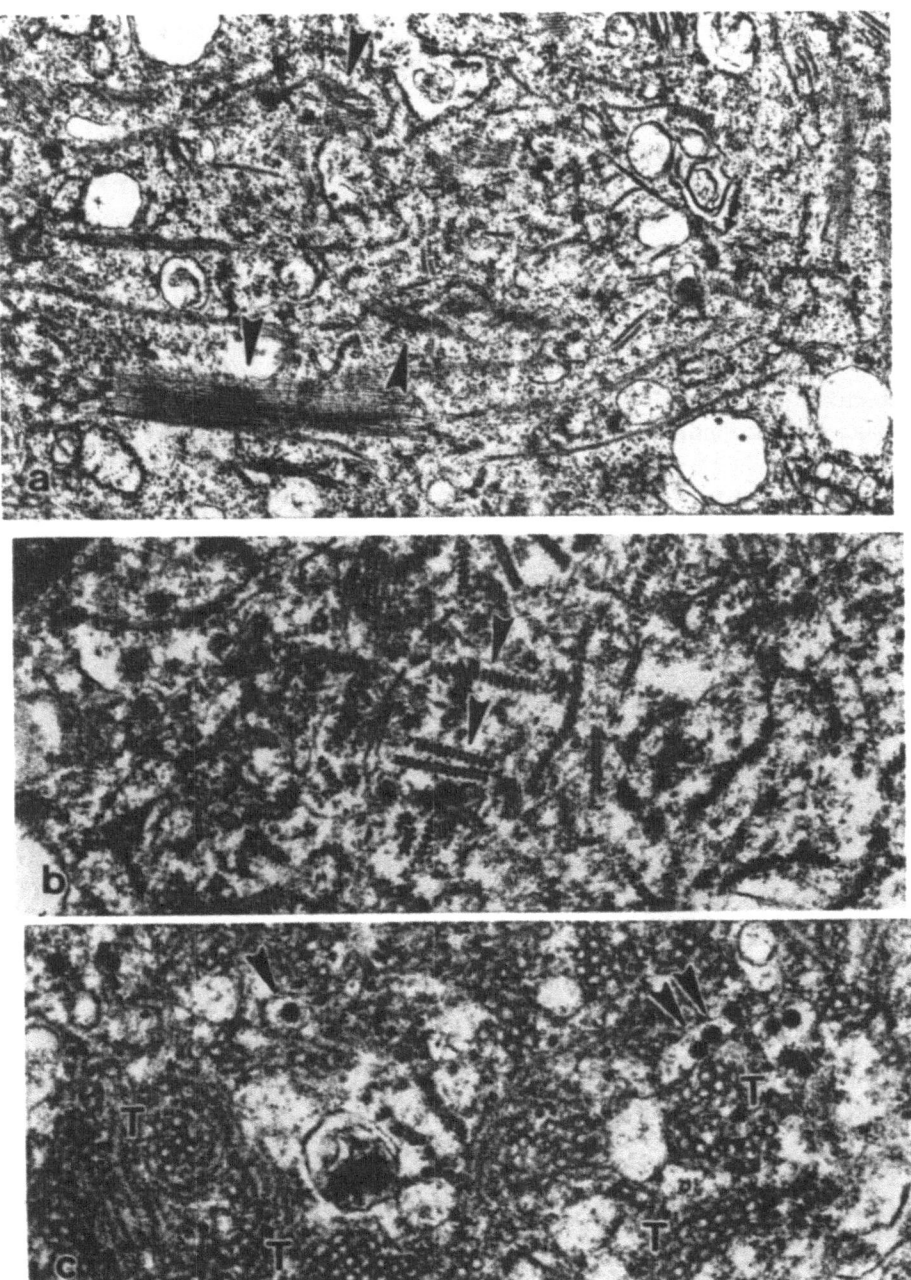

Fig. 8-16. Unique intracytoplasmic inclusions in cells infected with *Flaviviruses (a)*. In cells infected with SLE virus, the cytoplasmic matrix is filled with electron-dense, filamentous material in sheets or bundles (arrowheads) *(b)*. At higher magnification, fine, electron-dense cross-striations are seen on the filaments in a regular pattern which forms a lattice on the sheets and a beaded appearance on the filamentous bundles (arrowheads) . *(c)* shows cytoplasmic inclusions of tightly packed tubules *(T)* in a honeycomb pattern. The compositions of the filamentous and tubular inclusions are unknown.

Virions are seen in vacuoles (arrowheads). Magnifications: *(a)* ×20,000; *(b)* and *(c)* ×47,500

viruses in negatively stained preparations are only 7 nm in diameter (Fig. 8-11a and b), and are therefore probably different from the SHA. It is not clear whether the SHA is derived from virions or from cellular membranes which have been altered by the insertion of virus proteins.

A second subviral particle found in *Flavivirus* preparations from brain or serum of infected animals is a 4S "soluble" complement-fixing antigen (SCF) which is highly antigenic. The SCF apparently consists of non-structural virus antigens. The mechanisms of formation and release of this SCF are not yet understood.

VI. Organization of the Virion

The foregoing discussion of *Flavivirus* morphogenesis illustrates the mysteries which still surround the formation and replication of these medically important viruses. A model to illustrate present hypotheses about *Flavivirus* replication and assembly is shown in Fig. 8-17. A schematic model showing possible organization of *Flavivirus* structural proteins is shown in Fig. 8-18.

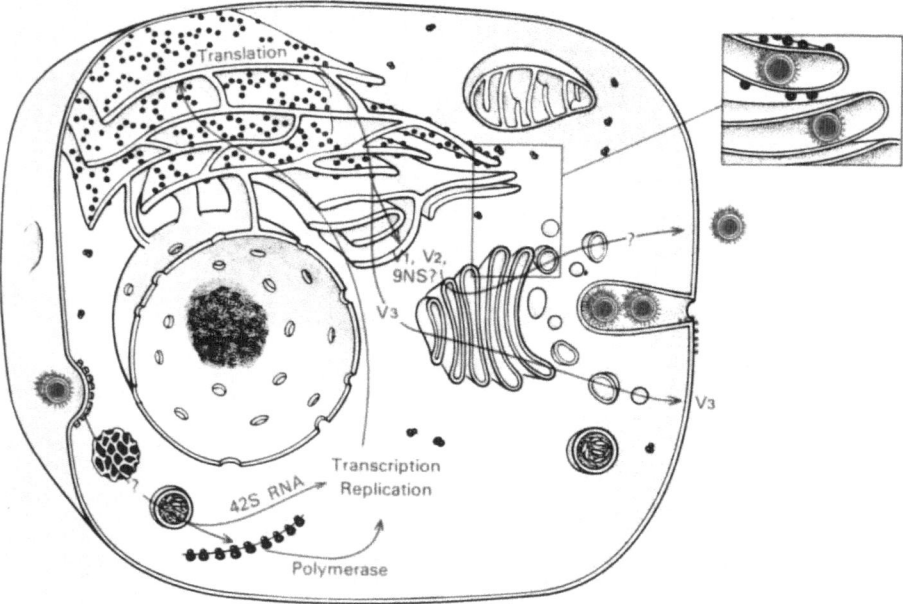

Fig. 8-17. A model summarizing current concepts of the replication, transcription and assembly of *Flaviviruses*. It is not clear whether virions are taken up by clathrin-coated vesicles. The virus genomic RNA acts as mRNA to direct the synthesis of RNA-dependent RNA polymerase. Only full-length negative and positive strands of RNA are found in *Flavivirus*-infected cells. Translation of all 3 structural and 9 non-structural proteins of *Flaviviruses* appears to occur on membrane-bound polysomes. Non-structural proteins appear to remain in the perinuclear area while the peplomeric glycoprotein V3 is transported through the Golgi to the plasma membrane. Late in infection, many of the virus proteins are found in association with the plasma membrane. No NCs have been observed in the cytoplasm of infected cells. Although enveloped virions are found in the lumen of the RER and Golgi (window) and on the plasma membrane, budding of virions has not been seen. Release of virions from intact cells may be *via* the cellular secretory apparatus

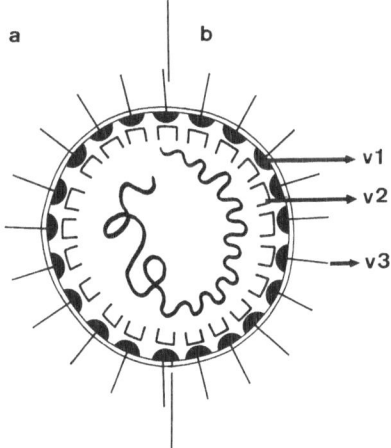

Fig. 8-18. Tentative model of the organization of a *Flavivirus*. The peplomers appear to consist of the V3 glycoprotein. A second, non-glycosylated envelope protein, *V1*, appears to lie beneath the lipid bilayer. The capsid protein, *V2*, with the genomic RNA, forms a NC with cubic symmetry. *(a)* indicates that the RNA may be coiled within the capsid. Alternatively, the RNA may interact in a regular manner with the capsid protein V2 *(b)*

A major advance in the study of *Flavivirus* will be the development of molecular clones of the virus genes which will permit genetic mapping and analysis of virus protein synthesis, processing and transport. Monoclonal antibodies against the virus structural and non-structural proteins will also help to characterize the intracellular transport and processing of these proteins. Immunoelectron microscopy and cytochemical techniques such as those applied to the *Alphaviruses* will contribute to our understanding of the morphogenesis of *Flaviviruses*.

Rubiviruses

Rubella virus, which causes a mild exanthem in children and adults and can induce severe congenital malformations following intrauterine infection, is the only member of the *Rubivirus* genus. It is included in the *Togaviridae* family because of its structural similarities to other family members (Brinton, 1980). The virions are spherical particles 50 to 75 nm in diameter, containing 33 nm cores with cubic symmetry. A fuzzy layer of surface projections covers the virus envelope (Bardelletti *et al.*, 1975; Murphy, 1980).

Rubella virus has a single-stranded RNA genome of MW $3-4 \times 10^6$. The genomic RNA is of positive polarity, since it is infectious. The virions contain at least 3 structural proteins: a NC protein, and two or three glycoproteins associated with the virus envelope (van Alstyne *et al.*, 1981; Ho-Terry and Cohen, 1982; Okerblom *et al.*, 1983).

Rubella virus matures by budding from intracellular membranes or, in some cells, from the plasma membrane (Murphy, 1980). It is unclear what factors determine the different budding sites of rubella virus. Like the *Flaviviruses*, no rubella virus NCs are observed free in the cytoplasm. Unlike the *Flaviviruses*, however, budding virions are readily observed in thin sections of infected cells.

Pestiviruses

The *Pestiviruses*, hog cholera virus and bovine diarrhea virus, show morphological and biochemical similarities to other togaviruses, although they are not arthropod-borne (Murphy, 1980; Brinton, 1980). The virions are spherical particles 40 to 60 nm in diameter, with an indistinct layer of surface projections 6 to 8 nm long. The cores released from the virions appear to have cubic symmetry and are 25–30 nm in diameter (Horzinek, 1973 b).

The formation of *Pestiviriuses* is similar to the formation of *Flaviviruses* in that budding virons are hard to detect, no free NCs are observed in the cytoplasm and virions are released into the lumen of the RER or Golgi apparatus.

Non-Arthropod-Borne Togaviruses

Some non-arthropod-borne viruses including lactic dehydrogenase virus and equine arteritis virus have been included in the togavirus group because of morphological similarities (Brinton, 1980; Murphy, 1980; Horzinek, 1981). However, these viruses are structurally somewhat different from other togaviruses (Murphy, 1980) and they do not cross react serologically with any of them. It appears likely that they may form a separate genus within the family of *Togaviridae*. Additional studies on the molecular biology and assembly of the non-arthropod borne togaviruses will be required to aid in their classification.

The virions of equine arteritis virus are 50 to 70 nm in diameter and contain hollow cores of 35 nm diameter. The virions appear to form by budding into intra-cytoplasmic membranes (Murphy, 1980).

Conclusion

The *Togaviridae* family is a collection of several different types of small enveloped RNA viruses. It is clear that the mechanism of virus budding must be considered separately for each of the togavirus genera. Only the *Alphaviruses* have been studied sufficiently at the molecular level to permit construction of a detailed model of virus budding. Future molecular biological studies will elucidate the mechanisms for assembly release of the other togaviruses.

9

Assembly of *Retroviridae*

I. Introduction

Retroviruses are widely distributed among vertebrate species, can be transmitted horizontally and vertically, require RNA-directed DNA synthesis for their replication, and are natural agents of oncogenesis in several species (Bishop, 1978). The virion is about 100 nm in size, contains a coiled strand of RNA complexed to proteins (NC), which forms a crescent structure during budding and a dense core after release, and has an envelope bearing clusters of spikes or knobs (Fig. 9-1). The most extensively studied retroviruses are of avian and murine origin, but retroviruses also exist in feline and bovine species, as well as in primates. The prototype avian retrovirus is avian sarcoma virus (ASV), whereas there are two prototypes of murine retroviruses, murine leukemia virus (MuLV), and mouse mammary tumor virus (MMTV). There are also retroviruses in sheep called "lentiviruses"; one of them is visna virus which causes a slow encephalitis in sheep (Haase, 1975). The prototype non-human primate virus is MPMV (Mason-Pfizer monkey virus). More recently, a new group of retroviruses has been isolated from human T cell leukemia (reviewed in Gallo and Wong-Staal, 1982). Since less information is available on the molecular structure and synthesis of primate retrovirus polypeptides, they will not be covered in the first section of this chapter. In most cases, retroviruses are not cytopathic and do not have dramatic effects on cellular metabolism.

We will first briefly review the molecular organization of avian and murine retroviruses, and then the assembly events and mechanisms. The reader is referred to extensive reviews on the subject for more details (Bishop, 1978; Varmus, 1982; Dickson *et al.*, 1982). As in the review of Bolognesi *et al.* (1978), we will attempt to present a unifying view of retrovirus assembly in spite of the well-known differences in the site of NC assembly and virion structure existing between subgroups of this large family.

II. Molecular Organization

Retroviruses have a diploid genome, *i.e.* two identical single-stranded, positive-sense RNA molecules, which are capped and polyadenylated, are found within each

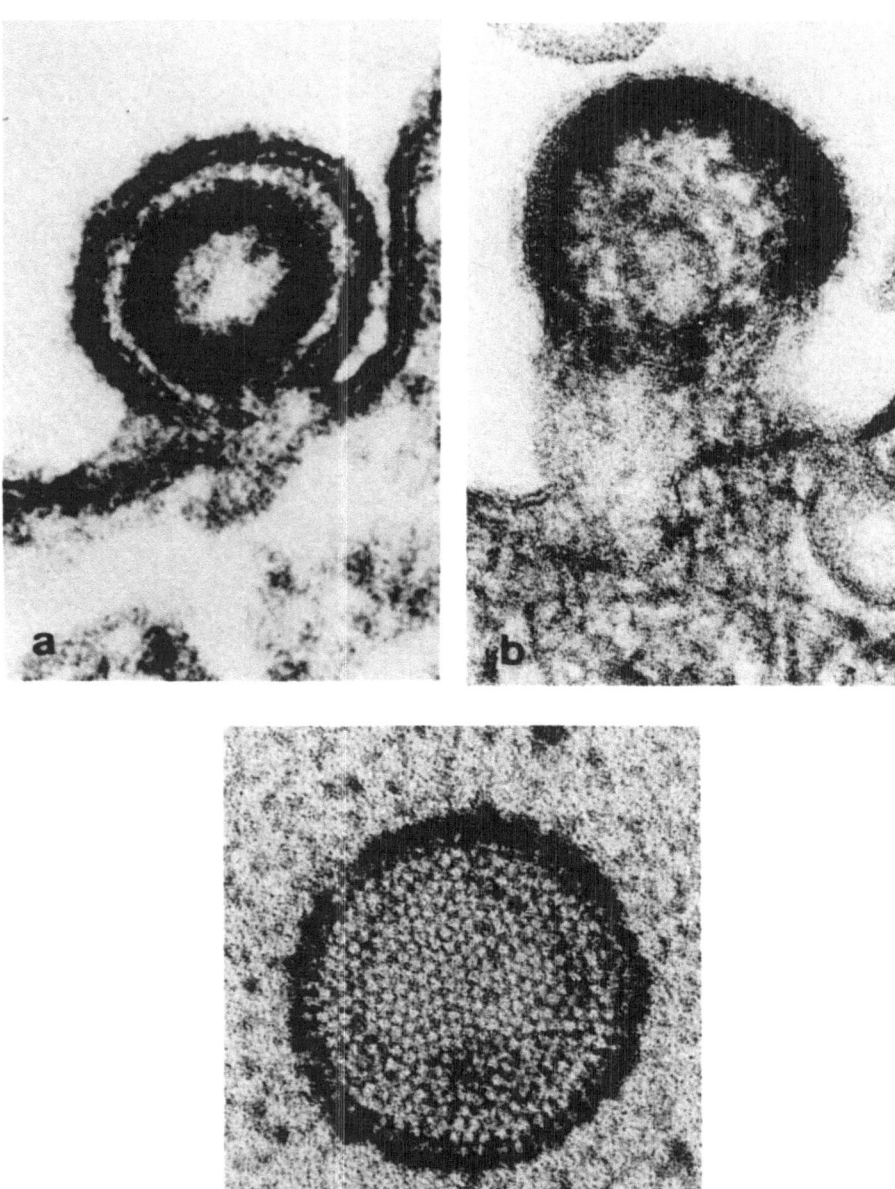

Fig. 9-1. Retrovirus particles budding (a and b) or released from the cell (c). *(a)* Cell infected with Friend leukemia virus. The outer and inner leaflets of the plasma membrane are resolved and in continuity with the leaflets of the virus envelope. The dense core is separated from the virus envelope by an electron-lucent layer. The core is forming a ring, because the virus is almost detached from the cell. *(b)* Visna virus particles budding from a sheep choroid plexus cell. In this case, the dense core is still growing and has the shape of a crescent closely apposed to the unit membrane which displays distinct spikes. *(c)* Mouse mammary tumor virion after negative staining with phosphotungstic acid at neutral pH. Distinct spikes are seen which are sometimes organized hexagonally. Magnifications: (a) ×232,000, (b) ×190,000, (c) ×285,000. [Courtesy of (a) Dr. E. de Harven from De Harven, 1974; reproduced with permission of Academic Press, New York; (b) Dubois-Dalcq *et al.,*1979 b; reproduced with permission of Academic Press, New York; (c) Dr. C. M. Calberg-Bacq]

virion. This is a unique phenomenon among the known animal viruses, and it facilitates formation of heterozygotes and genetic recombination. Each single-stranded RNA molecule has a short segment at its 5' end where hydrogen bonds link the two subunits together (Bender and Davidson, 1976). Viral RNA has redundancies (R) at both 5' and 3' ends. During the replicative cycle, parental RNA is transcribed by reverse transcriptase *(pol)* into virus-specific DNA (Fig. 9-2) (Varmus, 1982; Hughes, 1983). Synthesis of the first negative DNA strand is initiated by a host tRNA primer which is bound close to the 5' end of the genome RNA. Reverse transcription produces a linear form of virus DNA which is longer than the haploid subunit of the virus genome. This occurs by the transfer of a nascent DNA strand twice between templates and the fusion of unique sequences (U) at both ends of the genomic RNA (U3 and U5) during reverse transcription (Fig. 9-2). Unintegrated linear virus DNA molecules have long terminal repeats (LTRs) with the sequence U_3-R-U_5 at both ends. LTRs carry regulatory signals, such as a transcription initiation site and a poly(A) addition site. Before virus DNA is integrated into the host cell genome, it exists in the form of circular duplexes, containing either one or two LTRs. Integrated linear virus DNA (provirus) serves as the template for synthesis in the nucleus of virus genomic RNA and of virus messenger RNAs from which virus proteins are synthesized (Fig. 9-2).

The haploid subunit of single-stranded genomic RNA contains three genes necessary for replication of infectious virus. They are, from the 5' end, *gag*, which

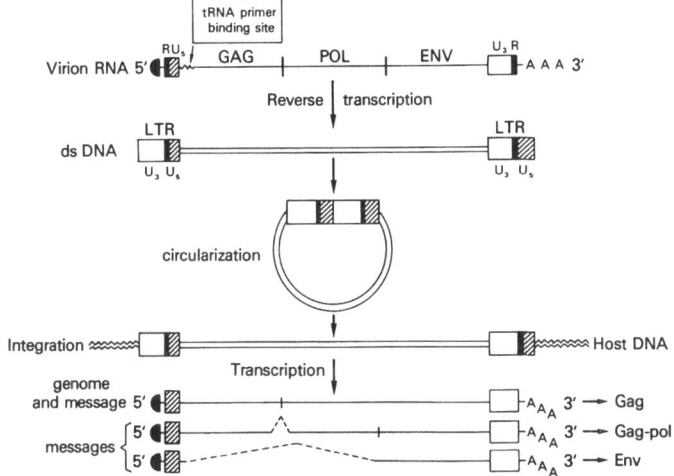

Fig. 9-2. Representation of retrovirus replication and transcription. The virion RNA is capped at its 5' end and polyadenylated at its 3' end and contains the three replicative genes, GAG, POL and ENV. There is a short sequence repeated at each terminus called R. A sequence unique to the 5' terminus is called U5 while a sequence unique to the 3' terminus is called U3. After reverse transcription, the double stranded DNA has at each end a long terminal repeat or LTR composed of U3, R, and U5. This DNA then becomes a supercoiled circular structure which contains one or two LTR sequences (a circle with two LTRs is represented). The provirus DNA then becomes integrated in the host DNA. Transcription of the provirus DNA results in the formation of new virus genomes and mRNAs. The messages molecules for the structural genes are indicated. The dashed lines in the mRNA diagrams indicate regions spliced out to create subgenomic mRNA molecules. In some cases, the *pol* mRNA is thought to contain a splice site near the *gag-pol* junction. Modified from Dickson *et al.* (1982) and Varmus (1982)

Table 9-1. *Proteins of Retroviridae: The gag Molecule and its Products*[1]

	Precursor		*gag 1* Lipid binding, hydrophobic (amino-terminal)	*gag 2* Phosphoprotein, binding to genome	*gag 3* Core shell protein	*gag 4* Basic, RNA-binding	*gag 5* Protease (carboxyl-terminal)
Avian C Prototype, ASV	p76		pp19	p10	p27	p12	p15
Murine C Prototype, MuLV	p65		p15	pp12	p30	p10	?
Murine B Prototype, MMTV	p77		p10	pp21	p28	p14	?

[1] From Dickson *et al.*, 1982.

codes for the core proteins antigens, *pol*, which codes for reverse transcriptase, and *env*, which codes for the envelope glycoproteins (Fig. 9-2). Both *gag* and *env* antigens may be group- or type-specific. The order of these three genes seems invariant among retroviruses. A cellular transforming gene may be added to these genes by a recombination event between transforming and non-transforming strains. More commonly, the acquisition of a host cell transforming gene is done at the expense of virus genetic information.

Virus Proteins

We will now summarize the structure of each protein encoded by the replicative genes and their putative functions. For the purpose of clarity, the *gag* and *env* products, which have similar properties in the prototype viruses ASV, MuLV and MMTV, have been given numbers (Tables 9-1, 9-2, and 9-3).

(1) *gag* (Table 9-1). The primary product of translation of the *gag* gene is a precursor polyprotein (p76^{gag} in ASV, p65^{gag} in MuLV, and p77^{gag} in MMTV), which are cleaved into four or five core proteins. In addition, in MuLV, there is another glycosylated *gag* polyprotein, gp80 *gag* (see section III, p. 158). The cleaved *gag* proteins of ASV, MuLV, and MMTV are listed in Table 9-1 and numbered 1 to 4 or 5 from the amino terminus to the carboxyl terminus. The fifth protein is a virus-encoded protease highly specific for p76 *gag*, which has been found only in avian retroviruses (Pepinski and Vogt, 1983). The *gag* polyprotein probably becomes phosphorylated while still in a precursor form (Bishop, 1978; Bolognesi *et al.*, 1978) and this phosphorylation increases the structural stability of the core (Durbin and Manning, 1982). Within the *gag* polyproteins, one can distinguish four to five domains between the amino- and carboxyl-termini. The first domain corresponds to the hydrophobic membrane protein, the second to the phosphoprotein, the third to the core shell protein, the fourth to the basic RNA-binding protein and the fifth to the virus protease (Barbacid and Aaronson, 1978; reviewed in Dickson *et al.*, 1982). The pp19 of ASV has both lipid-binding and genome-binding properties (*gag* 1).

Table 9-2. *Proteins of* Retroviridae: *The* gag-pol *Molecule and its Products*[1]

	Precursor in cytoplasm	Intermediates at plasma membrane and immature virion	Final product *pol* in mature virions
Avian C (prototype, ASV)	Pr180	Pr130	two subunits $\alpha = 58\,K$ $\beta = 92\,K^*$
Murine C (prototype, MuLV)	Pr180	Pr150, 140	p80
Murine B (prototype, MMTV)	Pr160	Pr130	p100

[1] From Dickson *et al.*, 1982.
* Phosphorylated.

(2) *pol* (Table 9-2). MW $=160$ to 180. An essential part of all retroviruses is the RNA-dependent DNA polymerase (reverse transcriptase) or *pol* (20 to 70 copies/ virus), which is also responsible for the specific incorporation of the primer tRNA necessary for the initiation of DNA synthesis (Fig. 9-2). The reverse transcriptase is apparently expressed by uninterrupted translation from *gag* and *pol* in the 38S mRNA. A splicing mechanism which removes the termination codon at the end of the *gag* gene generates a *gag-pol* mRNA that lacks the *gag* termination signals (Fig. 9-2). p180$^{gag\text{-}pol}$ is the product of that read-through translation and contains the antigenic determinants of both *gag* and *pol* proteins. This p180$^{gag\text{-}pol}$ cannot be a precursor to any of the mature *gag* gene products but it is the precursor of the active reverse transcriptase found in mature virions. In avian retroviruses, the mature enzyme is a bimolecular complex with two non-identical subunits, α and β, while the murine polymerases consist of only one polypeptide (Table 9-2). Synthesis of the precursor p180$^{gag\text{-}pol}$, at low levels probably serves to mediate the incorporation of a small number of polymerase molecules into the virion (Bolognesi, 1978). There are about 50 *gag* molecules for one *pol* molecule in the virion.

(3) *env*. The *env* product is a polyprotein precursor Pr92 in ASV; Pr80 to 90 in MuLV; gPr73 in MMTV, which yields the envelope glycoproteins gp85 and gp37 in ASV, gp70 and p15E in MuLV, and gp52 and gp36 in MMTV (Table 9-3). In both avian and murine retroviruses, the smaller of the two cleavage products is the one anchored in the cell membrane (spike anchor or *env* 2), while the larger one makes the body of the spike and knobs (surface spike or *env* 1) and is attached to *env* 2 by disulfide bonds (Fig. 9-3, Table 9-3). The smaller product is often more highly conserved, although many type-specific variations have been observed in MMTV-gp36 (Calberg-Bacq *et al.*, 1981). In avian retroviruses, gp85 and gp37 can form a disulfide-linked dimer or tetramer (Ewert and Halpern, 1982). In MuLV, gp70 is linked to p15E not only by disulfide bonds but also by noncovalent associations (Montelaro *et al.*, 1978). A model for the membrane orientation of p15E and its interactions with gp70 has been proposed (Fig. 9-3) (Pinter and Honnen, 1983). The

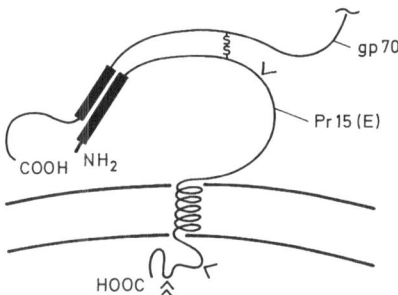

Fig. 9-3. Diagram of structural features of the two envelope proteins of MulV, gp70 and pl5E. Arrow-heads represent sites sensitive to trypsin cleavage. The double arrowheads represent the natural cleavage sites used in the late processing of prl5E to pl5E. The carboxyl-terminal non-polar region of pl5E is represented traversing the membrane in a helical conformation. The heavy lines represent non-polar regions at the aminoterminus of pl5E and carboxyl terminus of gp70, which may be noncovalently associated. In addition, gp70 and pl5E are linked by disulfide bonds. (Courtesy of Dr. A. Pinter, from Pinter and Honnen, 1983; reproduced with permission of ASM Publications, Washington, D.C.)

Table 9-3. *Proteins of* Retroviridae*: The* env *Molecule and its Products*[1]

	Precursor	env 1 Surface spike protein	env 2 Spike-anchoring protein
Avian C (ASV)	Pr92	gp85	gp37
Murine C (MuLV)	Pr80–90	gp70	p15E*
Murine B (MMTV)	gPr73	gp52	gp36

[1] From Dickson *et al.*, 1982.
* Nonglycosylated.

various domains of MuLV gp70 have been analyzed recently using monoclonal antibodies (Pinter *et al.*, 1982). These studies demonstrate that gp70 contains structurally distinct amino- and carboxy-terminal domains with multiple disulfide bonds within but not between these domains. In MMTV, three gp52 and three gp36 molecules make a trimeric structure and a rosette pattern resembling the prominent spikes of MMTV (Dickson *et al.*, 1982) (Fig. 9-1c). Using monoclonal antibodies to MMTV gp52, it was demonstrated that the three patterns of antigenic reactivity (type-, class, and group-specific) were related to individual determinants on the gp52 molecule (Massey *et al.*, 1980).

III. Intracellular Synthesis of Virus Components

We describe here the site of synthesis of the virus precursors, as well as their transport to the site of assembly and their cleavage or post-translational modification. The precursor, *gag*, is synthesized on free ribosomes from the 28S mRNA and is found diffusely throughout the cytoplasm of MuLV-infected cells when viewed by immunofluorescence (Satake and Luftig, 1983) (Fig. 9-4a). In contrast, the core shell protein, p28, of MMTV is detected in patches and inclusions in infected cells of the mouse mammary gland (Kozma *et al.*, 1979) (Fig. 9-5). However, one should realize that the detection of a *gag* product by immunological methods does not necessarily mean that this product has already been cleaved from its precursor.

Many observations suggest indeed that *gag* cleavage occurs mostly at the assembly site at the plasma membrane. For instance, in ASV infected cells, protease p15 (*gag* 5) is only detected at sites of virus assembly under the plasma membrane, suggesting that *gag* cleavage occurs only at the membrane (Vogt *et al.*, 1979). Avian cells contain a cellular protease which cleaves p76^{gag}, allowing the release of p15, which will then cleave the other products (Vogt *et al.*, 1982). Such a protease is not present in mammalian cells which, therefore, do not support good replication of avian retroviruses. Once the avian virus protease, p15, is activated, it cleaves the resulting *gag* intermediate at two to three sites (Fig. 9-6). It is thought that two cellular proteases might be involved in the cleavage of MuLV *gag* (Dickson *et al.*, 1982). Viral protein kinase phosphorylates p65^{gag} at the site of pp12 (*gag* 2) and triggers the cleavage into two intermediates (Yoshinaka and Luftig, 1982) (Fig. 9-6). There are at least four protease sites in p65^{gag} and one is very near the carboxyl terminus of p15 (*gag* 1), which may thus be a possible *in vivo* cleavage site (Yoshinaka and Luftig, 1981). There is also a protease in MMTV, but its identity with one of the known virus proteins has not been established (Dickson, 1982). Cleavage of p77[gag] is sequential, releasing first *gag* 1, then *gag* 2 and finally cleaving *gag* 3 and 4 (Fig. 9-6). An exception to the membrane cleavage of *gag* polyprotein is seen with visna virus: the core shell protein, p30, is detected in large amounts in the cell cytoplasm, suggesting that some cleavage of the *gag* precursor occurs inside the cell (Vigne *et al.*, 1982).

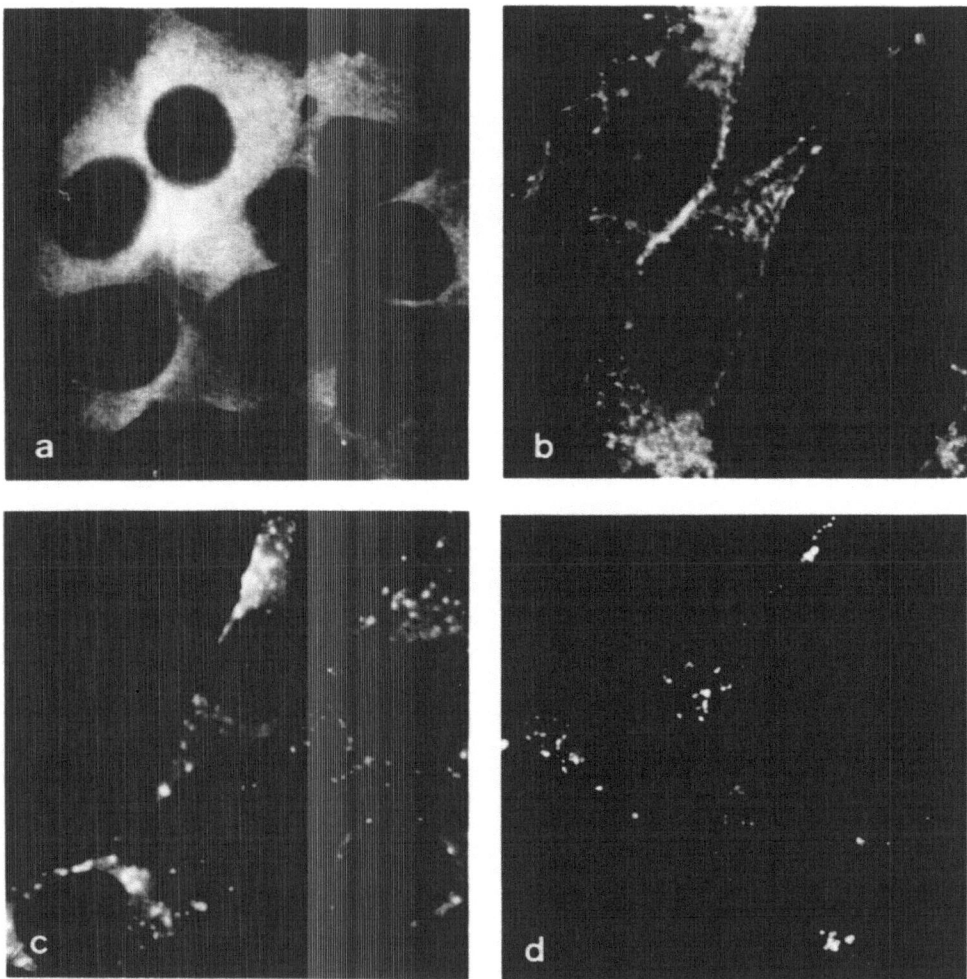

Fig. 9-4. Immunofluorescent staining of four proteins of MuLV after formaldehyde fixation. *(a)* Indirect labeling was performed using as the first antiserum anti-p30. *(b)* Labeling with anti-gp70. *(c)* Labeling with anti-p15. *(d)* Labeling with anti-p15E. The core shell protein, p30, is diffuse throughout the cytoplasm, while all the other proteins are present as fine dots and clusters close to the membrane. Colocalization of p15 and p15E at the membrane level was demonstrated in other experiments. Magnification: × 1290. (Courtesy of Dr. R. B. Luftig, from Satake and Luftig, 1983; reproduced with permission of Academic Press, New York)

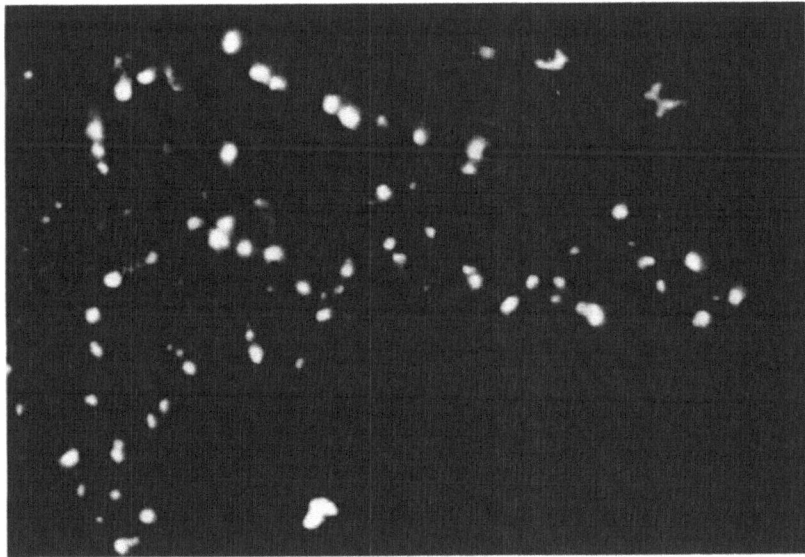

Fig. 9-5. Immunofluorescent localization of the core shell protein, p28, of MMTV in the mouse mammary gland in lactation. This *gag* gene product is clearly located in inclusions inside the epithelium. Magnification: × 360. (Courtesy of Dr. S. Kozma, from Kozma *et al.*, 1979; reproduced with permission of Cambridge University Press, England)

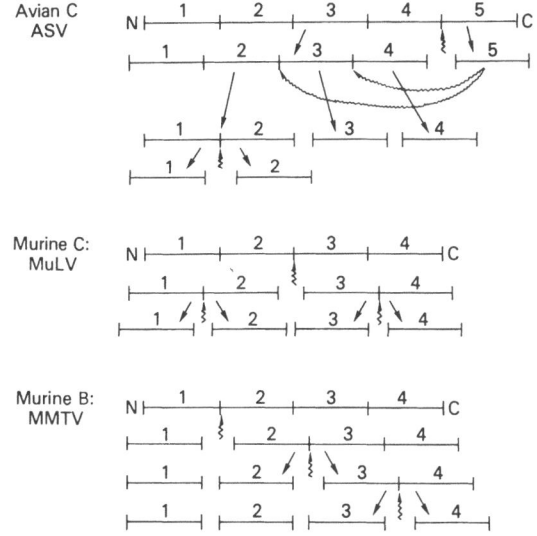

Fig. 9-6. Representation of *gag* cleavage occurring in the three prototype viruses ASV, MuLV, and MMTV. The *gag* products are named 1 to 5 from the amino terminus to the carboxyl terminus, as shown in Table 9-1. ASV has a virus-encoded protease, *gag* 5, which is cleaved by an unknown host cell protease and then activated to cleave the other part of *gag*, at least at two sites. *gag* 1 and 2 are further separated by an unknown protease. Enzyme cleavage sites are indicated by the zig-zag arrows. Cleavage of MuLV occurs in a more symmetrical way and two different proteases, one probably viral, the other probably cellular, appear to be involved. MMTV cleavage occurs in a cascade from the amino terminus toward the carboxyl terminus of the *gag* molecule; no virus-encoded proteases have been identified yet (modified from Dickson *et al.*, 1982)

In MuLV-infected cells, there is a second *gag* protein which is glycosylated (gp80 *gag*) and synthesized independently from p65 *gag*, probably because two different initiation sites in MuLV RNA give rise to the two polyproteins. Virus-specific 28S mRNA associated with free ribosomes might synthesize p65^{gag}, while membrane-bound 38S RNA synthesizes the precursor of gp80 *gag* (Schultz *et al.*, 1981; Dickson *et al.*, 1982). Glycosylated *gag* is not phosphorylated like p65^{gag}, but contains sugar residues and an additional peptide sequence at the amino terminus is not found in p65^{gag}. Such a sequence may function as signal peptide. Additional glycosylation may occur during transport through the Golgi to the cell surface, where the protein might be shed from the cell, probably as cleavage products. Association of gp80 *gag* with the cell surface and with extracellular matrix components has been described (Edwards *et al.*, 1982) and may alter the growth of the cell. Thus, this polyprotein is not present in the virion and is probably not involved in its assembly.

The read-through product *gag-pol* is also synthesized on free ribosomes and cleavage into intermediates occurs at the plasma membrane and in immature virions (Dickson *et al.*, 1982) (Table 9-2). However, the final cleavage necessary for transcriptase activity only occurs during virion maturation after release (see below). This may prevent the enzyme from making transcripts of cytoplasmic RNA inside the infected cells.

In contrast to *gag, env* is synthesized on membrane-bound ribosomes inserted in the rough ER membrane and cleaved within the cell. *env* is probably transported to budding sites in vesicles (see Chapter 1). In ASV, unglycosylated pr63^{env}, synthesized on membrane-bound ribosomes, has a hydrophobic signal sequence at the start of the molecule which is removed by proteolytic cleavage. Later, the uncleaved precursor Pr92^{env} (the glycosylated form) is found associated with the rough ER and membrane-bound polysomes but not with newly released virus (Hayman, 1978). The polypeptides gp85 and gp37 are the two cleavage products of avian *env* found in cell lysates (Bosch *et al.*, 1982). Thus, glycosylation and cleavage of *env* probably occur simultaneously in the rough ER and the Golgi apparatus. However, Klemenz and Diggelman (1979) claim that almost all cleavage of the glycoprotein precursor of ASV takes place in freshly budded virus particles. Processing of the particles could be a fast but very late process just before release. In that case, the cleavage point may be on the outer side of the envelope and be accessible only to a host membrane protease. It is possible that the site of cleavage varies with the host cell. In MuLV, only the cleaved mature forms of *env* (gp70 and p15E) are accessible to surface membrane probes (Witte *et al.*, 1977). These two cleavage products of *env* have a patchy membrane localization by immuno-fluorescence, shown in Fig. 9-4b and 9-4d (Satake and Luftig, 1983).

Fig. 9-7. Formation of cytoplasmic NCs (A particles) and budding of MMTV in the mammary gland in lactation. *(a)* Accumulation of numerous A particles with their typical ring-shaped cores in the cytoplasm at 15 days of lactation. *(b)* Production of virions at the cell surface at three days of lactation. One NC is apposed to a cytoplasmic vacuole (arrowhead), while the other is seen inside a budding virion (double arrowhead). Inset shows the mature form of the released virus with an eccentric core. Magnifications: (a) ×22,000, (b) ×60,000. (Courtesy of Dr. C. M. Calberg-Bacq)

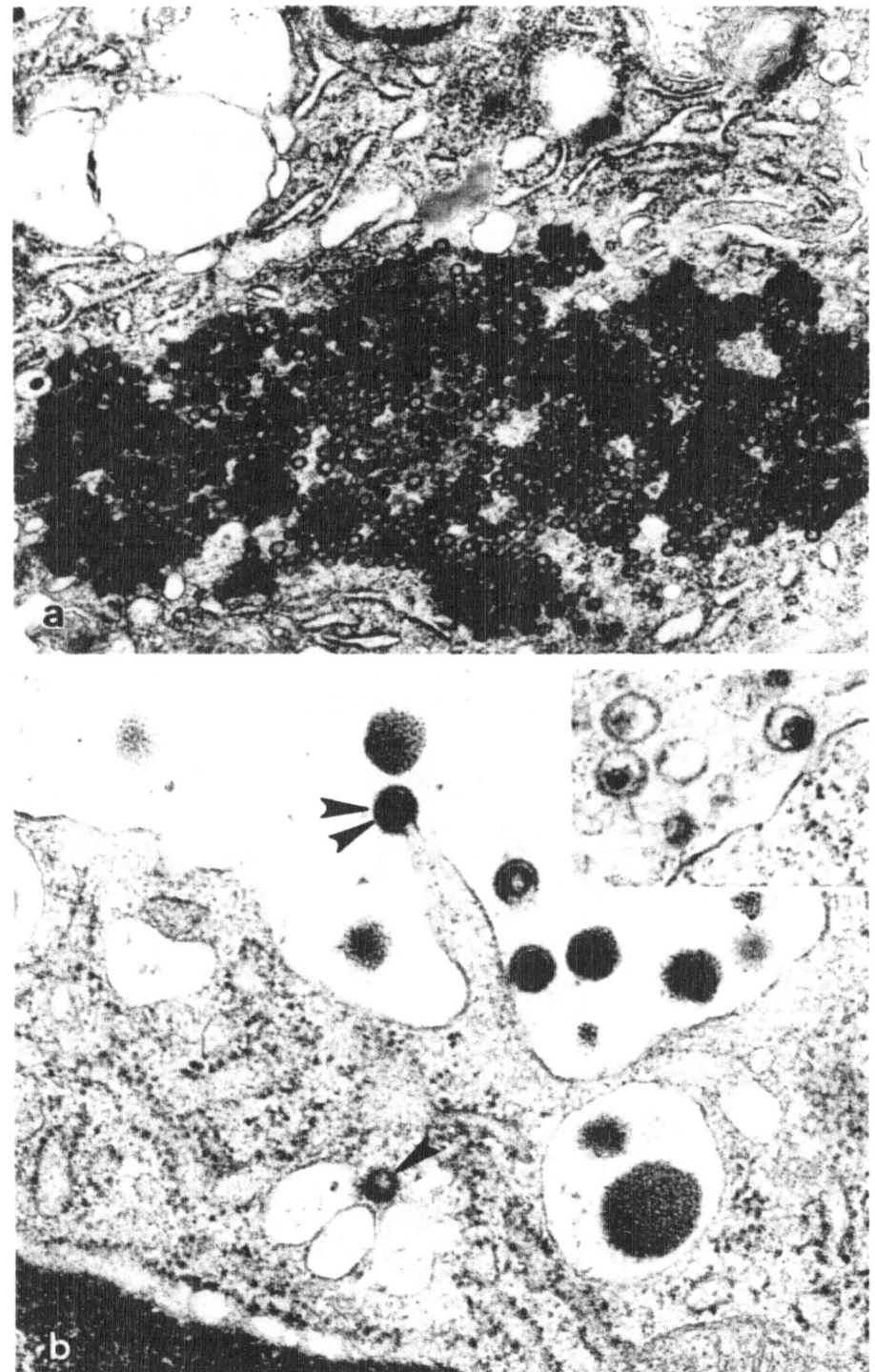

IV. Assembly of Virus Components

Retrovirus assembly has multiple aspects, depending on the types and families. One can distinguish two types: (1) viruses which assemble their NC before interacting with envelope components and budding (Fig. 9-7). These include MMTV (type B retroviruses) and some monkey retroviruses (type D retroviruses, prototype MPMV). (2) viruses which assemble their NC at the site of budding and simultaneously with budding. These are the type C retroviruses such as ASV, MuLV, other murine and avian retroviruses, as well as endogenous primate retroviruses other than MPMV. In these retroviruses, the interaction between *gag* and membrane might be necessary to trigger the final organization of the NC.

The specific case of the intracisternal A particles (IAP) should be mentioned here (reviewed in Schidlovsky, 1977; Kuff *et al.*, 1983). They are found in non-productive cells of early mouse embryos and many mouse tumors. In contrast to the type C viruses, these particles have their NCs closely apposed to their envelope, and subsequently bud into vacuoles. IAPs are released into cytoplasmic vacuoles but are never released in the extracellular space like type C viruses. They are not known to have biological activity or infectivity, although they contain polyadenylated genomic RNA species and DNA polymerase activity. New observations by Kuff *et al.* (1983) have shown that the DNAs of IAP genetic elements are incorporated into the $\varkappa$ light chain gene and thus appear to be movable elements in the mouse genome. Therefore, IAPs are not related to infectious type C virions which only occasionally bud into vacuoles in MuLV and murine sarcoma virus-infected cells (Schidlovsky, 1977) as well as from the cell surface (Fig. 9-8).

NC Assembly Occurs Independently from Budding

MMTV- and MPMV-infected cells show numerous spherical doughnut-shaped structures (70 nm) sometimes organized in inclusions within the cytoplasm (Fig. 9-7a) (Hageman *et al.*, 1981). Assembled NCs have been called intracytoplasmic A particles and form clusters close to cytoplasmic vacuoles. These inclusions can be stained by antibody against the *gag* core protein (see Fig. 9-5). MPMV NCs (or A particles) are similar, except that their outer ring has a more fuzzy appearance than in MMTV.

Earlier work had shown that intracytoplasmic A particles of MMTV contain a large polypeptide (MW: 70 K) precursor of the three major polypeptides found in the budding MMTV particles (Tanaka, 1977). It was later demonstrated that these particles contain $p77^{gag}$ precursor, RNA, and $p100^{pol}$ (reviewed by Nusse *et al.*, 1979), all of the components of the virion except *env*. In addition, when the MMTV *gag* precursor is not cleaved or phosphorylated, virus budding at the membrane cannot occur (Nusse *et al.*, 1979). These observations suggest that interactions between NC and membrane trigger *gag* cleavage and allow virus budding. Once NCs are assembled and form A particles, they migrate to the site of budding at the plasma membrane by an unknown mechanism (Fig. 9-7b).

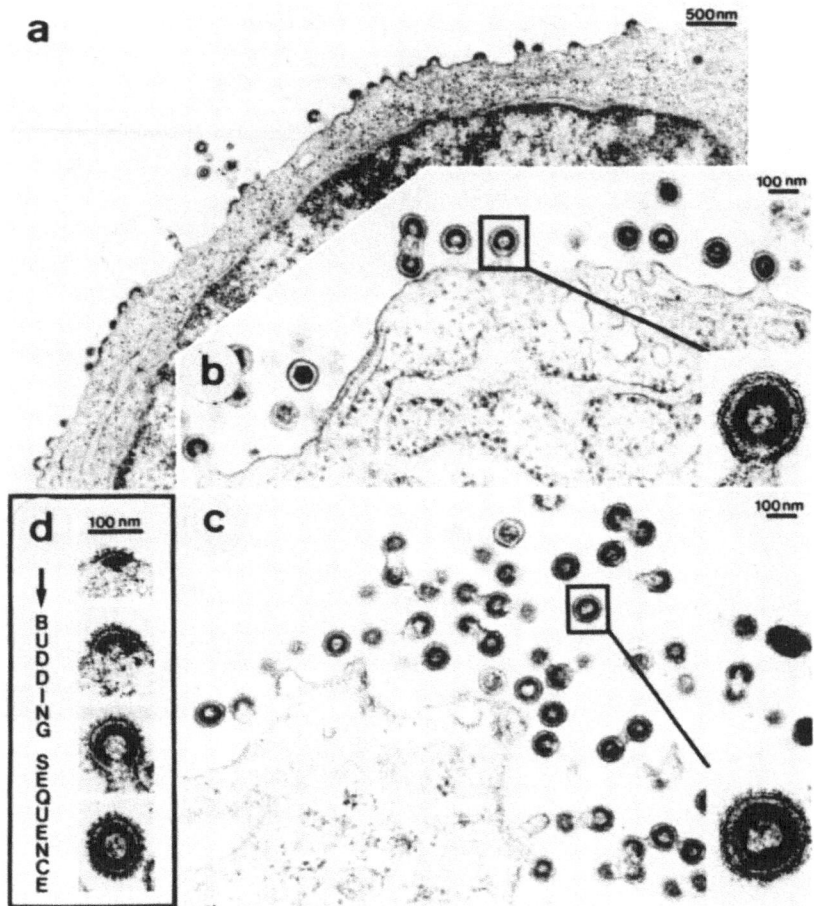

Fig. 9-8. This panel shows a typical leukemia cell producing Friend leukemia virus. *(a)* Numerous particles are seen budding from the cell surface. *(b)* Virions showing no surface projections, a result of treatment with propylene oxyde during embedding. *(c)* and *(d)* Surface projections are readily visible when no propylene oxide is used during embedding. *(d)* The budding sequence, resulting in a immature particle with a circular capsid. (Courtesy of Dr. H. Frank, 1982)

NC Assembly Is Coordinated with Budding

As mentioned earlier, retroviruses which assemble their cores directly at the budding sites are called type C. The budding site and the detailed morphology of the bud vary with the different types of viruses. Type C viruses usually bud at the plasma membrane (Fig. 9-8), but occasionally bud into ER vacuoles or cisterns. The NC then forms a half hollow sphere or crescent during maturation, and this crescent is closely apposed to the bud membrane in visna virus, while it is more distant from the envelope in ASV and MuLV (compare Figs. 9-8 and 9-9c).

The extent of budding, as well as the structural changes in the surface membrane of budding virions, are best seen by scanning electron microscopy and

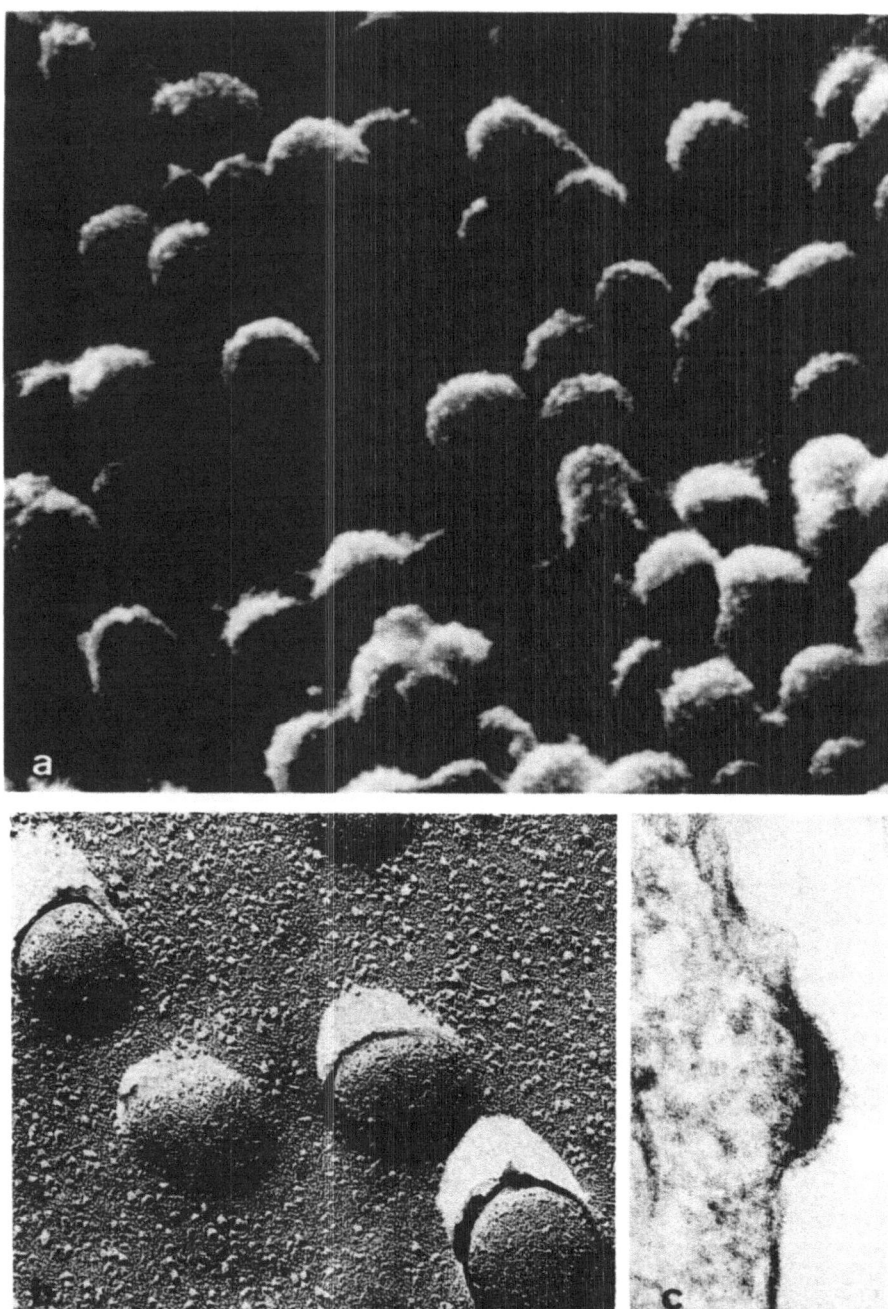

Fig. 9-9. Budding events in sheep choroid plexus cells infected with visna virus. *(a)* High resolution scanning electron microscopy reveals numerous virus buds with distinct knobs, probably corresponding to groups of spikes. *(b)* Early budding sites are detected after freeze-fracture of the membrane of an infected cell. The protoplasmic face is devoid of intramembrane particles (8–13 nm) at the site of budding. *(c)* Thin section of an early bud with its typical crescent-shaped NC directly under the envelope, which is covered with delicate spikes. Magnifications: (a) ×100,000, (b) ×120,000, (c) ×170,000. [(a) From Dubois-Dalcq *et al.,* 1979 a; reproduced with permission of Academic Press, New York]

freeze-fracture techniques (Demsey *et al.*, 1977), as seen in cells infected with visna virus (Fig. 9-9a, b) (Dubois-Dalcq *et al.*, 1976 b). The exclusion of the usual intramembrane particles from the virus bud may correspond to the exclusion of host membrane proteins. The surface knobs likely correspond to groups of spikes which are visible only where the NC is apposed to the membrane (Fig. 9-9c).

The molecular events underlying virus assembly and budding are not quite elucidated yet. Of the three genes common to all non-defective retroviruses, only the *gag* gene must be functional to allow encapsidation of genomic RNA and packaging and is sometimes called the "particle-making machine" (Vogt *et al.*, 1982; Dickson *et al.*, 1982). The *gag* molecule appears to contain "all the components to achieve a three-dimensional configuration that facilitates assembly" (Dickson *et al.*, 1982).

How is the virus genomic RNA chosen by gag polyprotein for packaging? Most likely the amino-terminal end of *gag* has some of the RNA-binding properties of its cleavage product. These proteins may find more binding sites on virus genomic RNA than on cellular or virus mRNA. There may be a specific site on the genome that interacts with a virion protein to direct RNA packaging specifically (Mann *et al.*, 1983). Retroviruses package full length RNA but not the spliced *env* mRNA (see Fig. 9-2), indicating that an essential signal might be in the region spliced out to form *env* mRNA. A mutant of Moloney MuLV with a deletion of 350 bases between the left long terminal repeat (LTR) and the start of the *gag* codon is also defective in packaging (Shank and Linial, 1980; Watanabe and Tenim, 1982; Mann *et al.*, 1983). So in this case, the key to packaging appears to be a site on the RNA rather than a region encoding a protein needed for packaging.

Interactions Between NC and Envelope Proteins

It has been suggested that *gag* molecules migrate to the cell membrane, bind to the virus genome, and cluster together at virus assembly sites before cleavage occurs (Naso *et al.*, 1982). This would lead to the formation of an electron-dense crescent under the membrane and a partially mature bud prior to cleavage of the *gag* proteins (Witte and Baltimore, 1978). However, membrane association of *gag* has not yet been demonstrated by immunofluorescence (Fig. 9-4a). Close interaction between *gag* molecules may trigger a change in *gag* conformation under the membrane, which may enhance binding to the genome of the phosphoproteins and small basic proteins of *gag*. Clustering of virus components is apparently inhibited in interferon-treated cells, because interferon alters membrane structure and function (see Chapter 1) and therefore might hamper the lateral mobility of *gag* and *env* molecules and their cleavage (Naso *et al.*, 1982; Sen and Pinter, 1983).

Virus Budding

As mentioned earlier, cleavage of *gag* appears to be a major event during budding. Cleavage of MuLV *gag* is slightly less efficient than cleavage of avian *gag*. Some, if not all, of the *gag* products may self-associate under the membrane and yield proteins organized in concentric shells (Pepinski *et al.,* 1980). The core shell protein (*gag* 3) has been localized in concentric dots in the NCs of budding visna virus by electron microscopic immunocytochemistry (Fig. 9-10a). In MuLV, the core shell protein may form higher-order homotypic multimers and the hexon subunits seen in isolated cores (Fig. 9-10b) (Nermut, 1972; Langue *et al.,* 1973). Probably, the core shell protein molecules have some specific associations with the NC. In addition, it is not rare to see radial strands between the outer shell and the envelope, which may correspond to binding sites between core and envelope (Yuen and Wong, 1977).

Generally, the less *gag* is cleaved, the more immature the particle. Observations on murine sarcoma virus (Gazdar MSV) suggest that budding and release may occur with an uncleaved *gag,* but the released virions have a dense crescent close to the membrane in contrast to the normal mature virions which have a central core (see below) (Pinter and de Harven, 1979). Cleavage of *gag* can be triggered *in vitro* and is associated with virion maturation (Yoshinaka and Luftig, 1977). A *ts* mutant of MuLV in which *gag* is not cleaved at the non-permissive temperature yields large amounts of this precursor in the cell and produces submembraneous electron-dense crescents which do not mature to budding particles (Yeger *et al.,* 1978). When these cells are switched to permissive temperature, synchronized budding begins in 2 to 4 min (Yeger *et al.,* 1978). Similarly, a *ts* mutant of ASV yields only atypical viruses with abnormal cores (Rohrschneider *et al.,* 1976). No core shell protein and hydrophobic phosphoprotein were found in these cores, suggesting that the first protein is necessary for core assembly, whereas the second is required for interaction with the membrane.

Once cleavage of *gag* has occurred, the hydrophobic proteins of the amino-terminal end of *gag* associate noncovalently with the spike anchoring protein of *env*. For instance, a fraction of pp19 molecules (ASV) associate with gp37, while p15 (MuLV) associates with p15E (Fig. 9-4c and d). Co-localization of these two proteins at the membrane level has been recently demonstrated by immunofluorescent staining with monoclonal antibodies in MuLV-infected cells (Sataka and Luftig, 1983). The polypeptide pp19 (*gag* 1) in avian retroviruses has been proposed as the mediator of packaging, since it has both lipid- and RNA-binding regions. These two functions are performed, respectively, by *gag* 1 and 2 in MuLV and MMTV. Thus the amino-terminal end product of *gag* may play a role analogous to the M protein of paramyxo-, myxo-, and rhabdoviruses (Simons and Garoff, 1980; Weiss and Bennett, 1980; Weiss, 1980). Studies on pseudotypes between the retrovirus avian leucosis virus (ALV) and the rhabdovirus VSV further support this concept. Indeed, a VSV mutant (*ts* 045), defective in G, the envelope glycoprotein, can acquire retrovirus glycoproteins, suggesting that the M protein of a rhabdovirus can interact efficiently with *env* products. Similarly, a VSV mutant in M (*ts* 31) allows functional assembly of rhabdovirus G into ALV, as if the amino-terminal product of *gag* (pp19) can interact with VSV rhabdovirus G, as well as with retrovirus glycoproteins. However, in this situation, only the retrovirus genome can be rescued, not the rhabdovirus one, probably because hydrophobic *gag* pp19, a phosphoprotein, cannot specifically interact with the VSV genome.

What exactly is the role of *env* glycoproteins in virus budding? Nondefective retroviruses have, in their genomes, information leading to infectious particles.

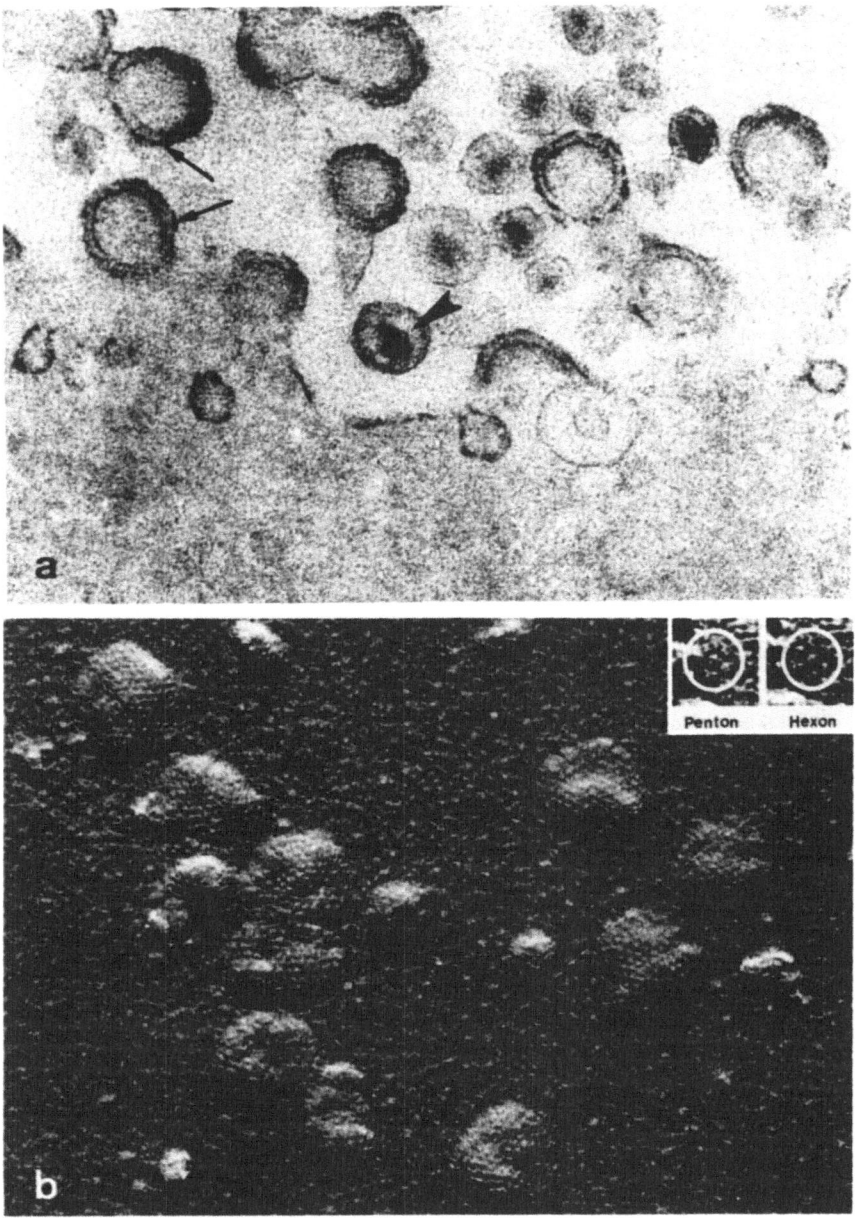

Fig. 9-10. Visualization of the core shell protein of retroviruses. *(a)* Immunoperoxidase labeling of the core shell protein (p30) of visna virus. The section is unstained and a black electron-opaque product corresponds to the localization of the protein. p30 is organized in dots, mostly in the outer part of the virus capsid in virus buds (small arrows). The detached mature viruses are smaller and have a central core, sometimes labeled by the antibody (arrowhead) (no counterstain). *(b)* Isolated cores of Friend leukemia virus after freeze-drying, shadowed with platinum carbon. The cores are covered with hexagonal structures, probably composed of core shell protein, organized in pentons and hexons as shown in the inset. Magnifications: (a) 80,000, (b) 125,000. [(a) from Dubois-Dalcq *et al.,* 1979 b; reproduced with permission of Academic Press, New York; (b) Courtesy of Dr. H. Frank]

They thus code for specific molecular interactions between *gag* and *env* components. The *gag* proteins are sufficient to produce a physical particle and *env* confers infectivity to the virion. Mutants deficient in glycoprotein assembly yield non-infectious particles at non-permissive temperatures (Pinter and de Harven, 1979). It appears that *env* proteins are not essential to virus budding and other markers in the membrane may direct the assembly of *gag* molecules in the *env* mutants (Dickson *et al.*, 1982). Reduced cleavage of the *env* precursor have been observed in ASV mutants with a defect in the glycoprotein (Hardwick and Hunter, 1981). The precursor is found in the virions and some anomalous cleavage occurs in the extracellular space; as a result, the surface spike protein gp85 falls off the envelope. Thus, correct cleavage is essential to make an infectious particle. Similarly, monensin treatment of MPMV-infected cells results in non-infectious budding viruses. Again, the *env* precursor is migrating, but not cleaved, in the monensin-treated cells, and therefore its cleavage products are not inserted into the virion (Chatterjee *et al.*, 1982). Mooren *et al.* (1981) showed that antibody cross-linking of the spike-anchoring protein p15 E inhibits budding of a MuLV *ts* mutant (*gag* cleavage lesion) at permissive temperatures, as if this protein was essential in initiation of budding. In contrast, most molecules of the surface spike glycoprotein gp70 can be cross-linked without inhibiting budding. Thus, although the integrity of the surface spike glycoprotein is not required for budding, that of the spike-anchoring protein may be necessary to this process.

The location of *env* proteins definitely determines the budding site of various retroviruses at the basolateral domains of polarized epithelial cells (Roth *et al.*, 1983 a). The positioning of the cleaved products of *env* in the plasma membrane may help *gag* to find its site of interaction with the spike-anchoring protein. Monensin treatment of MuLV-infected cells blocks entry of the glycoprotein gp70 into the Golgi apparatus (Srinivas *et al.*, 1982). In this case the glycoprotein does not migrate further to the cell surface and the virus buds into vacuoles instead. Thus the site of budding depends at least partly on the presence of *env* in a specific membrane domain.

Virus Release and Post-Release Maturation

Very little is known about the mechanism of virus release. A molecular event, however, has been correlated with release of MuLV (Green *et al.*, 1981 b). Removal of the carboxyl-terminal tail of the spike-anchoring protein p15E at the time of budding may trigger association of the *env* proteins, causing localized pinching of the membrane and virus budding.

Retrovirions undergo further maturation after release (Fig. 9-11c and d). This consists of a shift from a virus with a concentrically coiled internal NC and an electrolucent center, to a collapsed core situated in the middle of the virion (type C and type D, such as MPMV) or eccentrically (type B, such as MMTV). The central core of MMPV is sometimes triangular or tubular in shape. The reasons why mature virions have an eccentric core in some families is not clear; perhaps this may indicate a loss of binding between *gag* proteins in the core and *env* components so

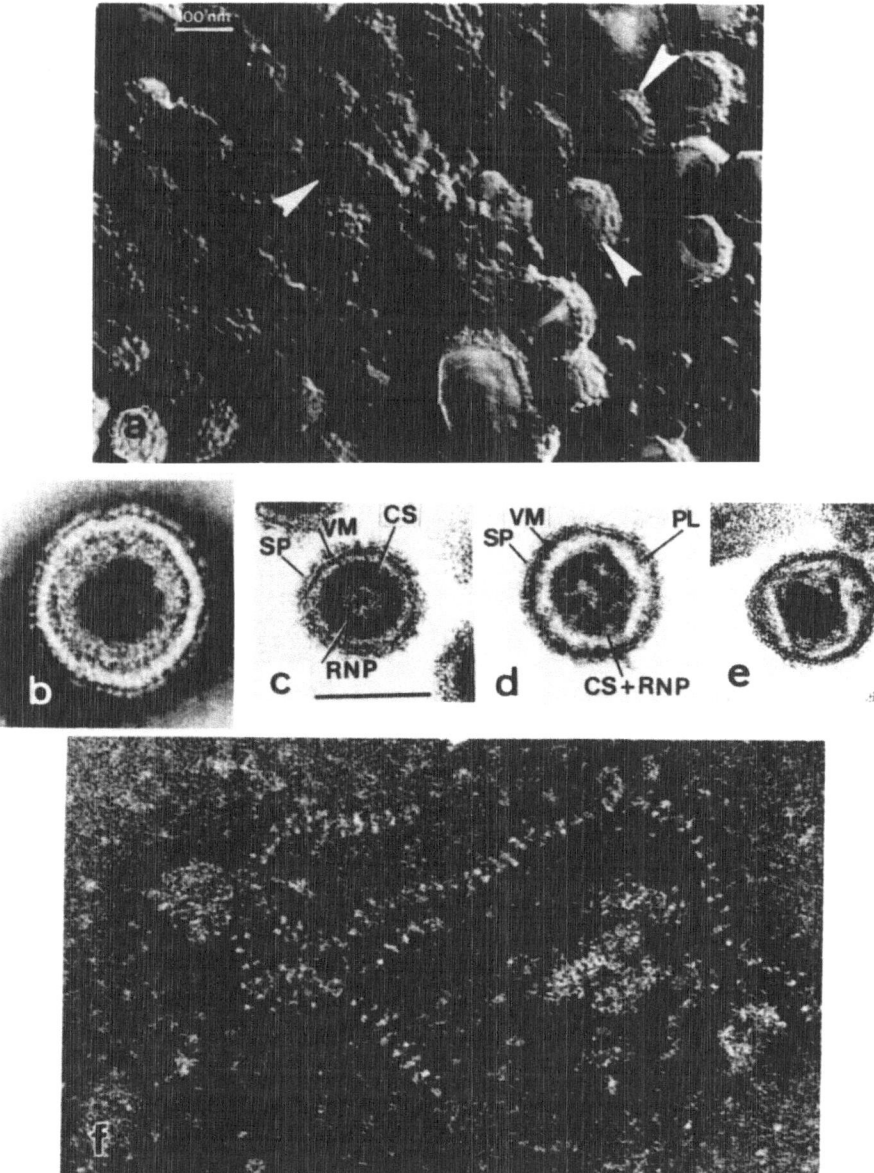

Fig. 9-11. Structure of immature and mature released retrovirions. *(a)* Clusters of Friend leukemia virus particles which have been freeze-dried. The knobs on their surface are very clear and, in some cases, part of the envelope has been lifted away, revealing the core shell under the envelope (white arrowheads). *(b)* Negatively stained immature MMTV after fixation with glutaraldehyde. Note the prominent spikes which appear to be radially connected with the capsid ring. *(c)* and *(d)* Thin sections of immature and mature Friend leukemia virus respectively. *SP* surface projections; *VM* virus membrane layer; *CS* core shell; *RNP* ribonucleoprotein (NC); *PL* protein layer. *(e)* Mature visna virion with a layer between the core and envelope. *(f)* Negatively stained NC with helical symmetry released from Friend leukemia virus after treatment at pH 3.5. Uranyl acetate staining. Magnifications: (b) × 227,000, (c) and (d): bar = 100 nm, (e) × 120,000, (f) × 220,000. [(a), (c), (d) and (f) courtesy of Dr. H. Frank, and (b) of Dr. C. M. Calberg-Bacq]

that the core is free to move in various positions inside the envelope. Alternatively, eccentric cores may have a point of attachment to the envelope (type B). On the other hand, viruses with a central core must have some way to regulate the central position of the NC (type C and type D). It is thought that immature particles are more ordered and stable structures than mature particles, since immature particles have a fuzzy skeleton material extending between core and envelope (Hageman *et al.*, 1981). It is only in the released particle that the polymerase becomes active after cleavage from the *gag-pol* precursor. This is why free virus shows an increase in reverse transcriptase activity and infectivity (Lu *et al.*, 1979).

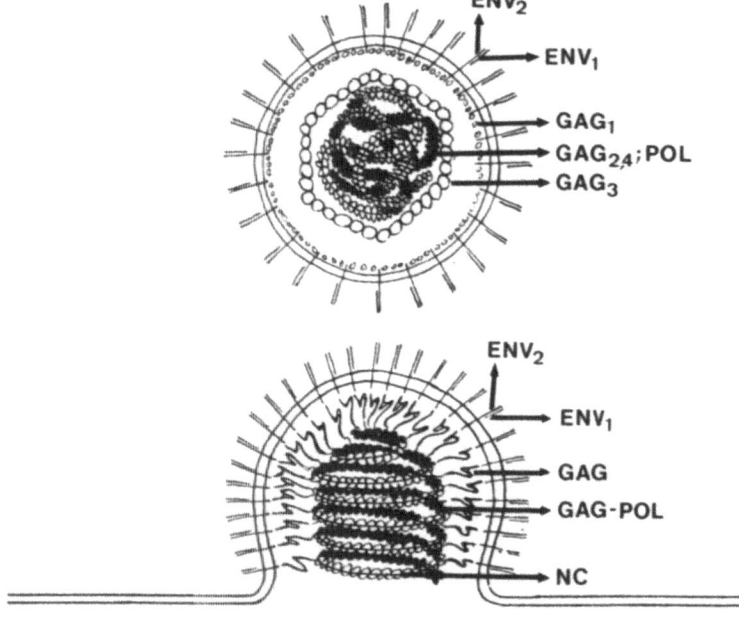

Fig. 9-12. Representation of a budding retrovirus C-type particle and the mature form released from the cells. The locations of the proteins are indicated. Numbers on *gag* and *env* proteins correspond to those given in Table 9-1 and 9-3. NC, nucleocapsid. The *gag* molecule in the budding particle may span the space between the NC and the inner envelope, but its shape is hypothetical. One possible organization of the coiling of the NC in the budding virus is shown. It takes into consideration that the center of the budding particle is clear. It is also possible that retrovirus NC forms a helix similar to that shown in the coronavirus scheme (see Chapter 7, Fig. 7-12)

Fig. 9-13. Representations of the events of retrovirus replication and assembly. *(a)* Maturation of a retrovirus C particle at the membrane. The processes of reverse transcription, integration and replication have been shown in Fig. 9-2. The three mRNAs synthesized in the nucleus are translated on the rough ER (ENV) or on polysomes (GAG and GAG-POL). ENV glycosylation is completed in the Golgi apparatus and cleavage occurs close to or at the cell surface. GAG is cleaved in its five products at the budding site, while POL is cleaved only in the released virion, which acquires a central core. *(b)* Typical maturation events occurring in retroviruses which assemble their NCs in the cytoplasm before budding, such as MMTV and MPMV. Differences in structure between the released virions in these two strains are indicated. B virions have an eccentric core, while D virions appear to have an intermediate layer between the envelope and the irregularly shaped NC. *(c)* Typical murine leukemic cell which can produce virus both at the tips of villi and inside vacuoles

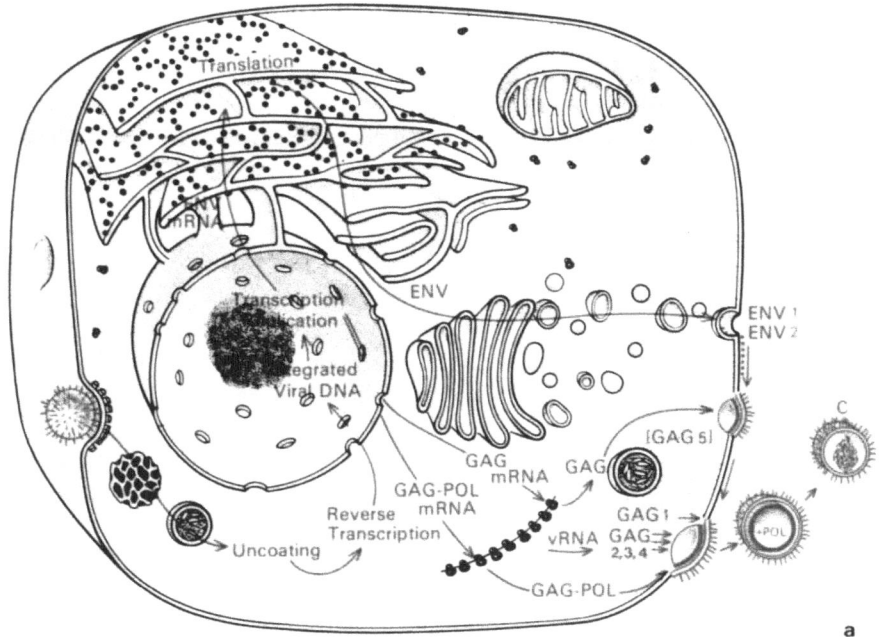

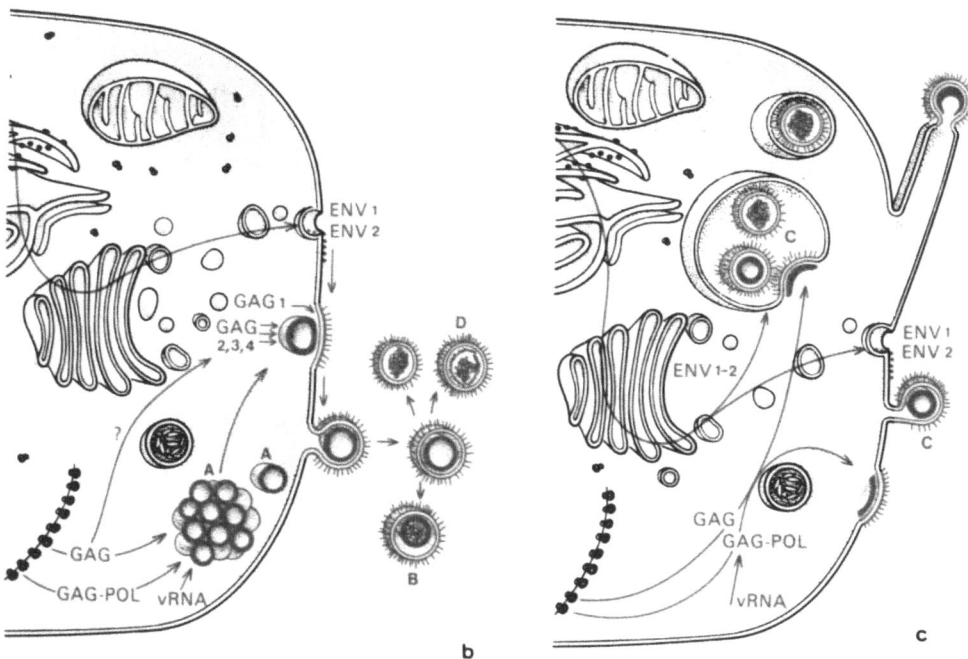

V. Organization of the Virion

The virion consists of a 100 nm enveloped particle, usually containing one core (Fig. 9-11d and e), although multiple or tubular cores may also occur. The core consists of a spiral filamentous strand, as proposed earlier by Sarkar *et al.* (1971). This strand corresponds to the virus NC which can be measured after uncoiling by detergent. It is 7–10 nm in diameter and about $1\,\mu$ long. Partial coiling can be maintained in isolation (Fig. 9-11f) (Frank *et al.*, 1978). Around the NC core is an outer shell of hexagonally arranged subunits (Nermut *et al.*, 1972) (Fig. 9-10b), probably made of core shell proteins. The exact relationship between these subunits and the coiled RNA is not known. Around the core and its shell is a perinucleoid electron-lucent space. The degree of electron opacity in that zone varies with the type of retrovirus. Virions sometimes have an intermediate layer between the core and the envelope (Fig. 9-11e). The envelope usually bears visible spikes. The longest spikes seen on retrovirus envelopes pertain to MMTV (100 Å) (Fig. 9-11b) (Calberg-Bacq *et al.*, 1976). One spike consists of a spherical knob 55 Å in diameter joined to the viral envelope by a stalk (40 Å long, 20 Å in diameter), whereas ASV spikes seem shorter (50 Å). In MuLV, one spike contains 4 to 6 gp70-p15E complexes (Schneider *et al.*, 1980) and knobs are seen on the surface after freeze-drying (Fig. 9-11a).

A model of the organization of immature and mature retroviruses has been proposed by Bolognesi *et al.* (1978) and later by Dickson *et al.* (1982). Our scheme, modified from these authors, is shown in Fig. 9-12. Fig. 9-13a, b, and c summarizes the events of replication and assembly of retroviruses in the cell.

10

Assembly of Rotaviruses

I. Introduction

The *Reoviridae* family includes three genera: *Reoviruses, Rotaviruses,* and *Orbiviruses* (Joklik, 1983). These are non-enveloped viruses which have an inner capsid and an outer protein shell with icosahedral symmetry. The viruses have segmented genomes of double-stranded RNA. Although mature infectious virions of *Reoviridae* are not enveloped, recent evidence has shown that an enveloped particle is an intermediate in the formation of *Rotavirus* virions (Estes *et al.,* 1983). For this reason, the assembly of *Rotaviruses* has been included in this book. Assembly of *Reoviruses* will not be considered, since these viruses assemble completely in the cytoplasm and do not have an intermediate budding form. *Orbiviruses* do have enveloped forms in infected cells, but relatively little is known about their formation (Verwoerd *et al.,* 1979).

Several comprehensive reviews on *Rotaviruses* should be consulted for additional detailed information (McNulty, 1978; Flewett and Woode, 1978; Wyatt *et al.,* 1978; Kapikian *et al.,* 1979; Holmes, 1983; Estes *et al.,* 1983).

Rotaviruses cause acute gastroenteritis in neonatal and young animals of many species (Wyatt *et al.,* 1978). The *Rotaviruses* which have been studied most intensively are simian rotavirus SA11, bovine rotavirus, murine rotavirus (epizootic diarrhea of infant mice, EDIM), and human *Rotaviruses. Rotaviruses* are frequently observed in feces of young animals and children with diarrhea, but until recently most *Rotaviruses* were very difficult to propagate *in vitro.* With the recognition that infectivity could be activated or enhanced by proteolytic enzymes and the identification of several cell types suitable for *Rotavirus* replication, it became possible to propagate many types of *Rotaviruses* in cultured cells (Almeida *et al.,* 1978; Barnett *et al.,* 1979; Wyatt *et al.,* 1980; Sato *et al.,* 1981).

The virions of *Rotaviruses* are non-enveloped particles about 70 nm in diameter (McNulty, 1978). In negatively stained preparations, they resemble a wheel with a central hub and short spokes radiating to a thin, sharply defined rim. This appearance suggested the name rotavirus, as *rota* is Latin for wheel (Flewett *et al.,* 1974) (Fig. 10-1).

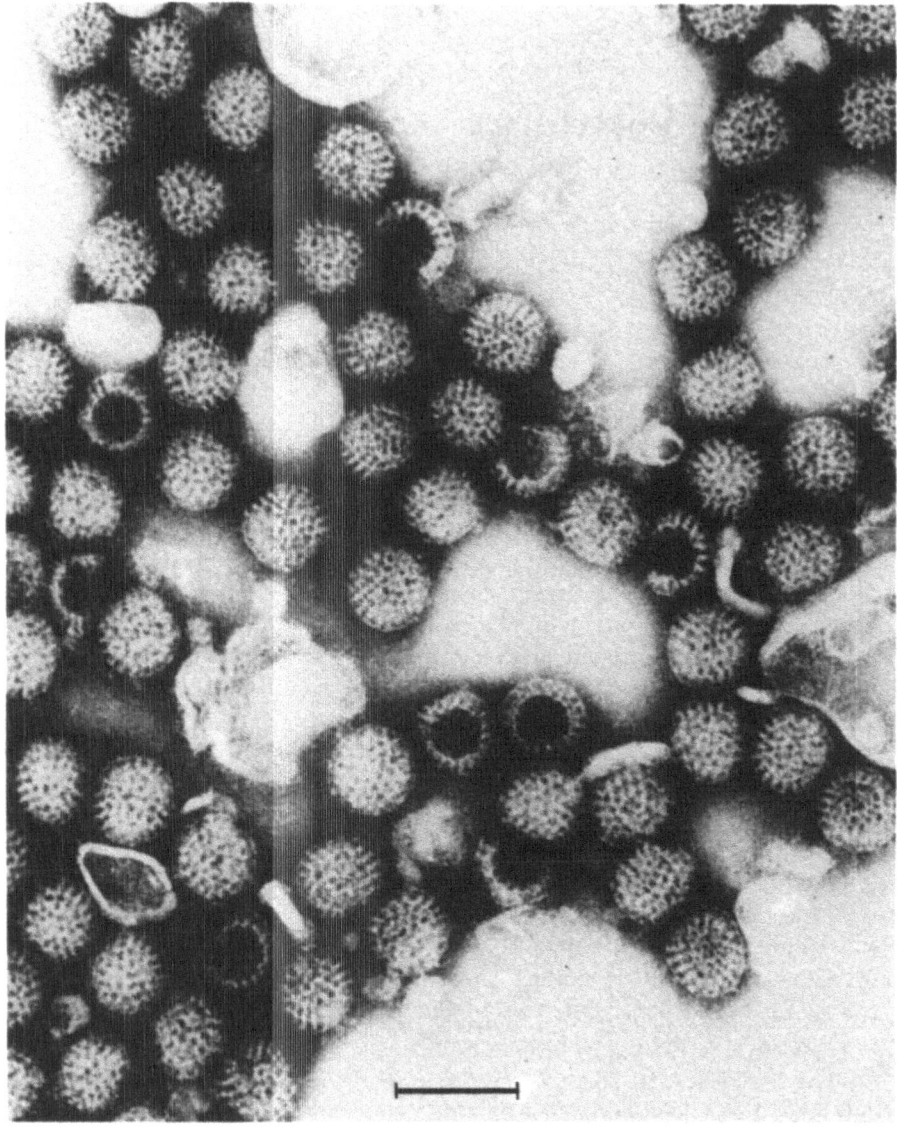

Fig. 10-1. Human rotavirus from a stool filtrate of a patient with gastroenteritis. Virions have a double-shelled capsid with icosahedral symmetry, hollow capsomers and a sharply defined edge. These mature infectious virions do not have an envelope. Magnification: ×175,000. (Courtesy of Dr. A. Z. Kapikian from Kapikian *et al.*, 1974; with permission of the American Association for the Advancement of Science)

Seroepidemiologic and virus isolation studies have shown that *Rotavirus* infections occur in all human populations. At least four major serotypes of human *Rotavirus* have been defined by plaque reduction assay (Kapikian *et al.*, 1974, 1979, 1983; Wyatt *et al.*, 1978, 1982; Estes *et al.*, 1983). Infection induces systemic and local immune responses, but individuals may be reinfected with *Rotaviruses*, sometimes resulting in inapparent infection (Kapikian *et al.*, 1974, 1983; McNulty, 1978;

Holmes, 1983). Serological cross-reactions have been identified between the inner capsid proteins of many different *Rotaviruses*, and outer capsid proteins show both type-specific (McNulty, 1978) and cross-reactive antigenic determinants (Offit *et al.*, 1983).

II. Molecular Organization

For details, the recent comprehensive reviews of Estes *et al.* (1983) and Holmes (1983) should be consulted.

Infection with *Rotaviruses* is initiated when virions bind to cell surface receptors on the tips of microvilli of mature enterocytes (Coelho *et al.*, 1981). Treatment of erythrocytes with neuraminidase prevents hemagglutination by *Rotaviruses*, so the erythrocyte receptor for the virus may contain sialic acid (Bastardo and Holmes, 1980; Holmes, 1983). Penetration of the cell by virions probably occurs by the endocytic pathway (Petrie *et al.*, 1981) and virions may be uncoated in endosomes to yield 50 nm subviral particles (Chasey, 1977; Petrie *et al.*, 1981).

Transcription of virus genes by a virion-associated RNA dependent RNA polymerase yields full length, single-stranded non-polyadenylated mRNA copies of each genome segment (Cohen, 1977; Cohen and Dubois, 1979; Bernstein and Hruska, 1981; McCrae and McCorquodale, 1983). The 3' ends of each segment of genomic or mRNA have a common conserved sequence of 8 nucleotides (McCrae and McCorquodale, 1983). This common sequence may mediate RNA polymerase recognition and it may also play a role in packaging the segments into the capsid. After virus protein synthesis has begun, double-stranded genomic RNA is synthesized in the cytoplasm (McCrea and Faulkner-Valle, 1981), where the entire replicative cycle takes place.

The genome of *Rotaviruses* consists of 11 segments of double-stranded RNA, ranging from 0.2 to 2.2×10^6 in MW and totalling 36 kb (Holmes, 1983; Estes *et al.*, 1983). The segments can be grouped in 4 size classes. The electrophoretic mobility of each RNA segment may differ slightly from one *Rotavirus* to another (Kalica *et al.*, 1978 b; Rodger *et al.*, 1981; Lourenco *et al.*, 1981). These differences have been used to distinguish *Rotavirus* strains or electropherotypes, to study genetic recombination or reassortment of *Rotavirus* genes and to assign gene products to each genome segment (Kalica *et al.*, 1981 a, b, 1983; Rodger and Holmes, 1979; Rodger *et al.*, 1981; Lourenco *et al.*, 1981; Greenberg *et al.*, 1981, 1983 c).

These gene assignments have been confirmed by *in vitro* translation studies using virus mRNAs or denatured double-stranded genomic RNA and monoclonal antibodies and by studies on genetic reassortment (Smith *et al.*, 1980; Dyall-Smith and Holmes, 1981; Arias *et al.*, 1982; McCrae and McCorquodale, 1982; Mason *et al.*, 1980, 1983; Estes *et al.*, 1983; Greenberg *et al.*, 1983 b; Ramig, 1982, 1983). Thus, the general pattern of *Rotavirus* gene-coding assignments is clear, although there is as yet no uniform nomenclature for the gene products and there is still controversy over the origin of some of the virus-specific polypeptides, which may be derived by processing of primary gene products (Estes *et al.*, 1983).

Table 10-1 summarizes the gene coding assignments of simian rotavirus SA11, with the known or putative functions of the virus-specific proteins. The terminology of Estes *et al.* (1983) for the proteins of SA11 simian rotavirus will be used in this chapter.

Table 10-1. *Proteins of Simian rotavirus SA11*[a]

Name	Apparent MW of gene product $\times 10^{-3}$	Genome Segment	Known or putative function
VP2	94	S2	Inner capsid protein
VP6	41	S6	Inner capsid protein Group and subgroup specific antigens
VP1	125	S1	Minor inner capsid protein
VP3	88	S4	Outer capsid protein Hemagglutinin Protease-enhanced infectivity
VP7	39	S9	Outer capsid glycoprotein Elicits type specific neutralizing antibody
VP9	26	S11	Outer capsid protein?
NS28[b]	28	S10	Non-structural glycoprotein Conversion from enveloped NC to mature virion
NS	53	S5	?
NS	34	S7	?
NS	35	S8	?
?	?	S3	?

[a] Adapted from Estes *et al.*, 1983.
[b] NS indicates non-structural polypeptides.

(1) *VP2* (MW: 91 K) is found in the inner capsid of double-shelled virions and in single-shelled particles. VP2 may be cleaved to form a 88 K or 84 K product, but it is not clear whether this cleavage is associated with enhanced virus infectivity (Estes *et al.*, 1981).

(2) *VP6* (MW: 41 K) forms 80% of the inner capsid protein. It can be released from single-shelled particles by treatment with chaotropic agents (Bican *et al.*, 1982). In the virion, VP6 may be cross-linked by disulfide bonds into trimers and may comprise the visible capsomers (Bastardo *et al.*, 1981; Esparza and Gil, 1981; Holmes, 1983). VP6 carries common group-specific antigenic determinants and subgroup determinants (Greenberg *et al.*, 1983 a, c; Holmes, 1983).

(3) *VP1* (MW: 125 K) is the largest inner capsid protein and is present in only small amounts.

(4) *VP7* (MW: 38 K) is a glycoprotein in the outer protein shell (Rodger *et al.*, 1977). It elicits neutralizing antibody (Killen and Dimmock, 1982; Bastardo *et al.*, 1981). It contains N-linked high-mannose oligosaccharides but no complex oligosaccharides (McCrae and Faulkner-Valle, 1981; Ericson *et al.*, 1982, 1983). Several

forms of VP7 can appear in purified virions, which may reflect heterogeneity of protein processing (Novo and Esparza, 1981; Estes *et al.*, 1982).

(5) *VP3* (MW: 88 K) is an outer capsid protein which can be cleaved to yield VP5* (60 k) and VP8* (28 k), enhancing virus infectivity (Barnett *et al.*, 1979; Graham and Estes, 1980; Espejo *et al.*, 1981; Estes *et al.*, 1981). VP3 is the hemagglutinin of certain animal *Rotaviruses* (Kalica *et al.*, 1978 a, 1983).

(6) *VP9* (MW: 26 K) has been reported to be a minor outer shell protein, but it is not found consistently by all investigators (Arias *et al.*, 1982).

(7) *NS28* (MW: 28 K) is a nonstructural glycoprotein encoded in genome segment S10. It contains a noncleavable signal sequence (Ericson *et al.*, 1983) and it appears to play a role in the maturation of virions (Petrie *et al.*, 1983). The functions of the other nonstructural proteins of *Rotaviruses* are not known.

III. Translation, Transport and Processing of Virus Proteins

Each of the mRNA segments appears to be translated to form a single polypeptide species which may then undergo processing. Identification of the origin of each of the virus-specific polypeptides of different *Rotaviruses* has been a major effort for several years (summarized in Estes *et al.*, 1983). Synthesis of most of the virus proteins occurs on free polysomes, but synthesis of the two virus glycoproteins, VP7 and NS28, occurs on membrane-bound ribosomes, as shown by immu-

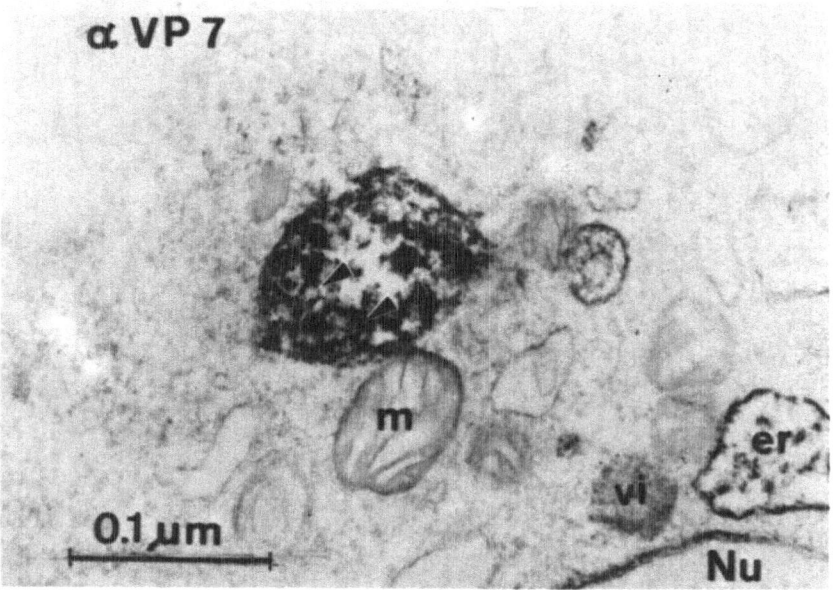

Fig. 10-2. Immunoperoxidase labeling of the endoplasmic reticulum and virus particles within ER cisternae by antibody directed against the VP7 glycoprotein, the major outer capsid glycoprotein. The nucleus *(Nu)*, mitochondria *(m)*, and viroplasm *(vi)* are unstained, while the ER and the nuclear envelope contain the VP7 virus antigen. Similar labeling with antibody against inner capsid proteins shows that these are found free in the cytoplasm and in the viroplasm. Magnification: ×27,500. (Courtesy of Dr. B. Petrie from Petrie *et al.*, 1982; with permission from the Cambridge University Press)

noelectron microscopy (Chasey, 1980; Petrie *et al.*, 1982; Fig. 10-2). Oligosaccharides of VP7 are susceptible to cleavage by endoglycosidase H and glycosylation can be inhibited with tunicamycin, showing that these are high-mannose, N-linked oligosaccharides (Arias *et al.*, 1982; Ericson *et al.*, 1982, 1983). This suggests that the virus glycoprotein is not transported through the Golgi apparatus where further trimming would occur, and correlates with the observation that virions are formed in the RER (Altenburg *et al.*, 1980) and with the immunoelectron microscopic data (Petrie *et al.*, 1982).

Proteolytic cleavage of several primary gene products of *Rotaviruses* can occur either within the cell or in released virions. The biological significance of most of these proteolytic cleavages is not known. However it is clear that cleavage of the outer capsid protein, VP3, on the virion by trypsin or other proteases results in enhanced virus infectivity (Graham and Estes, 1980; Greenberg *et al.*, 1981, 1983 c; Kalica *et al.*, 1983). As noted above, this observation has been of major importance in the cultivation *in vitro* of many *Rotaviruses* (Almeida *et al.*, 1978; Wyatt *et al.*, 1980, 1982; Sato *et al.*, 1981).

Synthesis of genomic RNA and non-glycosylated proteins probably occurs near intracytoplasmic, electron-dense inclusions called "viroplasm" (Estes *et al.*, 1983; Fig. 10-3). Immunoelectron microscopy has revealed inner capsid proteins VP6 and VP2 in this location (Petrie *et al.*, 1982).

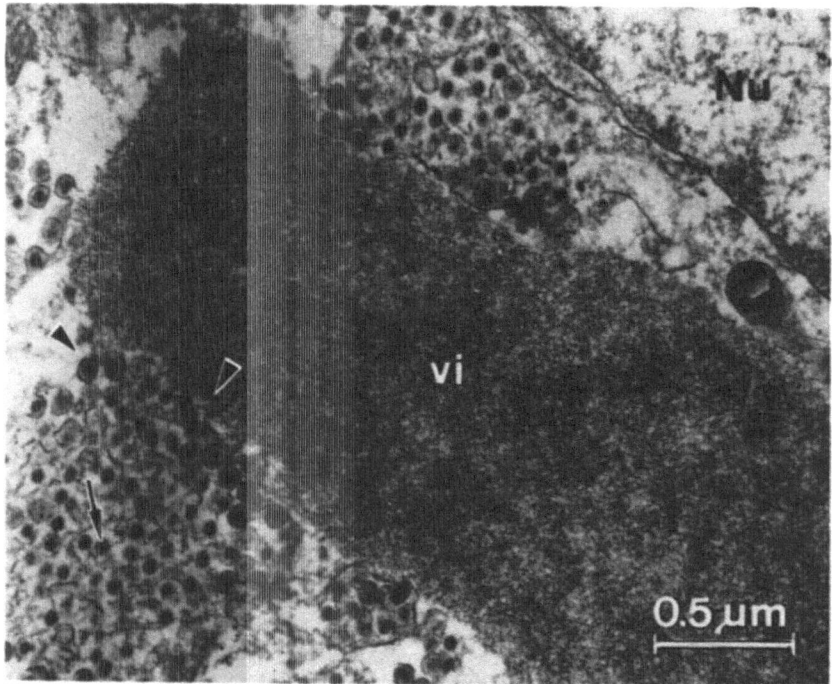

Fig. 10-3. Simian rotavirus SA11-infected cell contains a large viroplasm *(vi)* in the cytoplasm adjacent to the nucleus *(Nu)*. Enveloped forms of virus (arrowheads) are seen budding from the periphery of the viroplasm into the ER. In the lumen of the ER, mature virions which lack envelopes are observed (arrow). (Courtesy of Dr. B. Petrie)

IV. Assembly of Virus Components

NC Assembly

The inner and outer protein shells of rotaviruses form sequentially and by very different mechanisms. The inner capsids, containing the genomic RNA segments, are the equivalent of NCs of other viruses such as togaviruses (Chapter 8). These 55–65 nm NCs are sometimes referred to as single-shelled particles (Petrie *et al.*, 1982). They assemble in the cytoplasm at the periphery of the viroplasm (Pearson and McNulty, 1979; Altenberg *et al.*, 1980) by condensation of inner capsid proteins around 38 nm "cores" containing virion RNA (Estes *et al.*, 1983). This important step in *Rotavirus* maturation has not been studied extensively. It is not clear how one copy of each of the 11 genome segments is assembled into the core. Since recombination of genome segments occurs at high frequency in cells simultaneously infected with two *Rotaviruses,* it is clear that the genome segments do not necessarily remain associated throughout replication. There must be some as yet unexplained mechanism for assembly of a full complement of genomic segments into a single virion. This might be achieved by recognition of specific nucleotide sequences within each segment.

The 55–65 nm NC has cubic symmetry and is covered with ring-like capsomers (Stannard and Schoub, 1977; Esparza and Gil, 1978; Roseto *et al.*, 1979; Kogasaka *et al.*, 1979; Petrie *et al.*, 1981). The capsomers on the NC are probably formed by trimeric aggregates of the VP6 inner capsid protein (Bastardo *et al.*, 1981; Novo and Esparza, 1981; Palmer and Martin, 1982).

Virus Budding

The NC acquires a transient envelope, or pseudoenvelope, by budding at RER membranes adjacent to the viroplasm (Adams and Kraft, 1967; Banfield *et al.*, 1968; McNulty *et al.*, 1976; Pearson and McNulty, 1979; Altenburg *et al.*, 1980; Tektoff *et al.*, 1980; Suzuki *et al.*, 1981; Petrie *et al.*, 1981; Fig. 10-3). These RER membranes have been shown by cell fractionation and immunoelectron microscopy to contain outer capsid proteins (Soler *et al.*, 1982; Chasey, 1980; Petrie *et al.*, 1982). Enveloped capsids are about 75 nm in diameter. Similar enveloped particles have been observed in cells infected with *Orbiviruses* (Schnagl and Holmes, 1975; Verwoerd *et al.*, 1979), but no enveloped forms are seen with *Reoviruses,* which assemble completely within the cytoplasmic matrix. Enveloped forms of *Rotavirus* are rarely observed outside of cells and have been rather difficult to purify. Normally, the enveloped forms are seen only at the periphery within distended RER vesicles and the lumen of the RER is filled with double-shelled, non-enveloped virions (Holmes *et al.*, 1975; Chasey, 1977; Altenburg *et al.*, 1980; Petrie *et al.*, 1981).

Modification After Budding

Further maturation of *Rotaviruses* occurs within the cisternae of the RER where enveloped particles are converted into mature, double-shelled virions by a process that has not yet been elucidated. The fate of the lipids of the temporary envelope during this process is not known. The conversion of enveloped to double-shelled rotaviruses appears to depend upon the non-structural glycoprotein, NS28. When rotavirus-infected cells are treated with tunicamycin, glycosylation of VP7 and NS28 is inhibited (Sabara *et al.*, 1982; Petrie *et al.*, 1983). Conversion of enveloped to double-shelled virions is inhibited by tunicamycin, so that large numbers of enveloped particles accumulate in the RER (Petrie *et al.*, 1983; Fig. 10-4). A variant of SA11 which makes only non-glycosylated VP7 was observed to convert enveloped particles to double-shelled virions in a normal manner. This suggests that glycosylation of VP7 is not required for post-budding maturation. Therefore, the tunicamycin-induced defect in virus maturation is probably caused by failure to glycosylate NS28, the non-structural glycoprotein (Petrie *et al.*, 1983). The p20 non-glycosylated precursor of NS28 formed in the presence of tunicamycin does not show enhanced degradation by cellular proteases, but its function in virus maturation is inhibited. Further studies of this important virus protein may elucidate the mechanism of conversion of the enveloped forms to mature virions.

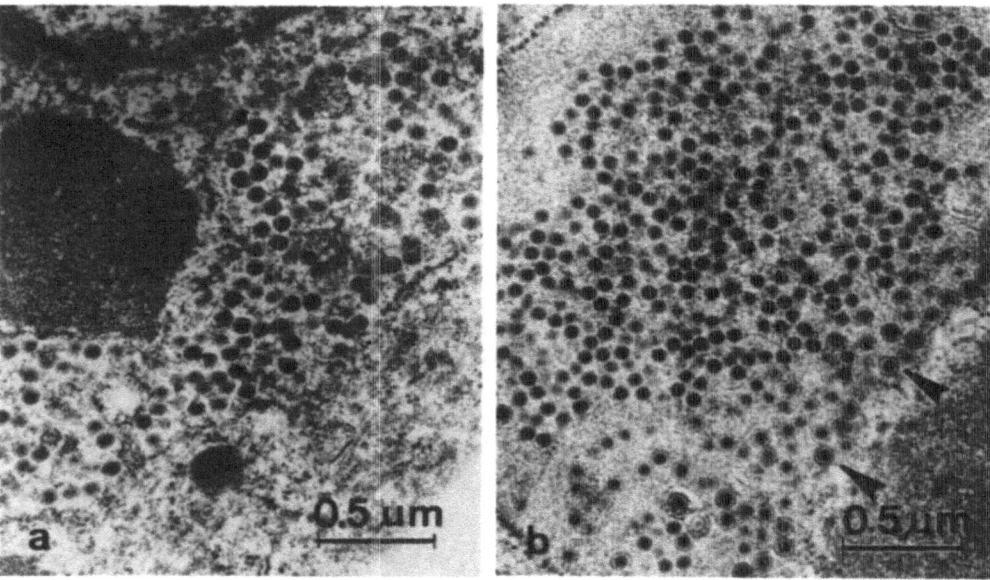

Fig. 10-4. Effect of tunicamycin on the morphogenesis of simian rotavirus SA11. *(a)* shows infected cells grown in the presence of 1.0 μg/ml of tunicamycin. All virus particles in the ER are enveloped. *(b)* shows that, without tunicamycin, enveloped particles are observed only at the periphery of the rough ER (arrowheads), while mature virions in the lumen of the rough ER are not enveloped. Thus, tunicamycin, which prevents glycosylation of the structural glycoprotein VP7 and the non-structural glyco-protein NS28, does not prevent virus budding but does prevent subsequent maturation of the virions associated with loss of the envelope. Magnification: × 35,000. (Courtesy of Dr. B. Petrie)

Two theories have been advanced to explain how *Rotaviruses* acquire the outer capsid proteins in the RER (Holmes, 1983; Estes *et al.*, 1983). The theory most in accord with the assembly of other enveloped RNA viruses discussed in this book suggests that outer capsid proteins, at least one of which is glycosylated, are inserted into RER membranes either co-translationally or post-translationally, and form an essential part of a transient virus envelope. This envelope would then be altered by an unknown action of the NS28 non-structural glycoprotein to form the outer capsid. During this process, lipids of the envelope would have to be removed, and changes in the composition and/or conformation of the outer capsid proteins might occur. The alternative theory suggests that the transient envelope is required only to deliver the inner capsid into the lumen of the RER, after which the envelope is removed and outer capsid proteins which are present in high concentrations in the lumen of the RER condense upon the inner core to form mature, double-shelled virions. Indeed, VP7 may be secreted into the lumen of the RER, since no cytoplasmic domain of this glycoprotein has been detected in protease digestion experiments with microsomal membranes. VP7 has been localized on the cisternal side of RER membranes by immunoelectron microscopy (Petrie et al., 1982). Further study is clearly needed to elucidate this unique step in the assembly of an icosahedral virus.

Release of Virions from Infected Cells

It is not clear how *Rotaviruses* are released from the RER of infected cells. Most infectious virus remains cell-associated even after cell death. Release has been postulated to occur following cell lysis or by exocytosis prior to cell lysis (Holmes, 1983; Estes et al., 1983).

Defective Assembly

Rotaviruses have been observed to bud into the nuclear envelope and into mitochondria (Altenburg et al., 1980; Tektoff et al., 1980). Synthesis of coreless, probably defective virus particles has been observed in some non-permissive cell types (McNulty et al., 1978; Payne et al., 1981). Convoluted smooth membranes are sometimes seen in association with the viroplasm, but their function is not known (Altenburg et al., 1980).

In *Rotavirus*-infected cells and in feces of animals infected with *Rotaviruses*, long rigid tubules are sometimes observed (Chasey and Labram, 1983; Fig. 10-5). These rotatubes have been divided into two classes on the basis of surface structure and diameter. Rotatube 1 is 80 nm in diameter and has a hexagonal surface lattice, whereas rotatube 2 is only 40 nm in diameter and has a delicate surface lattice with a different configuration. Tubules like these occur in RER cisternae of rotavirus-infected cells, and occasionally extend into the nucleus (Banfield et al., 1968; Rodriguez-Toro, 1980; Tektoff et al., 1980; Suzuki et al., 1981). These rotatubes are believed to be composed of inner capsid proteins (Kimura, 1981; Chasey and Labram, 1983).

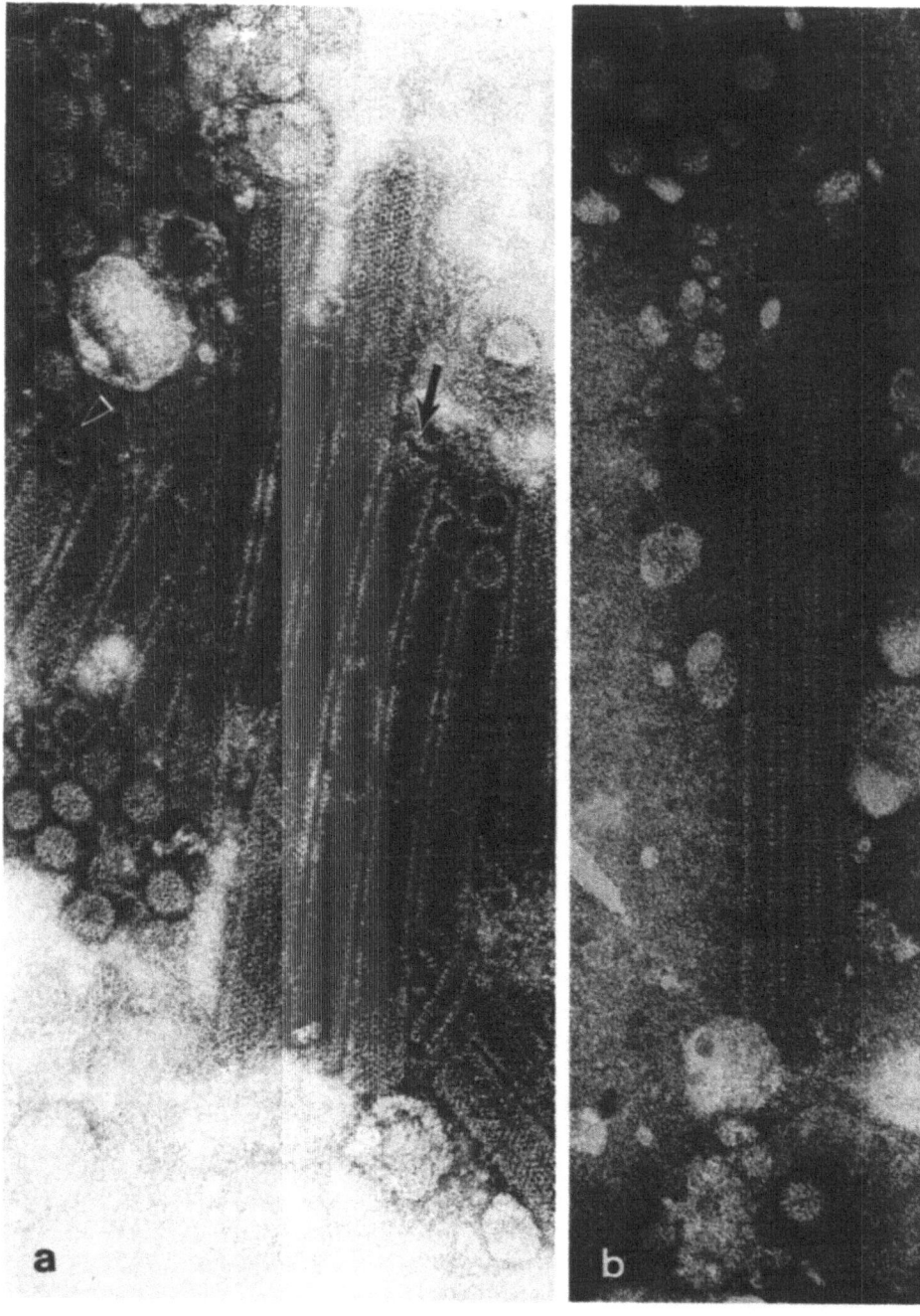

Fig. 10-5. Tubular assemblies found in bovine feces during naturally occurring bovine rotavirus infection. *(a)* double-shelled virions are shown in the upper left and a single shelled particle is shown by an
arrowhead. The arrow indicates a rotatube 1, a tube approximately 80 nm in diameter with a surface
lattice of hexagonally arranged subunits. *(b)* shows rotatube 2, half the diameter of rotatube 1. The
surface of rotatube 2 shows a different lattice from rotatube 1. Rotatubes are probably composed of
capsid proteins. Magnifications: (A) × 96,000; (B) × 100,000. (Courtesy of Dr. D. Chasey, British
Crown Copyright, 1984)

V. Organization of the Virion

A hypothetical scheme summarizing the events in *Rotavirus* replication is shown in Fig. 10-6. Chasey (1977) identified 5 different particle types in *Rotavirus*-infected cells. The relationships of the particles seen in negatively stained preparations with those in thin sections were demonstrated by Petrie *et al.* (1981). These particle types probably represent the partially uncoated virion in endosomes, the core, the NC, the enveloped NC, and the mature virion. Many important questions remain about the molecular interactions between different *Rotavirus* structural proteins and the 11 segments of genomic RNA and about the assembly of *Rotaviruses*.

The structure of the released virions has been studied most extensively. The location of the virus structural proteins in the inner and outer capsids has been

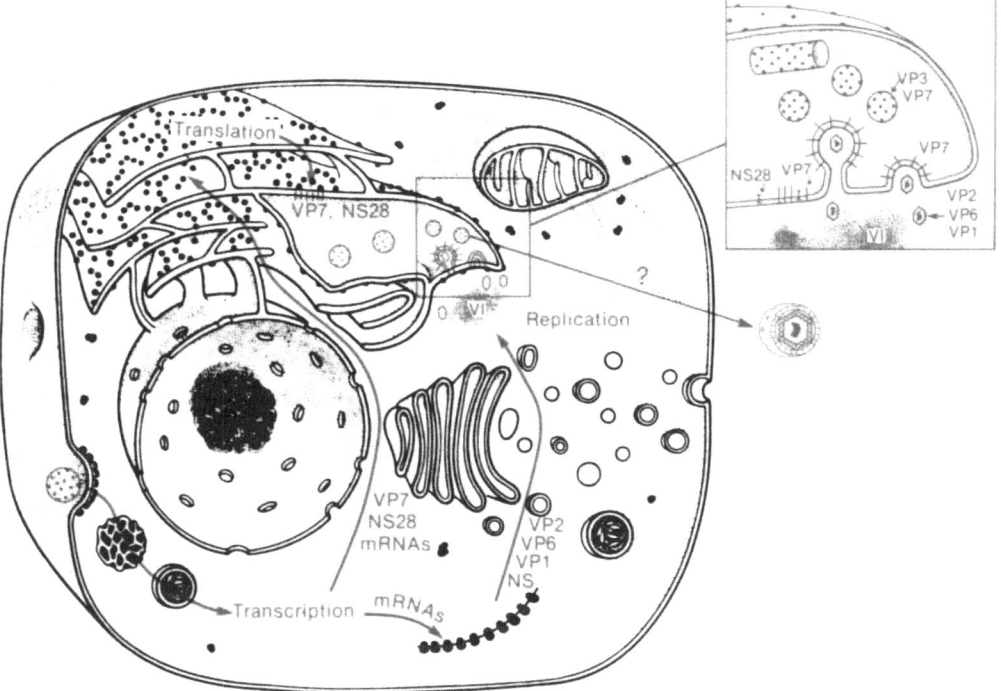

Fig. 10-6. A model of the replication of simian rotavirus SA11. Virions enter by adsorptive endocytosis and are partially uncoated in endosomes. Transcription of the 11 genome segments by a virion-associated RNA-dependent RNA polymerase results in the formation of 11 mRNA species which are translated on free or membrane-bound polysomes to form virus structural and non-structural proteins. A viroplasm *(VI)* of inner capsid proteins and genomic RNA forms in the cytoplasm, and single-shelled particles assemble at its periphery. These particles acquire an envelope by budding into the RER. This model shows outer capsid glycoprotein VP7 inserted into the transient virus envelope. Enveloped forms are released from the RER membrane and converted into double-shelled, non-enveloped virions by a process which may require a function of non-structural glycoprotein NS28. Alternatively, the virus envelope is removed and outer capsid proteins from a pool in the lumen of the RER assemble around the inner capsids. It is not yet clear whether virions are released from the cell by lysis or by exo-cytosis. (After Estes *et al.*, 1983, and Holmes, 1983)

described above. The double-shelled virion has a diameter of 65 to 75 nm, which is smaller than the 70–90 nm diameter of the enveloped particles (Holmes, 1983). The capsomers of the outer shell are shaped like push-pins and are masked by a thin layer of glycoprotein (Palmer and Martin, 1982), so that in negatively stained preparations the virion has a smooth surface, which appears to be pierced by regularly spaced holes. Models of virion structure have been proposed by Roseto *et al.* (1979) and Kogasaka *et al.* (1979).

References

Abelson, H. F., Smith, G. H., Hoffman, A., Rowe, W. P.: Use of enzyme-labeled antibody for EM localization of lymphocytic choriomeningitis. J. Natl. Cancer Inst. *42*, 497–515 (1969).

Abraham, G., Pattnaik, A. K.: Early RNA synthesis in Bunyamwera virus-infected cells. J. gen. Virol. *64*, 1277–1290 (1983).

Acheson, N. H., Tamm, I.: Replication of Semliki Forest virus: an electron microscopic study. Virology *32*, 128–143 (1967).

Ada, G. L., Yap, K. L.: The measurement of haemagglutinin and matrix protein present on the surface of influenza virus infected P815 mastocytoma cells. J. gen. Virol. *42*, 541–553 (1979).

Adams, R. H., Brown, D. T.: Inhibition of Sindbis virus maturation after treatment of infected cells with trypsin. J. Virol. *41*, 692–702 (1982).

Adams, W. R., Kraft, L. M.: Electron microscopic study of the intestinal epithelium of mice infected with the agent of epizootic diarrhea of infant mice (EDIM) virus. Am. J. Pathol. *51*, 39–60 (1967).

Air, G. M.: Nucleotide sequence coding for the "signal peptide" and N terminus of the hemagglutinin from Asian (H₂N₂) strain of influenza virus. Virology *97*, 468–472 (1979).

Akashi, H., Bishop, D. H. L.: Comparison of the sequences and coding of La Crosse and Snowshoe hare bunyaviruses S RNA species. J. Virol. *45*, 1155–1158 (1983).

Allen, R. D., Travis, J. L., Allen, N., Yilmaz, H.: Video-enhanced contrast polarization (AVEC-POL) microscopy: a new method applied to the detection of birefringence in the motile reticulopodial network of Allogromia laticollaris. Cell Motility *1*, 275–289 (1981).

Almeida, J. D., Tyrrell, D. L. J.: The morphology of three previously uncharacterized human respiratory viruses that grow in organ culture. J. gen. Virol. *1*, 175–178 (1967).

Almeida, J. D., Hall, T., Banatvala, J. E., Totterdell, B. M., Chrystie, I. L.: The effect of trypsin on the growth of rotavirus. J. gen. Virol. *40*, 213–218 (1978).

Alonso, F. V., Compans, R. W.: Differential effect of monensin on enveloped viruses that form at distinct plasma membrane domains. J. Cell Biol. *89*, 700–705 (1981).

Alonso, F. V., Roth, M. G., Melsen, L. R., Srinivas, R. V., Compans, R. W.: Inhibition of viral glycoprotein transport in MDCK cells. In: Proceedings Electron Microscopy Society of America (Bailey, G. W., ed.), 34–37. 1982.

Altenburg, B. C., Graham, D. Y., Estes, M. K.: Ultrastructural study of rotavirus replication in cultured cells. J. gen. Virol. *46*, 75–85 (1980).

Altsteil, L. D., Landsberger, F. R.: Lipid-protein interactions between the surface glycoprotein of vesicular stomatitis virus and the lipid bilayer. Virology *115*, 1–9 (1981).

Amos, L. A.: Structure of Microtubules, In: Microtubules (Roberts, K., Hyams, J. S., eds.), 1–64. London-New York-Toronto-Sydney-San Francisco: Academic Press 1979.

Aoki, H., Maeno, K., Tsurumi, T., Takeura, S., Shibata, M., Hamaguchi, M., Nagai, Y., Sugiura, Y.: Analysis of the inhibitory effect of canavanine on the replication of influenza RI/5+ virus. II. Interaction of M protein with the plasma membrane. Microbiol. Immunol. *25*, 1279–1289 (1981).

Arias, C. F., Lopez, S., Espejo, R. T.: Gene protein products of SA11 simian rotavirus genome. J. Virol. *41*, 42–50 (1982).

Arias, C., Bell, J. R., Lenches, E. M., Strauss, E. G., Strauss, J. H.: Sequence analysis of two mutants of Sindbis virus defective in the intracellular transport of their glycoproteins. J. Mol. Biol. *168*, 87–102 (1983).

Arnheiter, H., Dubois-Dalcq, M., Schubert, M., Davis, N., Patton, J., Lazzarini, R.: Microinjection of monoclonal antibodies to vesicular stomatitis virus nucleocapsid protein into host cells: effect on virus replication. In: The Molecular Biology of Negative Strand Viruses (Bishop, D. H. L., Compans, R. W., eds.), 393–398. New York: Academic Press.

Auperin, D., Bishop, D. H. L.: Molecular cloning and nucleotide sequence of cDNA encoding Pichinde arenavirus nucleocapsid, N protein. Ann. Meeting of the American Society for Virology, East Lansing, Mich., July 1983.

Auperin, D., Dimock, K., Cash, P., Rawls, W. E., Leung, W.-C., Bishop, D. H. L.: Analyses of the genomes of prototype Pichinde arenaviruses and a virulent derivative of Pichinde Munchique: evidence for sequence conservation at the 3' termini of their viral RNA species. Virology *116*, 363–367 (1982).

Bächi, T.: Intramembrane structural differentiation in Sendai virus maturation. Virology *106*, 41–49 (1980).

Ball, L. A.: Gene order. In: Rhabdoviruses (Bishop, D. H. L., ed.), Vol. II, 61–74. Boca Raton, Fla.: CRC Press 1980.

Baltimore, D.: Expression of animal virus genomes. Bacteriol. Rev. *35*, 235–241 (1971).

Banfield, W. G., Kasnic, G., Blackwell, J. H.: Further observations on the virus of epizootic diarrhea of infant mice. An electron microscopic study. Virology *36*, 411–421 (1968).

Barbacid, M., Aaronson, S. A.: Membrane properties of the *gag* gene-coded p15 protein of mouse type-C RNA tumor viruses. J. Biol. Chemistry *253*, 1408–1414 (1978).

Bardelletti, G., Kessler, N., Aymard-Henry, N.: Morphology, biochemical analysis and neuraminidase activity of rubella virus. Arch. Virol. *49*, 175–186 (1975).

Baric, R. S., Stohlman, S. A., Lai, M. M. C.: Characterization of replicative intermediate RNA of mouse hepatitis virus: presence of leader RNA sequences on nascent chains. J. Virol. *48*, 633–640 (1983).

Barnett, B. B., Spendlove, R. S., Clark, M. L.: Effect of enzymes on rotavirus infection. J. Clin. Microbiol. *10*, 111–113 (1979).

Baruch, E.: Electron microscopic study of spinal cord of mice infected with Yellow Fever virus. J. Ultrastr. Res. *9*, 209–224 (1963).

Bastardo, J. W., Holmes, I. H.: Attachment of SA11 rotavirus to erythrocyte receptors. Infect. Immun. *29*, 1134–1140 (1980).

Bastardo, J. W., McKimm-Breschkin, J. L., Sonza, S., Mercer, L. D., Holmes, I. H.: Preparation and characterization of antisera to electrophoretically purified SA11 virus polypeptides. Infect. Immun. *34*, 641–647 (1981).

Becht, H.: Cytoplasmic synthesis of an arginine rich nuclear component during infection with an influenza virus. J. Virol. *7*, 204–207 (1971).

Becht, H., Hämmerling, A., Rott, R.: Undisturbed release of influenza virus in the presence of univalent antineuraminidase antibody. Virology *46*, 337–343 (1971).

Becker, W. B., McIntosh, K., Dees, J. H., Chanock, R. M.: Morphogenesis of avian infectious bronchitis virus and a related human virus (strain 229E). J. Virol. *1*, 1019–1027 (1967).

Bell, J. W., Garry, R. F., Waite, M. R. F.: Effect of low-NaCl medium on the envelope glycoproteins of sindbis virus. J. Virol. *25*, 764–769 (1978).

Bellini, W. J., Silver, G. D., McFarlin, D. E.: Biosynthesis of measles virus hemagglutinin in persistently infected cells. Arch. Virol. *75*, 87–101 (1983).

Bellini, W. J., Englund, G., Richardson, C. D., Hogan, N., Rozenblatt, S., Myers, C. A., Lazzarini, R. A.: Molecular cloning of the phosphoprotein gene of measles virus. In: The Molecular Biology of Negative Strand Viruses (Bishop, D. H. L., Compans, R. W., eds.), 359–364. New York: Academic Press.

Bender, W., Davidson, N.: Mapping of poly(A) sequences in the electron microscope reveals unusual structures of type I oncornavirus RNA molecules. Cell *7*, 595–607 (1976).

Bergmann, J. E., Tokuyasu, K. T., Singer, J. S.: Passage of an integral membrane protein, the vesicular stomatitis virus glycoprotein, through the Golgi apparatus en route to the plasma membrane. Proc. Natl. Acad. Sci. U.S.A. 78, 1746–1750 (1981).

Berkaloff, A.: Etude au microscope électronique de la morphogenèse de la particule de virus Sendai. J. Microscopie 2, 633–638 (1963).

Bernstein, J. M., Hruska, J. F.: Characterization of RNA polymerase products of Nebraska calf diarrhea virus and SA11 rotavirus. J. Virol. 37, 1071–1074 (1981).

Bhatt, P. N., Johnson, E. A., Smith, A. L., Jacoby, R. O.: Genetic resistance to lethal flaviviral encephalitis. III. Replication of Banzi virus in vitro and in vivo in tissues of congenic susceptible and resistant mice. Arch. Virol. 69, 273–286 (1981).

Bican, P., Cohen, J., Carpilienne, A., Scherrer, R.: Purification and characterization of bovine rotavirus cores. J. Virol. 43, 1113–1117 (1982).

Biddison, W. E., Doherty, P. C., Webster, R. G.: Antibody to influenza virus matrix protein detects a common antigen on the surface of cells infected with Type A influenza viruses. J. Exp. Med. 146, 690–697 (1977).

Bishop, D. H. L., Roy, P.: Kinetics of RNA synthesis by vesicular stomatitis virus particles. J. Mol. Biol. 57, 513–527 (1971).

Bishop, D. H. L., Shope, R. E.: Bunyaviridae. In: Comprehensive Virology (Fraenkel-Conrat, H., Wagner, R. R., eds.), Vol. 14. New York: Plenum Press 1979.

Bishop, D. H. L., Calisher, C. H., Casals, J., Chumakov, M. P., Gaidamovich, S. Y., Hannoun, C., Lvov, D. K., Marshall, I. D., Oker-Blom, N., Pettersson, R. F., Porterfield, J. S., Russell, P. K., Shore, R. E., Westaway, E. G.: Bunyaviridae. Intervirology 14, 125–143 (1980).

Bishop, D. H. L., Clerx, J. P. M., Clerx-Van Haaster, C. M., Robeson, G., Rozhon, E. J., Ushijima, H., Veerisetty, V.: Molecular and genetic properties of members of the Bunyaviridae. In: The Replication of Negative Strand Viruses (Bishop, D. H. L., Compans, R. W., eds.), 135–145. New York: Elsevier/North-Holland 1981.

Bishop, D. H. L., Ihara, T., Eshita, Y., Reid, E.: The coding strategies of bunyavirus RNA species. In: The Molecular Biology of Negative Strand Viruses (Bishop, D. H. L., Compans, R. W., eds.). New York: Academic Press. In Press.

Bishop, J. M.: Retroviruses. Ann. Rev. Biochem. 47, 35–88 (1978).

Blinzinger, K.: Comparative electron microscopic studies of several experimental group B arbovirus infections of the murine CNS (CEE virus, Zimmern virus, Yellow fever virus). Ann. Inst. Pasteur 123, 497–519 (1972).

Blinzinger, K., Müller, W., Anzil, A. P.: Microhelics in the endoplasmic reticulum of murine neurons infected with a group B arbovirus. Arch. ges. Virusforsch. 35, 194–202 (1971).

Blobel, G.: Intracellular protein topogenesis. Proc. Natl. Acad. Sci. U.S.A. 77, 1496–1500 (1980).

Blok, J., Air, G. M., Laver, W. G., Ward, C. W., Lilley, G. G., Woods, E. F., Roxburgh, C. M., Inglis, A. S.: Studies on the size, chemical composition, and partial sequence of the neuraminidase (NA) from type A influenza viruses show that the N-terminal region of the NA is not processed and serves to anchor the NA in the membrane. Virology 119, 109–121 (1982).

Blough, H. A., Tiffany, J. M.: Theoretical aspects of structure and assembly of viral envelopes. In: Current Topics in Microbiology and Immunology (Brody, J. A., ed.), Vol. 70, 1–30. Berlin-Heidelberg-New York: Springer 1975.

Blumberg, B. M., Kolakofsky, D.: Intracellular vesicular stomatitis virus leader RNAs are found in nucleocapsid structures. J. Virol. 40, 568–576 (1981).

Blumberg, B. M., Leppert, M., Kolakofsky, D.: Interaction of VSV leader RNA and nucleocapsid protein may control VSV genome replication. Cell 23, 837–845 (1981).

Blumberg, B. M., Giorgi, C., Kolakofsky, D.: N Protein of vesicular stomatitis virus selectively encapsidates leader RNA in vitro. Cell 32, 559–567 (1983).

Boege, U., Wengler, G., Wengler, G., Wittmann-Liebold, B.: Primary structures of the core proteins of the alphaviruses Semliki Forest virus and Sindbis virus. Virology 113, 293–303 (1981).

Bohn, W., Rutter, G., Mannweiler, K.: Production of monoclonal antibodies to measles virus proteins by immunization of mice with heated and detergent-treated antigens. Virology 116, 368–371 (1982).

Bohn, W., Rutter, G., Hohenberg, H., Mannweiler, K.: Inhibition of measles virus budding by phenothiazines. Virology 130, 44–45 (1983).

Bolognesi, D. P., Montelaro, R. C., Frank, H., Schafer, W.: Assembly of type C oncornaviruses: a Model. Science *199*, 183–186 (1978).

Bonatti, S., Cancedda, R., Blobel, G.: Membrane biogenesis: *in vitro* cleavage, core glycosylation and integration into microsomal membranes of Sindbis virus glycoproteins. J. Cell Biol. *80*, 219–224 (1979).

Bosch, J. V., Schwarz, R. T., Zeimiecki, A., Friss, R. A.: Oligosaccharide modifications and the site of processing of gPr92^{env}, the precursor of the viral glycoproteins of Rous sarcoma virus. Virology *119*, 122–132 (1982).

Bouloy, M., Hannoun, C.: Studies on Lumbo virus replication. I. RNA-dependent RNA polymerase associated with virions. Virology *69*, 258–264 (1976).

Bouloy, M., Plotch, S. J., Krug, R. M.: Globin mRNAs are primers for the transcription of influenza viral RNA *in vitro*. Proc. Natl. Acad. Sci. U.S.A. *75*, 4886–4890 (1978).

Bowen, H. A., Lyles, D. C.: Structure of Sendai viral proteins in plasma membranes of virus-infected cells. J. Virol. *37*, 1079–1082 (1981).

Braam, J., Ulmanen, I., Krug, R. M.: Molecular model of a eukaryotic transcription complex: functions and movements of influenza P proteins during capped RNA-primed transcription. Cell *34*, 609–618 (1983).

Bracha, M., Schlesinger, M. J.: Altered E$_2$ glycoprotein of Sindbis virus and its use in complementation studies. J. Virol. *26*, 126–135 (1978).

Braciale, T. J.: Immunologic recognition of influenza virus-infected cells. II. Expression of influenza A matrix protein on the infected cell surface and its role in recognition by cross-reactive cytotoxic T cells. J. Exp. Med. *146*, 673–689 (1977).

Brand, C. M., Skehel, J. J.: Crystalline antigen from the influenza virus envelope. Nature New Biology *238*, 145–147 (1972).

Brayton, P. R., Lai, M. M. C., Patton, C. D., Stohlman, S. S.: Characterization of two RNA polymerase activities induced by mouse hepatitis virus. J. Virol. *42*, 847–853 (1982).

Breitenfeld, P. M., Schafer, W.: The formation of fowl plague virus antigens in infected cells, as studied with fluorescent antibodies. Virology *4*, 328–345 (1975).

Bridger, J. C., Woode, G. N., Meyling, A.: Isolation of coronaviruses from neonatal calf diarrhoea in Great Britain and Denmark. Vet. Microbiol. *3*, 101–113 (1978).

Briedis, D. J., Conti, G., Munn, E. A., Mahy, B. W. J.: Migration of influenza virus-specific polypeptides from cytoplasm to nucleus of infected cells. Virology *111*, 154–164 (1981).

Brinton, M. A.: Non-arbo togaviruses. In: The Togaviruses (Schlesinger, R. W., ed.), 623–666. New York: Academic Press 1980.

Brown, D. T.: Assembly of alphaviruses. In: The Togaviruses (Schlesinger, R. W., ed.), 473–501. New York: Academic Press 1980.

Brown, D. T., Riedel, B.: Morphogenesis of vesicular stomatitis virus: electron microscope observations with freeze-fracture techniques. J. Virol. *21*, 601–609 (1977).

Brown, D. T., Waite, M. R. F., Pfefferkorn, E. R.: Morphology and morphogenesis of Sindbis virus as seen with freeze-etching techniques. J. Virol. *10*, 524–536 (1972).

Brown, F., Crick, J.: Natural history of the rhabdoviruses of vertebrates and invertebrates. In: Rhabdoviruses (Bishop, D. H. L., ed.), Vol. I, 1–22. Boca Raton, Fla.: CRC Press 1979.

Bruns, M., Peralta, L. M., Lehmann-Grübe, F.: Lymphocytic choriomeningitis virus. III. Structural proteins of the virion. J. gen. Virol. *64*, 599–611 (1983).

Bucher, D., Kharitonenkov, J., Zakomirdin, I. G., Grigoriev, B. V., Klimenko, S. M., Davis, J. F.: Incorporation of influenza M-protein into liposomes. J. Virol. *36*, 586–590 (1980).

Buchmeier, M. J., Oldstone, M. B. A.: Identity of the viral protein responsible for serologic cross reactivity among the Tacaribe complex arenaviruses. In: Negative Strand Viruses and the Host Cell (Mahy, B. W. J., Barry, R. D., eds.), 91–97. New York-San Francisco-London: Academic Press 1978.

Buchmeier, M. J., Oldstone, M. B. A.: Protein structure of lymphocytic choriomeningitis virus: evidence for a cell-associated precursor of the virion glycopeptides. Virology *99*, 111–120 (1979).

Buchmeier, M. J., Oldstone, M. B. A.: Molecular studies of LCM virus induced immunopathology: development and characterization of monoclonal antibodies to LCM virus. In: The Replication of Negative Strand Viruses (Bishop, D. H. L., Compans, R. W., eds.), 71–78. New York-Amsterdam-Oxford: Elsevier/North-Holland 1981.

Buchmeier, M. J., Elder, J. H., Oldstone, M. B. A.: Protein structure of lymphocytic choriomeningitis virus: identification of the virus structural and cell associated polypeptides. Virology *89*, 133–145 (1978).

Buchmeier, M. J., Welsh, R. M., Dutko, F. J., Oldstone, M. B. A.: The virology and immunobiology of lymphocytic choriomeningitis virus infection. In: Advances in Immunology (Dixon, F. J., Kunkel, H. G., eds.), Vol. 30, 275–331. New York-London-Toronto-Sydney-San Francisco: Academic Press 1980.

Buchmeier, M. J., Lewicki, H. A., Tomori, O., Oldstone, M. B. A.: Monoclonal antibodies to lymphocytic choriomeningitis and Pichinde viruses: generation, characterization, and cross-reactivity with other arenaviruses. Virology *113*, 73–85 (1981).

Bukrinskaya, A., Starov, A., Issayeva, C.: Abortive infection of Ehrlich tumor cells by influenza virus: virus-specific RNA, protein synthesis and transport. Intervirology *16*, 43–48 (1981).

Butler, P. J., Klug, A.: The assembly of a virus. In: Molecules to Living Cells. Reading from Scientific American, 116–121. San Francisco: Freeman 1980.

Büechi, M., Bächi, T.: Immunofluorescence and electron microscopy of the cytoplasmic surface of the human erythrocyte membrane and its interaction with Sendai virus. J. Cell Biol. *83*, 338–347 (1979).

Büechi, M., Bächi, T.: Microscopy of internal structures of Sendai virus associated with the cytoplasmic surface of host membranes. Virology *120*, 349–359 (1982).

Calberg-Bacq, C.-M., Francois, C., Kozma, S., Gosselin, L., Osterrieth, P. M., Rentier-Delrue, F.: On the organization of the mouse mammary tumor virus envelope. Sixth European Congress on Electron Microscopy, Jerusalem 1976.

Calberg-Bacq, C.-M., Francois, C., Kozma, S., Osterrieth, P. M., Teramoto, Y. A.: Immunological characterization of a mammary tumour virus from Swiss mice: multiple epitopes associated with the viral gene products. J. gen. Virol. *57*, 75–83 (1981).

Caliguiri, L. A., Holmes, K. V.: Host-dependent restriction of influenza virus maturation. Virology *92*, 15–30 (1979).

Capone, J., Toneguzzo, F., Ghosh, H. P.: Synthesis and assembly of membrane glycoproteins. Membrane anchoring COOH-terminal domain of vesicular stomatitis virus envelope glycoprotein G contains fatty acids. J. Biological Chem. *257*, 16–19 (1982).

Cardiff, R. D., Russ, S. B., Brandt, W. E., Russell, P. K.: Cytological localization of dengue-2 antigens: an immunological study with ultrastructural correlation. Infect. Immun. *7*, 809–816 (1973).

Carroll, A. R., Wagner, R. R.: Role of the membrane (M) protein in endogenous inhibition of *in vitro* transcription by vesicular stomatitis virus. J. Virol. *29*, 134–142 (1979).

Cartwright, B.: Effect of Concanavalin A on vesicular stomatitis virus maturation. J. gen. Virol. *34*, 249–256 (1977).

Cartwright, B., Smale, C. J., Brown, F., Hull, R.: Model for vesicular stomatitis virus. J. Virol. *10*, 256–260 (1972).

Cartwright, B., Talbot, P., Brown, F.: The proteins of biologically active sub-units of vesicular stomatitis virus. J. gen. Virol. *7*, 267–272 (1970).

Caul, E. O., Egglestone, S. I.: Further studies on human enteric coronaviruses. Arch. Virol. *54*, 107–117 (1977).

Caul, E. O., Ashley, C. R., Ferguson, M., Egglestone, S. I.: Preliminary studies on the isolation of coronavirus 229E nucleocapsids. FEMS Microbiology Lett. *5*, 101–105 (1979).

Cavanagh, D.: Structural polypeptides of coronavirus IBV. J. gen. Virol. *53*, 93–103 (1981).

Chanas, A. C., Young, P. R., Ellis, D. S., Mann, G., Stanford, S. S., Howard, C. R.: Evaluation of plaque size reduction as a method for the detection of Pichinde virus antibody. Arch. Virol. *65*, 157–167 (1980).

Chasey, D.: Different particle types in tissue culture and intestinal epithelium infected with rotavirus. J. gen. Virol. *37*, 443–451 (1977).

Chasey, D.: Investigation of immunoperoxidase labeled rotavirus in tissue culture by light and electron microscopy. J. gen. Virol. *50*, 195–200 (1980).

Chasey, D., Alexander, D. J.: Morphogenesis of Avian infectious bronchitis virus in primary chick kidney cells. Arch. Virol. *52*, 101–111 (1976).

Chasey, D., Labram, J.: Electron microscopy of tubular assemblies associated with naturally occurring bovine rotavirus. J. gen. Virol. *64*, 863–872 (1983).

Chatis, P. A., Morrison, T. G.: Mutational changes in the vesicular stomatitis virus glycoprotein affect the requirement of carbohydrate in morphogenesis. J. Virol. *37*, 307–316 (1981).

Chatterjee, S., Bradac, J. A., Hunter, E.: Effect of monensin on Mason-Pfizer montery virus glycoprotein synthesis. J. Virol. *44*, 1003–1012 (1982).

Cheley, S., Anderson, R.: Cellular synthesis and modification of murine hepatitis virus polypeptides. J. gen. Virol. *54*, 301–311 (1981).

Cheley, S., Anderson, R., Cupples, M. J., LeeChan, E. C. M., Morris, L. V.: Intracellular murine hepatitis virus-specific RNAs contain common sequences. Virology *112*, 596–604 (1981).

Choppin, P. W., Compans, R. W.: Phenotypic mixing of envelope proteins of the parainfluenza virus SV5 and vesicular stomatitis virus. J. Virol. *5*, 609–616 (1970).

Choppin, P. W., Compans, R. W.: Reproduction of Paramyxoviruses. In: Comprehensive Virology (Fraenkel-Conrat, H., Wagner, R. R., eds.), Vol. 4, 95–178. New York: Plenum Press 1975.

Clegg, J. C. S.: Sequential translation of capsid and membrane protein genes of alphaviruses. Nature *254*, 454–455 (1975).

Clerx, J. P. M., Fuller, F., Bishop, D. L. H.: Tick-borne viruses structurally similar to Orthomyxoviruses. Virology *127*, 205–219 (1983).

Clinton, G. M., Little, S. P., Hagen, F. S., Huang, A. S.: The matrix (M) protein of vesicular stomatitis virus regulates transcription. Cell *15*, 1455–1462 (1978).

Clinton, G. M., Burge, B. W., Huang, A. S.: Phosphoproteins of vesicular stomatitis virus: identity and interconversion of phosphorylated forms. Virology *99*, 84–94 (1979).

Coelho, K. I. R., Bryden, A. S., Hall, C., Flewett, T. H.: Pathology of rotavirus infection in suckling mice: A study by conventional histology, immunofluorescence, ultra thin sections and scanning electron microscopy. Ultrastr. Pathol. *2*, 59–80 (1981).

Cohen, J.: Ribonucleic antipolymerase activity associated with purified calf rotavirus. J. gen. Virol. *36*, 395–402 (1977).

Cohen, J., Dubois, P.: Cell free transcription and translation of rotavirus RNA. Biochem. Biophys. Res. Commun. *88*, 791–796 (1979).

Collins, A. R., Knobler, R. L., Powell, H., Buchmeier, M. J.: Monoclonal antibodies to murine hepatitis virus 4 (strain JHM) define the viral glycoprotein responsible for attachment and cell-cell fusion. Virology *119*, 358–371 (1982).

Collins, P. L., Hightower, L. E., Ball, L. A.: Transcriptional map for Newcastle disease virus. J. Virol. *35*, 682–693 (1980).

Collins, P. L., Wertz, G. T. W., Ball, L. A., Hightower, L. E.: Translation of the separated messenger RNAs of Newcastle disease virus. In: The Replication of Negative Strand Viruses (Bishop, D. H. L., Compans, R. W., eds.), 537–543. New York: Elsevier/North-Holland 1981.

Compans, R. W.: Influenza virus proteins. II. Association with components of the cytoplasm. Virology *51*, 56–70 (1973).

Compans, R. W., Caliguiri, L. A.: Isolation and properties of an RNA polymerase from influenza virus-infected cells. J. Virol. *11*, 441–448 (1973).

Compans, R. W., Dimmock, N. J.: An electron microscopic study of single-cycle infection of chick embryo fibroblasts by influenza virus. Virology *39*, 449–515 (1969).

Compans, R. W., Klenk, H.-D.: Viral membranes. In: Comprehensive Virology (Fraenkel-Conrat, H., Wagner, R. R., eds.), Vol. 13, 293–407. New York: Plenum Press 1979.

Compans, R. W., Pinter, A.: Incorporation of sulfate into influenza virus glycoproteins. Virology *66*, 151–160 (1975).

Compans, R. W., Holmes, K. V., Dales, S., Choppin, P. W.: An electron microscopic study of moderate and virulent virus-cell interactions of the parainfluenza virus SV5. Virology *30*, 411–426 (1966).

Compans, R. W., Klenk, H.-D., Caliguiri, L. A., Choppin, P. W.: Influenza virus proteins. I. Analysis of polypeptides of the virion and identification of spike glycoproteins. Virology *42*, 880–889 (1970 a).

Compans, R. W., Dimmock, N. J., Meier-Ewert, H.: An electron microscopic study of the influenza virus-infected cell. In: The Biology of Large RNA Viruses (Barry, R. D., Mahy, B. W. J., eds.), 87–108. New York: Academic Press 1970 b.

Compans, R. W., Content, J., Duesberg, P. H.: Structure of the ribonucleoprotein of influenza virus. J. Virol. *10*, 795–801 (1972 a).

Compans, R. W., Mountcastle, W. E., Choppin, P. W.: The sense of the helix of paramyxovirus nucleocapsids. J. Mol. Biol. *65*, 167–169 (1972 b).

Compans, R. W., Boersma, D. P., Cash, P., Clerx, J. P. M., Gimenez, H. B., Kirk, W. E., Peters, C. J., Vezza, A. C., Bishop, D. H. L.: Molecular and genetic studies of Tacaribe, Pichinde, and lymphocytic choriomeningitis viruses. In: The Replication of Negative Strand Viruses (Bishop, D. H. L., Compans, R. W., eds.), 31–42. New York-Amsterdam-Oxford: Elsevier/North-Holland 1981.

Cox, N. J., Kendal, A. P.: Presence of a segmented single stranded genome in influenza C virus. Virology 74, 239–241 (1976).

Dahlberg, J. E., Obijeski, J. F., Korb, J.: Electron microscopy of the segmented RNA genome of the La Crosse virus: absence of circular molecules. J. Virol. 22, 203–209 (1977).

Dalrymple, J. M., Schlesinger, S., Russell, P. K.: Antigenic characterization of two Sindbis envelope glycoproteins separated by isoelectric focusing. Virology 69, 93–103 (1976).

Dalton, A. J., Haguenau, F.: Ultrastructure of Animal Viruses and Bacteriophages: An Atlas. New York-London: Academic Press 1973.

Dalton, A. J., Rowe, W. P., Smith, G. H., Wilsnack, R. E., Pugh, W. E.: Morphological and cytochemical studies on lymphocytic choriomeningitis virus. J. Virol. 2, 1465–1478 (1968).

Damsky, C. H., Sheffield, J. B., Tuszynski, G. P., Warren, L.: Is there a role for actin in virus budding? J. Cell Biol. 75, 593–605 (1977).

David-Ferriera, J. F., Manaker, R. A.: An electron microscope study of the development of a mouse hepatitis virus in tissue culture cells. J. Cell Biol. 24, 57–78 (1965).

Davies, H. A., Dourmashkin, R. R., Macnaughton, M. R.: Ribonucleoprotein of Avian infectious bronchitis virus. J. gen. Virol. 53, 67–74 (1981).

De, B. P., Thornton, G.j B., Luk, D., Banerjee, A. K.: Purified matrix protein of vesicular stomatitis virus blocks viral transcription in vitro. Proc. Natl. Acad. Sci. U.S.A. 79, 7137–7141 (1982).

de Harven, E.: Remarks on the ultrastructure of type A, B, and C virus particles. In: Advances in Virus Research, Vol. 19, 221–264. New York-San Francisco-London: Academic Press 1974.

Demsey, A., Steere, R. L., Brandt, W. E., Veltri, B. J.: Morphology and development of Dengue-2 virus employing the freeze fracture and thin-section techniques. J. Ultrastr. Res. 46, 103–116 (1974).

Demsey, A., Kawka, D., Stackpole, C. W.: Application of freeze-drying intact cells to studies of murine oncornavirus morphogenesis. J. Virol. 21, 358–367 (1977).

Dennis, D. E., Brian, D. A.: Coronavirus cell-associated RNA-dependent RNA polymerase. In: Biochemistry and Biology of Coronaviruses (ter Meulen, V., Siddell, S., Wege, H., eds.), 155–170. New York: Plenum Press 1981.

Dennis, D. E., Brian, D. A.: RNA-dependent RNA polymerase activity in coronavirus-infected cells. J. Virol. 42, 153–164 (1982).

Desselberger, V., Racaniello, V. R., Zazra, J. J., Palese, P.: The 3' and 5'-terminal sequences of influenza A, B and C virus RNA segments are highly conserved and show partial inverted complementarity. Gene 8, 315–328 (1980).

Deubel, V., Digoutte, J. P., Mattei, X., Pandare, D.: Morphogenesis of Yellow Fever virus in Aedes aegypti cultured cells. II. An ultrastructural study. J. Trop. Med. Hyg. 30, 1071–1077 (1981).

Dickson, C., Eisenmann, R., Fan, H., Hunter, E., Teich, N.: Protein biosynthesis and assembly. In: The Molecular Biology of Tumor Viruses (Weiss, R. A., Teich, N., Varmus, H. E., Coffin, J. M., eds.), RNA Tumor Viruses, 2nd ed., 513–648. Cold Spring Harbor, N.Y.: Cold Spring Harbor Laboratory 1982.

Dickson, R. B., Willingham, M. C., Pastan, I.: α_2-macroglobulin adsorbed to colloidal gold: a new probe in the study of receptor-mediated endocytosis. J. Cell Biol. 89, 29–34 (1981).

Dietzschold, B., Wunner, W. H., Wiktor, T. J., Lopes, A. D., Lafon, M., Smith, C. L., Koprowski, H.: Characterization of an antigenic determinant of the glycoprotein that correlates with pathogenicity of rabies virus. Proc. Natl. Acad. Sci. U.S.A. 80, 70–74 (1983).

Dimmock, N. J., Cook, R. F., Bean, W. J., Wignall, J. M.: Cellular and viral control processes affect the expression of matrix protein during influenza virus infection of avian erythrocytes. In: The Replication of Negative Strand Viruses (Bishop, D. H. L., Compans, R. W., eds.), 345–351. New York: Elsevier/North-Holland 1981.

Doughri, A. M., Storz, J.: Light and ultrastructural pathologic changes in intestinal coronavirus infection of newborn calves. Zentralbl. Vet. Med. B 24, 367–387 (1977).

Doughri, A. M., Storz, J., Hajer, J., Fernando, H. S.: Morphology and morphogenesis of a coronavirus infected intestinal epithelial cells of newborn calves. Exp. Mol. Pathol. 25, 355–370 (1976).

Dowdle, W. R., Downie, J. C., Laver, W. G.: Inhibition of virus release by antibodies to surface antigens of influenza viruses. J. Virol. *13,* 269–275 (1974).

Dowling, P. C., Giorgi, C., Roux, L., Dethlefsen, L. A., Galantowicz, M. E., Blumberg, B. M., Kolakofsky, D.: Molecular cloning of the 3′ proximal third of the Sendai virus genome. Proc. Natl. Acad. Sci. U.S.A. *80,* 5213–5216 (1983).

Dubois-Dalcq, M., Reese, T. S.: Structural changes in the membrane of Vero cells infected with a paramyxovirus. J. Cell Biol. *67,* 551–565 (1975).

Dubois-Dalcq, M., Barbosa, L. H., Hamilton, R., Sever, J. L.: Comparison between productive and latent subacute sclerosing panencephalitis viral infection *in vitro.* Lab. Invest. *30,* 241–250 (1974).

Dubois-Dalcq, M., Reese, T. S., Murphy, M., Fuccillo, D.: Defective bud formation in human cells chronically infected with SSPE. J. Virol. *19,* 579–593 (1976 a).

Dubois-Dalcq, M., Reese, T. S., Narayan, O.: Membrane changes associated with assembly of visna virus. Virology *74,* 520–530 (1976 b).

Dubois-Dalcq, M., Narayan, O., Griffin, D. E.: Cell surface changes associated with mutation of visna virus in antibody-treated cell cultures. Virology *92,* 353–366 (1979 a).

Dubois-Dalcq, M., Rodriguez, M., Reese, T. S.: Structural changes in the membrane of cells infected with scrapie and other neurotropic viruses. In: Slow Transmissible Diseases of the Nervous System (Prusiner, S. B., Hadlow, W. J., eds.), Vol. II, 123–145. New York-London-Sydney-Toronto-San Francisco: Academic Press 1979 b.

Dubois-Dalcq, M. E., Doller, E. W., Haspel, M. V., Holmes, K. V.: Cell tropism and expression of mouse hepatitis viruses (MHV) in mouse spinal cord cultures. Virology *119,* 317–331 (1982).

Ducatelle, R., Coussement, W., Pensaert, M. B., DeBouck, P., Hoorens, J.: *In vivo* morphogenesis of a new porcine enteric coronavirus CV777. Arch. Virol. *68,* 35–44 (1981).

Durbin, R. K., Manning, J. S.: The core of murine leukemia virus requires phosphate for structural stability. Virology *116,* 31–39 (1982).

Dutko, F. J., Kennedy, S. I. T., Oldstone, M. B. A.: Genome structure of lymphocytic choriomeningitis virus: cohesive complementary termini? In: The Replication of Negative Strand Viruses (Bishop, D. H. L., Compans, R. W., eds.), 43–50. New York-Amsterdam-Oxford: Elsevier/North-Holland 1981.

Dyall-Smith, M. L., Holmes, I. H.: Gene-coding assignments to rotavirus double-stranded RNA segments 10 and 11. J. Virol. *38,* 1099–1103 (1981).

East, J. L., Kingsbury, D. W.: Mumps virus replication in chick embryo lung cells: properties of RNA species in virions and infected cells. J. Virol. *8,* 161–173 (1971).

Edwards, S. G., Lin, Y. C., Fan, H.: Association of murine leukemia virus *gag* antigen with extracellular matrices in productively infected mouse cells. Virology *116,* 306–317 (1982).

Edwards, J., Mann, E., Brown, D. T.: Conformational changes in Sindbis virus envelope proteins accompanying exposure to low pH. J. Virol. *45,* 1090–1097 (1983).

Elder, K. T., Bye, J. M., Skehel, J. J., Waterfield, M. D., Smith, A. E.: *In vitro* synthesis, glycosylation and membrane insertion of influenza virus hemagglutinin. Virology *95,* 343–350 (1979).

Emerson, S. U.: Reconstitution studies detect a single polymerase entry site on the vesicular stomatitis virus genome. Cell *31,* 635–642 (1982).

Emerson, S. U., Yu, Y. H.: Both NS and L proteins are required for *in vitro* RNA synthesis by vesicular stomatitis virus. J. Virol. *15,* 1348–1356 (1975).

Enzmann, P.-J., Weiland, F.: Studies on the morphology of alphaviruses. Virology *95,* 501–510 (1979).

Ericson, B. L., Graham, D. Y., Mason, B. B., Estes, M. K.: Identification, synthesis and modifications of simian rotavirus SA11 polypeptides in infected cells. J. Virol. *42,* 825–829 (1982).

Ericson, B. L., Graham, D. Y., Mason, B. B., Hanssen, H. H., Estes, M. K.: Two types of glycoprotein precursors are produced by the simian rotavirus SA11. Virology *127,* 320–332 (1983).

Erwin, C., Brown, D. T.: Intracellular distribution of Sindbis virus membrane proteins in BHK-21 cells infected with wild-type virus and maturation-defective mutants. J. Virol. *36,* 775–786 (1980).

Erwin, C., Brown, D. T.: Requirement of cell nucleus for Sindbis virus replication in cultured Aedes albopictus cells. J. Virol. *45,* 792–799 (1983).

Esparza, J., Gil, F.: A study of the ultrastructure of human rotavirus. Virology *91,* 141–150 (1978).

Espejo, R. T., Lopez, S., Carlos, A.: Structural polypeptides of simian rotavirus SA11 and the effect of trypsin. J. Virol. *37,* 156–160 (1981).

Estes, M. K., Graham, D. Y., Mason, B. B.: Proteolytic enhancement of rotavirus infectivity: molecular mechanisms. J. Virol. *39*, 879–888 (1981).

Estes, M. K., Graham, D. Y., Ramig, R. F., Ericson, B. L.: Heterogeneity in the structural glycoprotein (VP7) of simian rotavirus SA11. Virology *122*, 8–14 (1982).

Estes, M. K., Palmer, E. L., Obijeski, J. F.: Rotaviruses: A review. In: Current Topics in Microbiology and Immunology (Cooper, M., Hofshneider, P. H., Koprowski, H., Melchers, F., Rott, R., Schweiger, H., Vogt, P. K., eds.), Vol. 105, 123–184. Berlin-Heidelberg-New York: Springer 1983.

Etchison, J. R., Summers, D. F.: Structure, synthesis, and function of the vesicular stomatitis virus glycoprotein. In: Rhabdoviruses (Bishop, D. H. L., ed.), Vol. 1, 151–160. Boca Raton, Fla.: CRC Press 1979.

Etkind, P. R., Cross, R. K., Lamb, R. A., Merz, D. C., Choppin, P. W.: *In vitro* synthesis of structural and non-structural proteins of Sendai and SV5 viruses. Virology *100*, 22–33 (1980).

Ewert, D. L., Halpern, M. S.: Avian endogenous retroviral envelope glycoprotein is assembled in two structural complexes of gp85 and gp37 subunits. Virology *122*, 506–509 (1982).

Famulari, N. G., Fleissner, E.: Kinetics of utilization of Sendai virus RNA and proteins in the process of virus assembly. J. Virol. *17*, 605–613 (1976).

Farber, F. E., Rawls, W. E.: Isolation of ribosome-like structures from Pichinde virus. J. gen. Virol. *26*, 21–31 (1975).

Farquhar, M. G., Palade, G. E.: The Golgi apparatus (complex)–(1954–1981)–from artifact to center stage. J. Cell Biol. *91*, 77s–103s (1981).

Faulkner, G., Dubois-Dalcq, M., Hooghe-Peters, E., McFarland, H. F., Lazzarini, R. A.: Defective interfering particles modulate VSV infection of dissociated neuron cultures. Cell *17*, 979–991 (1979).

Fekadu, M., Chandler, F. W., Harrison, A. K.: Pathogenesis of rabies in dogs inoculated with an Ehtiopian rabies virus strain. Immunofluorescence, histologic and ultrastructural studies of the central nervous system. Arch. Virol. *71*, 109–126 (1982).

Fields, S., Winter, G., Brownlee, G. G.: Structure of the neuraminidase gene in human influenza virus A/PR/8/34. Nature *290*, 213–217 (1981).

Finch, J. T., Gibbs, A. J.: Observations on the structure of the nucleocapsid of some paramyxoviruses. J. gen. Virol. *6*, 141–150 (1970).

Flawith, J. W. F., Dimmock, N. J.: Newly synthesized influenza virus proteins are transported from the nucleus. J. gen. Virol. *45*, 527–531 (1979).

Flewett, T. H., Woode, G. W.: The rotaviruses. Arch. Virol. *57*, 1–23 (1978).

Flewett, T. H., Bryden, A. S., Davies, H., Woode, G. N., Bridger, J. C., Derrick, J. M.: Relation between viruses from acute gastroenteritis of children and newborn calves. Lancet *2*, 61–63 (1974).

Frank, H., Schwarz, H., Graf, T., Schäfer, W.: Properties of mouse leukemia viruses. XV. Electron microscopic studies on the organization of Friend leukemia virus and other mammalian C-type viruses. Z. Naturforsch. *33c*, 124–138 (1978).

Francki, R. I. B., Randles, J. W.: Rhabdoviruses infecting plants. In: Rhabdoviruses (Bishop, D. H. L., ed.), Vol. III, 135–166. Boca Raton, Fla.: CRC Press 1980.

Friedman, R. M.: Antiviral activity of interferon. Bacteriol. Rev. *41*, 543–567 (1977).

Fuller, F., Bishop, D. H. L.: Identification of virus coded non-structural polypeptides in Bunyavirus infected cells. J. Virol. *41*, 643–648 (1982).

Gallione, C. J., Greene, J. R., Iverson, L. E., Rose, J. K.: Nucleotide sequences of the mRNA's encoding the vesicular stomatitis virus N and NS proteins. J. Virol. *39*, 529–535 (1981).

Gallo, R. C., Wong-Staal, F.: Retroviruses as etiological agents of some animal and human leukemias and lymphomas and as tools for elucidating the molecular mechanism of leukemogenesis. Blood *60*, 545–557 (1982).

Gard, G. P., Vezza, A. C., Bishop, D. H., Compans, R. W.: Structural proteins of Tacaribe and Tamiami virions. Virology *83*, 84–95 (1977).

Garoff, H.: Structure and assembly of the Semliki Forest virus membrane. Biochem. Soc. Transact. *7*, 301–306 (1979).

Garoff, H., Söderlund, H.: The amphiphilic membrane glycoproteins of Semliki Forest virus are attached to the lipid bilayer by their COOH-terminal ends. J. Mol. Biol. *124*, 545–549 (1978).

Garoff, H., Frischauf, A.-M., Simons, K., Lehrach, H., Delius, H.: The capsid protein of Semliki Forest virus has clusters of basic amino acids and prolines in its amino terminal region. Proc. Natl. Acad. Sci. U.S.A. *77*, 6376–6380 (1980 a).

Garoff, H., Frischauf, A.-M., Simons, K., Lehrach, H., Delius, H.: Nucleotide sequence of cDNA coding for Semliki Forest virus membrane glycoproteins. Nature *288*, 236–241 (1980 b).

Garoff, H., Kondor-Koch, C., Riedel, H.: Structure and assembly of alphaviruses. Curr. Top. Microbiol. Immunol. *99*, 1–50 (1982).

Garten, W., Kohama, T., Klenk, H.-D.: Proteolytic activation of the haemagglutinin-neuraminidase of Newcastle disease virus involves loss of a glycopeptide. J. gen. Virol. *51*, 207–211 (1980).

Garwes, D.J.: Structure and physicochemical properties of coronaviruses. In: Viral Enteritis in Humans and Animals (Bricout, F., Scherrer, R., eds.), Vol. 90, 141–162. Les Editions de L'Institut National de la Santé et de la Recherche Médicale 1980.

Garwes, D.J., Pocock, D.H.: The polypeptide structure of transmissible gastroenteritis virus. J. gen. Virol. *29*, 25–34 (1975).

Garwes, D.J., Pocock, D.H., Pike, B.V.: Isolation of subviral components from transmissible gastroenteritis virus. J. gen. Virol. *32*, 283–294 (1976).

Garwes, D.J., Lucas, M.H., Higgins, D.A., Pike, B.V., Cartwright, S.F.: Antigenicity of structural components from porcine transmissible gastroenteritis virus. Vet. Microbiol. *3*, 179–190 (1978/79).

Gentsch, J.R., Bishop, D.H.L.: Small viral RNA segment of bunyaviruses codes for viral nucleocapsid protein. J. Virol. *28*, 417–419 (1978).

Gentsch, J.R., Bishop, D.H.L.: M viral RNA segment of bunyaviruses codes for two glycoproteins, G_1 and G_2. J. Virol. *30*, 767–770 (1979).

Gentsch, J.R., Bishop, D.H.L., Obijeski, J.F.: The virus particle' nucleic acids and proteins of four bunyaviruses. J. gen. Virol. *34*, 257–268 (1977).

Genty, N., Bussereau, F.: Is cytoskeleton involved in vesicular stomatitis virus reproduction? J. Virol. *34*, 777–781 (1980).

Georges, J.C., Guedenet, J.C.: Etude ultrastructurale de l'organisation de la nucléocapside an cours de la morphogenèse du Myxovirus Parainfluenzae 2 en cultures cellulaires. J. Microscopie *19*, 221–230 (1974).

Gething, M.J., Sambrook, J.: Cell surface expression of influenza hemagglutinin from a cloned DNA copy of the RNA gene. Nature *293*, 620–625 (1981).

Gething, M.J., White, J.M., Waterfield, M.D.: Purification of the fusion protein of Sendai virus: analysis of the NH_2 terminal sequence generated during precursor activation. Proc. Natl. Acad. Sci. U.S.A. *76*, 2737–2740 (1978).

Gibson, R., Leavitt, R., Kornfeld, S., Schlesinger, S.: Synthesis and infectivity of vesicular stomatitis virus containing nonglycosylated G protein. Cell *13*, 671–679 (1978).

Gibson, R., Schlesinger, S., Kornfeld, S.: The nonglycosylated glycoprotein of vesicular stomatitis virus is temperature-sensitive and undergoes intracellular aggregation at elevated temperatures. J. Biol. Chem. *254*, 3600–3607 (1979).

Giorgi, C., Blumberg, B.M. and D. Kopakofsky: Sendai virus contains overlapping genes expressed from a single mRNA. Cell. *35*, 829–836 (1983).

Giuffre, R.M., Tovell, D.R., Kay, C.M., Tyrrell, D.L.J.: Evidence for an interaction between the membrane protein of a Paramyxovirus and actin. J. Virol. *42*, 963–968 (1982).

Glazier, K., Raghow, R., Kingsbury, D.W.: Regulation of Sendai virus transcription: evidence for a single promotor *in vivo*. J. Virol. *21*, 863–871 (1971).

Gomatos, P., Kääriäinen, L., Keränen, S., Ranki, M., Sawicki, D.L.: Semliki Forest virus replication complex capable of synthesizing 42S and 26S nascent RNA chains. J. gen. Virol. *49*, 61–69 (1980).

Graham, D.Y., Estes, M.K.: Proteolytic enhancement of rotavirus infectivity: biological mechanisms. Virology *101*, 432–439 (1980).

Granoff, A.: Heterozygosis and phenotypic mixing with Newcastle disease virus. In: Basic Mechanisms in Animal Virus Biology, Vol. 27. New York: Cold Spring Harbor Symposia on Quantitative Biology 1962.

Green, J., Griffiths, G., Louvard, D., Quinn, P., Warren, G.: Passage of viral membrane proteins through the Golgi complex. J. Mol. Biol. *152*, 663–698 (1981 a).

Green, N., Shinnic, T.M., Witte, W., Ponticelli, A., Sutcliffe, J.G., Lerner, R.A.: Sequence-specific antibodies show that maturation of Moloney leukemia virus envelope polyprotein involves removal of a COOH-terminal peptide. Proc. Natl. Sci. U.S.A. *78*, 6023–6027 (1981 b).

Greenberg, H. B., Kalica, A. R., Wyatt, R. G., Jones, R. W., Kapikian, A. Z., Chanock, R. M.: Rescue of noncultivatable human rotavirus by gene reassortment during mixed infection with *ts* mutants of a cultivatable bovine rotavirus. Proc. Natl. Acad. Sci. U.S.A. *78*, 420—424 (1981).

Greenberg, H. B., McAuliffe, V., Valdesuso, J., Wyatt, R. G., Flores, J., Kalica, A. R., Hoshino, Y., Singh, N.: Serological analysis of the subgroup protein of rotavirus, using monoclonal antibodies. Infect. Immun. *39*, 91—99 (1983 a).

Greenberg, H. B., Valdesuso, J., van Wyke, K., Midthun, K., Walsh, M., McAuliffe, V., Wyatt, R. G., Kalica, A. R., Flores, J., Hoshino, Y.: Production and preliminary characterization of monoclonal antibodies directed at two surface proteins of rhesus rotavirus. J. Virol. *47*, 267—275 (1983 b).

Greenberg, H. B., Flores, J., Kalica, A. R., Wyatt, R. G., Jones, R.: Gene coding assignments for growth restriction, neutralization and subgroup specificities of the W and DS-1 strains of human rotavirus. J. gen. Virol. *64*, 313—324 (1983 c).

Gregoriades, A.: The membrane protein of influenza virus: extraction from virus and infected cell with acidic chloroform-methanol. Virology *54*, 369—383 (1973).

Gregoriades, A.: Influenza virus-induced proteins in nuclei and cytoplasm of infected cells. Virology *79*, 449—454 (1977).

Gregoriades, A.: Interaction of influenza M protein with viral lipid and phosphatidyl-choline vesicles. J. Virol. *36*, 470—479 (1980).

Greig, A. S., Johnson, C. M., Bouillant, A. M. P.: Encephalitis of swine caused by haemagglutinating virus. VI. Morphology of the virus. Res. Vet. Sci. *12*, 305—307 (1971).

Griffin, J. A., Compans, R. W.: Effect of cytochalasin B on the maturation of enveloped viruses. J. Exp. Med. *150*, 379—391 (1979).

Griffiths, G., Brands, R., Burke, B., Louvard, D., Warren, G.: Viral membrane proteins acquire galactose in *trans* Golgi cisternae during intracellular transport. J. Cell Biol. *95*, 781—792 (1982).

Griffiths, G., Quinn, P., Warren, G.: Dissection of the Golgi complex. I. Monensin inhibits the transport of viral membrane proteins from medial to *trans* Golgi cisternae in baby hamster kidney cells infected with Semliki Forest virus. J. Cell Biol. *96*, 835—850 (1983).

Grimley, P. M., Berezesky, I. K., Friedman, R. M.: Cytoplasmic structures associated with an arbovirus infection: loci of viral ribonucleic acid synthesis. J. Virol. *2*, 1326—1338 (1968).

Grimley, P. M., Friedman, R. M.: Arboviral infection of voluntary striatal muscle. J. Infect. Dis. *122*, 45—52 (1970 a).

Grimley, P. M., Friedman, R. M.: Development of Semliki Forest virus in mouse brain: an electron microscopic study. Exp. Molec. Path. *12*, 1—12 (1970 b).

Grimley, P. M., Henson, D. E.: Electron microscopy in virus infections. In: Diagnostic Electron Microscopy (Trump, B. F., Jones, R. T., eds.), Vol. 4. New York: J. Wiley 1983.

Grimley, P. M., Levin, J. G., Berezesky, I. K., Friedman, R. M.: Specific membranous structures associated with the replication of group A arbovirus. J. Virol. *10*, 492—503 (1972).

Grimley, P. M., Berezesky, I. K., Levin, J. G.: Morphogenesis of C-type and A-type particles in cells infected by an arbovirus. J. Natl. Cancer Inst. *50*, 275—279 (1973).

Haase, A. T.: The slow infection caused by visna virus. Curr. Topics Microbiol. Immunol. *72*, 102—156 (1975).

Hackett, C. J., Askonas, B. A., Webster, R. G., van Wyke, K.: Quantitation of influenza virus antigens on infected target cells and their recognition by cross-reactive cytotoxic T cells. J. Exp. Med. *151*, 1014—1025 (1980).

Hageman, Ph. C., Calafat, J., Hilgers, J.: The biology of the mouse mammary tumor viruses. In: Mammary Tumors in the Mouse (Hilgers, J., Sluyser, M., eds.), 392—463. Amsterdam-New York-Oxford: Elsevier/North-Holland. Biomedical Press 1982.

Hall, W. W., Choppin, P. W.: Evidence for lack of synthesis of M polypeptide of measles virus in brain cells in subacute sclerosing panencephalitis. Virology *99*, 443—447 (1979).

Hall, W. W., Lamb, R. A., Choppin, P. W.: Measles and subacute sclerosing panencephalitis virus proteins: lack of antibody to the M protein in patients with subacute sclerosing panencephalitis. Proc. Natl. Acad. Sci. U.S.A. *76*, 2047—2051 (1979).

Hardwick, J. M., Bussell, R. H.: Glycoproteins of measles virus under reducing and non-reducing conditions. J. Virol. *25*, 687—692 (1978).

Hardwick, J. M., Hunter, E.: Rous sarcoma virus mutant LA3382 is defective in virion glycoprotein assembly. J. Virol. *40*, 752–761 (1981).

Harmison, G., Meier, E., Schubert, M.: The polymerase gene of VSV. In: The Molecular Biology of Negative Strand Viruses (Bishop, D. H. L., Compans, R. W., eds.), 35–40. New York: Academic Press.

Harnish, D. G., Leung, W. C., Rawls, W. E.: Characterization of polypeptides immunoprecipitable from Pichinde virus-infected BHK-21 cells. J. Virol. *38*, 840–848 (1981).

Harnish, D. G., Dimock, K., Bishop, D. H. L., Rawls, W. E.: Gene mapping in Pichinde virus: assignment of viral polypeptides to genomic L and S RNAs. J. Virol. *46*, 638–641 (1983).

Harrison, S. C., Kirckhausen, T.: Clathrin, cages and coated vesicles. Cell *33*, 650–652 (1983).

Hasony, H. J., Macnaughton, M. R.: Serological relationships of the subcomponents of human coronavirus strain 229E and mouse hepatitis virus strain 3. J. gen. Virol. *58*, 449–452 (1982).

Haspel, M. V., Lampert, P. W., Oldstone, M. B.: Temperature-sensitive mutants of mouse hepatitis virus produce a high incidence of demyelination. Proc. Natl. Acad. Sci. U.S.A. *75*, 4033–4036 (1978).

Hay, A. J.: Studies on the formation of the influenza virus envelope. Virology *60*, 398–419 (1974).

Hay, A. J., Skehel, J. J.: Studies on the synthesis of influenza virus proteins. In: Negative Strand Viruses (Barry, R. D., Mahy, B. W., eds.). New York: Academic Press 1974.

Hay, A. J., Skehel, J. J., McCauley, J.: Characterization of influenza virus RNA complete transcripts. Virology *116*, 517–522 (1982).

Hayman, M.: Synthesis and processing of avian sarcoma virus glycoprotein. Virology *85*, 475–486 (1978).

Heggeness, M. H., Scheid, A., Choppin, P. W.: Conformation of the helical nucleocapsids of paramyxoviruses and vesicular stomatitis virus: reversible coiling and uncoiling induced by changes in salt concentration. Proc. Natl. Acad. Sci. U.S.A. *77*, 2631–2635 (1980).

Heggeness, M. H., Smith, P. R., Choppin, P. W.: *In vitro* assembly of the nonglycosylated membrane protein (M) of Sendai virus. Proc. Natl. Acad. Sci. U.S.A. *79*, 6232–6236 (1982 a).

Heggeness, M. H., Smith, P. R., Ulmanen, I., Krug, R. M., Choppin, P. W.: Studies on the helical nucleocapsid of influenza virus. Virology *118*, 466–470 (1982 b).

Helenius, A., Kartenbeck, J.: The effects of octylglucoside on the Semliki Forest virus membrane. Evidence for spike protein-nucleocapsid interaction. Eur. J. Biochem. *106*, 613–618 (1980).

Helenius, A., von Bonsdorff, C.-H.: Semliki Forest virus membrane proteins. Preparation and characterization of spike complexes soluble in detergent-free medium. Biochim. Biophys. Acta. *436*, 845–899 (1976).

Helenius, A., Marsh, M.: Endocytosis of enveloped animal viruses. Ciba. Sym. *92*, 59–69 (1982).

Helenius, A., Kartenbeck, J., Simons, K., Fries, E.: On the entry of Semliki Forest virus into BHK 21-F cells. J. Cell Biol. *84*, 404–420 (1980).

Helenius, A., Savas, M., Simons, K.: Asymmetric and symmetric membrane reconstitution by detergent elimination. Studies with Semliki Forest spike glycoprotein and penicillinase from the membrane of bacillus licheniformis. Eur. J. Biochem. *116*, 27–35 (1981).

Helenius, A., Marsh, M., White, J.: Inhibition of Semliki Forest virus penetration by lysosomotropic weak bases. J. gen. Virol. *58*, 47–61 (1982).

Herrler, G., Nagele, A., Meier-Ewert, H., Brown, A. S., Compans, R. W.: Isolation and structural analysis of influenza C virion glycoproteins. Virology *113*, 439–451 (1981).

Herz, C., Stavnezer, E., Krug, R. M., Gurney, T., jr.: Influenza virus, an RNA virus, synthesizes its messenger RNA in the nucleus of infected cells. Cell *26*, 391–400 (1981).

Hewitt, J. A.: On the influence of polyvalent ligands on membrane curvature. J. Theoretical Biol. *64*, 455–472 (1977).

Hewitt, J. A., Nermut, M. V.: A morphological study of the M-protein of Sendai virus. J. gen. Virol. *34*, 127–136 (1977).

Hewlett, M. J., Pettersson, R. F., Baltimore, D.: Circular forms of Uukuniemi virion RNA: an electron microscopic study. J. Virol. *21*, 1085–1093 (1977).

Hill, V. M., Marnell, L., Summers, D. F.: *In vitro* replication and assembly of vesicular stomatitis virus nucleocapsid. Virology *113*, 109–118 (1981).

Hirokawa, N., Heuser, J.: The inside and outside of *gap*-junction membranes visualized by deep etching. Cell *30*, 395–406 (1982).

Hirst, J. K.: Genetic recombination with NDV, polio viruses, and influenza. Cold Spring Harbor Symposia on Quantitative Biology 27, 303—309 (1971).

Hirst, J. K., Pons, M.: Mechanisms of influenza virus recombination. II. Virus aggregation and its effect on plaque formation by so called noninfective virus. Virology 56, 620—631 (1973).

Hodes, D. S., Schnitzer, T. J., Kalica, A. R., Camargo, E., Chanock, R. M.: Inhibition of respiratory syncytial, parainfluenza 3 and measles viruses by 2-deoxy-D-glucose. Virology 63, 201—208 (1975).

Hogan, N., Rickaert, F., Dubois-Dalcq, M., Bellini, W. J., McFarlin, D.: Early appearance and co-localization of individual measles virus using double-label fluorescent antibody techniques. In: The Molecular Biology of Negative Strand Viruses (Bishop, D. H. L., Compans, R. W., eds.), 421—426. New York: Academic Press.

Holmes, I. H.: Morphological similarity of Bunyamwera supergroup viruses. Virology 43, 708—712 (1971).

Holmes, I. H.: Rotaviruses. In: The Reoviridae (Joklik, W. K., ed.). New York: Plenum Press 1983.

Holmes, I. H., Ruck, B. J., Bishop, R. F., Davidson, G. P.: Infantile enteritis virus: morphogenesis and morphology. J. Virol. 16, 937—943 (1975).

Holmes, K. V., Behnke, J. N.: Evolution of a coronavirus during persistent infection in vitro. In: Biochemistry and Biology of Coronaviruses (ter Meulen, J., Siddell, S., Wege, H., eds.), 287—299. New York: Plenum Press 1981.

Holmes, K. V., Doller, E. W., Sturman, L. S.: Tunicamycin resistant glycosylation of a coronavirus glycoprotein: demonstration of a novel type of viral glycoprotein. Virology 115, 334—344 (1981 a).

Holmes, K. V., Doller, E. W., Behnke, J. N.: Analysis of the functions of coronavirus glycoproteins by differential inhibition of synthesis with tunicamycin. In: Biochemistry and Biology of Coronaviruses (ter Meulen, V., Siddell, S., Wege, H., eds.), 133—142. New York: Plenum Press 1981 b.

Holmes, K. V., Frana, M. F., Robbins, S. G., Sturman, L. S.: Coronavirus maturation. In: Coronaviruses: Molecular Biology and Pathogenesis (Rottier, P. J. M., van der Zeijst, B. A. M., Horzinek, M., eds.), 37—52. New York: Plenum Press 1984.

Homma, M., Ohuchi, M.: Trypsin action on the growth of Sendai virus in tissue culture cells. III. Structural difference of Sendai viruses grown in eggs and tissue culture cells. J. Virol. 12, 1457—1465 (1973).

Horisberger, M. A.: Large P proteins of influenza A viruses are composed of one acidic and two basic polypeptides. Virology 107, 302—305 (1980).

Horzinek, M. C.: Comparative aspects of togaviruses. J. gen. Virol. 20, 87—103 (1973 a).

Horzinek, M. C.: The structure of togaviruses. Prog. Med. Virol. 16, 109—156 (1973 b).

Horzinek, M. C.: Non-arthropod-borne togaviruses, 54—63. New York: Academic Press 1981.

Horzinek, M. C., Mussgay, M.: Studies on the nucleocapsid structure of group A arboviruses. J. Virol. 4, 514—520 (1969).

Hosaka, Y., Shimizu, K.: Cell fusion by Sendai virus. In: Virus Infection and the Cell Surface (Poste, G., Nicolson, D., eds.), 129—155. Amsterdam: North-Holland 1977.

Howard, C. R., Simpson, D. I. H.: The biology of the arenaviruses. J. gen. Virol. 51, 1—14 (1980).

Howard, C. R., Buchmeier, M. J.: A protein kinase activity in lymphocytic choriomeningitis virus and identification of the phosphorylated product using monoclonal antibody. Virology 126, 538—547 (1983).

Howe, C., Coward, J. E., Fenger, T. W.: Viral invasion: morphological, biochemical and biophysical aspects. In: Comprehensive Virology (Fraenkel-Conrat, H., Wagner, R., eds.), Vol. 16, 1—71. New York: Plenum Press 1980.

Ho-Terry, L., Cohen, A.: Rubella virion polypeptides: characterization by polyacrylamide gel electrophoresis, isoelectric focusing and peptide mapping. Arch. Virol. 72, 47—54 (1982).

Hsu, C. H., Kingsbury, D. W.: Contribution of oligosaccharide sulfation to the charge heterogeneity of a viral glycoprotein. J. Biol. Chem. 257, 9033—9038 (1982 a).

Hsu, C. H., Kingsbury, D. W.: Topography of phosphate residues in Sendai virus proteins. Virology 120, 225—234 (1982 b).

Hsu, C. H., Kingsbury, D. W., Murti, K. G.: Assembly of vesicular stomatitis virus nucleocapsids in vivo: a kinetic analysis. J. Virol. 32, 304—313 (1979 a).

Hsu, M. C., Scheid, A., Choppin, P. W.: Reconstitution of membranes with individual paramyxovirus glycoproteins and phospholipids in cholate solution. Virology 95, 476—491 (1979 b).

Hsu, C. H., Morgan, E. M., Kingsbury, D. W.: Site-specific phosphorylation regulates the transcriptive activity of vesicular stomatitis virus NS protein. J. Virol. *43*, 104–112 (1982).

Huang, R. T. C., Rott, R., Klenk, H.-D.: Influenza viruses cause hemolysis and fusion of cells. Virology *110*, 243–247 (1981).

Huang, Y., Collins, P. L., Wertz, G. W.: Respiratory syncitial virus proteins: Analysis of structural and non-structural proteins by peptide mapping and translation *in vitro* using individual hybrid-selected mRNAs. In: The Molecular Biology of Negative Strand Viruses (Bishop, D. H. L., Compans, R. W., eds.). New York: Academic Press. In Press.

Hughes, S.: Synthesis, integration and transcription of the retroviral provirus. In: Current Topics in Microbiology and Immunology (Vogt, P. K., Koprowski, H., eds.), Vol. 103, 33–49. Berlin-Heidelberg-New York: Springer 1983.

Inglis, S. C., Carroll, A. R., Lamb, R. A., Mahy, B. W. J.: Polypeptides specified by the influenza virus genome. Virology *74*, 489–520 (1976).

Ishida, N., Homma, M.: Sendai virus. Adv. Virus Res. *123*, 349–383 (1978).

Jackson, D. A., Caton, A. J., McCready, S. J., Cook, P. R.: Influenza virus RNA is synthesized at fixed sites in the nucleus. Nature *296*, 366–368 (1982).

Jacobs, B. L., Penhoet, E. E.: Assembly of vesicular stomatitis virus: distribution of the glycoprotein on the surface of infected cells. J. Virol. *44*, 1047–1055 (1982).

Jay, F. T., Dawood, M. R., Friedman, R. M.: Interferon induces the production of membrane deficient and infectivity defective vesicular stomatitis virions through interference in the virion assembly process. J. gen. Virol. *64*, 707–712 (1983).

Johnson, D. C., Schlesinger, M. J.: Vesicular stomatitis virus and Sindbis virus glycoprotein transport to the cell surface is inhibited by ionophores. Virology *103*, 407–424 (1980).

Johnson, D. C., Schlesinger, M. J., Elson, E. L.: Fluorescence photobleaching recovery measurements reveal differences in envelopment of Sindbis and vesicular stomatitis virus. Cell *23*, 423–431 (1981).

Johnson, K. P., Norrby, E., Swoveland, P., Carrigan, D. R.: Expression of five viral antigens in cells infected with wild-type and SSPE strains of measles virus: correlation with cytopathic effects and productivity of infections. Arch. Virol. *73*, 255–262 (1982).

Joklik, W. K.: The Reoviridae. New York: Plenum Press 1983.

Kääriäinen, L., Söderlund, H.: Structure and replication of alpahviruses. Curt. Top. Microbiol. Immunol. *82*, 15–69 (1978).

Kääriäinen, L., Hashimoto, K., Saraste, J., Virtanen, I., Penttinen, K.: Monensin and FCCP inhibit the intracellular transport of alphavirus membrane glycoproteins. J. Cell Biol. *87*, 783–791 (1980).

Kalica, A. R., James, H. D., Kapikian, A. Z.: Hemagglutination by simian rotavirus. J. Clin. Microbiol. *7*, 314–315 (1978 a).

Kalica, A. R., Sereno, M. M., Wyatt, R. G., Mebus, C. A., Chanock, R. M., Kapikian, A. Z.: Comparison of human and animal rotavirus strains by gel electrophoresis of viral RNA. Virology *87*, 247–255 (1978 b).

Kalica, A. R., Greenberg, H. B., Espejo, R. T., Flores, J., Wyatt, R. G., Kapikian, A. Z., Chanock, R. M.: Distinctive ribonucleic acid patterns of human rotavirus subgroups 1 and 2. Infect. Immun. *33*, 958–961 (1981).

Kalica, A. R., Flores, J., Greenberg, H. B.: Identification of the rotaviral gene that codes for hemagglutination and protease-enhanced plaque formation. Virology *125*, 194–205 (1983).

Kaluza, G., Scholtissek, C., Rott, R.: Inhibition of the multiplication of enveloped RNA viruses by glucosamine and 2-deoxy-D-glucose. J. gen. Virol. *14*, 251–259 (1972).

Kamata, T., Watanabe, Y.: Role for nucleocapsid phosphorylation in the transcription of influenza virus genome. Nature *267*, 460–462 (1977).

Kapikian, A. Z., Kim, H. W., Wyatt, R. G., Rodriguez, W. J., Ross, S., Parrott, R. H., Chanock, R. M.: Reovirus like agent in stools: association with infantile diarrhea and development of serologic tests. Science *185*, 1049–1053 (1974).

Kapikian, A. Z., Yolken, R. H., Greenberg, H. B., Wyatt, R. G., Kalica, A. R., Chanock, R. M., Kim, H. W.: Gastroenteritis virus. In: Diagnostic Procedures for Viral, Rickettsial, and Chlamydial Infections (Lennette, E. H., Schmidt, N. J., eds.), 5th ed., 927–995. Washington, D.C.: American Public Health Association 1979.

Kapikian, A. Z., Wyatt, R. G., Levine, M. M., Yolken, R. H., Van Kirk, D. H., Dolin, R., Greenberg, H. B., Chanock, R. M.: Oral administration of human rotavirus to volunteers: Induction of illness and correlation of resistance. J. Infect. Dis. *147*, 95–106 (1983).

Karnovsky, M. J., Kleinfeld, A. M., Hoover, R. L., Klausner, R. D.: The concept of lipid domains in membranes. J. Cell. Biol. *94*, 1–6 (1982).

Kelly, D. C., Dimmock, N. J.: Fowl plague virus replication in mammalian cells-avian erythrocyte heterokaryons: studies concerning the actinomycin D and ultraviolet light-sensitive phase in influenza virus replication. Virology *61*, 210–222 (1974).

Kelly, D. C., Avery, R. J., Dimmock, N. J.: Failure of an influenza virus to initiate infection in enucleate BHK cells. J. Virol. *13*, 1155–1161 (1974).

Kendal, A. P., Galphin, J. C., Palmer, E. L.: Replication of influenza virus at elevated temperatures: production of virus-like particles with reduced matrix protein content. Virology *76*, 186–195 (1977).

Kennedy, D. A., Johnson-Lussenburg, C. M.: Isolation and morphology of the internal component of human coronavirus, strain 229E. Intervirology *6*, 197–206 (1975/76)

Kennedy, S. I. T.: Synthesis of alphavirus RNA. In: The Togaviruses (Schlesinger, R. W., ed.), 351–369. New York: Academic Press 1980.

Keppler, D. O. R., Rudigier, J. F. M., Bischoff, E., Decker, K. F. A.: The trapping of uridine phosphates by D-galactosamine, D-glucosamine, and 2-deoxy-D-galactose: a study on the mechanism of galactosamine hepatitis. Eur. J. Biochem. *17*, 246–253 (1970).

Kilbourne, E. D.: Inhibition of influenza virus multiplication with a glucose antimetabolite (2-deoxy-D-glucose). Nature *183*, 271–272 (1959).

Killen, H. M., Dimmock, N. J.: Identification of a neutralization-specific antigen of a calf rotavirus. J. gen. Virol. *62*, 297–311 (1982).

Kim, J., Hama, K., Miyake, Y., Okada, Y.: Transformation of intramembrane particles of HVJ (Sendai virus) envelopes from an invisible to visible form on aging of virions. Virology *95*, 523–535 (1979).

Kimura, T.: Immuno-electron microscopy study on the antigenicity of tubular structures associated with human rotavirus. Infect. Immun. *33*, 611–615 (1981).

King, B., Brian, D. A.: Bovine coronavirus structural proteins. J. Virol. *42*, 700–707 (1982).

Kingsbury, D. W.: Paramyxovirus replication. Curr. Top. Microbiol. *59*, 1–33 (1973).

Kingsbury, D. W.: The molecular biology of paramyxoviruses. Med. Microbiol. Immunol. *160*, 73–83 (1974).

Kingsbury, D. W.: Paramyxoviruses. In: The Molecular Biology of Animal Viruses (Nayak, D. P., ed.), Vol. 1, 349–382. New York: Marcel Dekker 1977.

Kingsbury, D. W., Portner, A.: On the genesis of incomplete Sendai virions. Virology *42*, 872–879 (1970).

Kingsbury, D. W., Bratt, M. A., Choppin, P. W., Hanson, R. P., Hosaka, Y., ter Meulen, V., Norrby, E., Plowright, W., Ropp, R., Hunnes, W.: Paramyxoviridae. Intervirology *10*, 137–152 (1978).

Kingsbury, D. W., Hsu, C. H., Morgan, E. M.: A role for NS protein phosphorylation in VSV transcription. In: The Replication of Negative Strand Viruses (Bishop, D. H. L., Compans, R. W., eds.), 821–827. New York-Amsterdam-Oxford: Elsevier/North-Holland 1981.

Kingsford, L., Emerson, S. U.: Transcriptional activities of different phosphorylated species of NS protein purified from vesicular stomatitis virus and cytoplasm of infected cells. J. Virol. *33*, 1097–1105 (1980).

Klemenz, R., Diggelmann, H.: Extracellular cleavage of the glycoprotein precursor of Rous sarcoma virus. J. Virol. *29*, 285–292 (1979).

Klenk, H.-D.: Viral envelopes and their relationship to cellular membrane. Curr. Top. Microbiol. Immunol. *68*, 29–58 (1974).

Klenk, H.-D.: Host cell membranes in virus replication. In: Mammalian Cell Membranes (Jamieson, G. A., Robinson, D. M., eds.), 260–283. London-Boston: Butterworths 1977.

Klenk, H.-D., Scholtissek, C., Rott, R.: Inhibition of glycoprotein biosynthesis of influenza virus by D-glucosamine and 2-deoxy-D-glucose. Virology *49*, 723–734 (1972).

Klenk, H.-D., Rott, R., Orlich, M., Blodorn, J.: Activation of influenza A viruses by trypsin treatment. Virology *68*, 426–439 (1975).

Knipe, D. M., Baltimore, D., Lodish, H. F.: Separate pathways of maturation of the major structural proteins of vesicular stomatitis virus. J. Virol. *21*, 1128–1139 (1977 a).

Knipe, D. M., Baltimore, D., Lodish, H. F.: Maturation of viral proteins in cells infected with temperature-sensitive mutants of vesicular stomatitis virus. J. Virol. *21*, 1149–1158 (1977 b).

Knipe, D., Lodish, H., Baltimore, D.: Analysis of the defects of temperature-sensitive mutants of vesicular stomatitis virus: intracellular degradation of specific viral proteins. J. Virol. *21*, 1140–1148 (1977 c).

Kogasaka, R., Akihara, M., Horino, K., Chiba, S., Nakao, T.: A morphological study of human rotavirus. Arch. Virol. *61*, 41–48 (1979).

Kolakofsky, D., Boy de la Tour, E., Delius, H.: Molecular weight determination of Sendai and Newcastle disease virus RNA. J. Virol. *13*, 261–268 (1974).

Kolakofsky, D., Bruschi, A.: Antigenomes in Sendai virions and Sendai virus-infected cells. Virology *66*, 185–191 (1975).

Kondor-Koch, C., Riedel, H., Söderberg, K., Garoff, H.: Expression of the structural proteins of Semliki Forest virus from cloned cDNA microinjected into the nucleus of baby hamster kidney cells. Proc. Natl. Acad. Sci. U.S.A. *79*, 4525–4529 (1982).

Koolen, M. J. M., Osterhaus, A. D. M. E., van Steenis, G., Horzinek, M. C., van der Zeijst, B. A. M.: Temperature-sensitive mutants of mouse hepatitis virus strain A59: Isolation, characterization, and neuropathogenic properties. Virology *125*, 393–402 (1983).

Kornfeld, S., Li, E., Tabas, I.: The synthesis of complex-type oligosaccharides. II. Characterization of the processing intermediates in the synthesis of complex oligosaccharide units of the vesicular stomatitis virus G Protein. J. Biol. Chem. *253*, 7771–7778 (1978).

Kozma, S., Calberg-Bacq, C.-M., Francois, C., Osterrieth, P. M.: Detection of virus antigens in Swiss albino mice infected by milk-borne mouse mammary tumour virus: the effect of age, sex and reproductive status. I. Localization by immunofluorescence of four antigens in mammary tissues and other organs. J. gen. Virol. *45*, 27–40 (1979).

Krug, R. M., Morgan, M. A., Shatkin, A. J.: Influenza viral mRNA contains internal N^6-methyladenosine and 5′-terminal 7-methylguanosine in cap structures. J. Virol. *20*, 45–53 (1976).

Kuff, E. L., Feenstra, A., Lueders, K., Smith, L., Hawley, R., Hozumi, N., Shulman, M.: Intracisternal A-particle genes as movable elements in the mouse genome. Proc. Natl. Acad. Sci. U.S.A. *80*, 1992–1996 (1983).

Kurilla, M. G., Keene, J. D.: The leader RNA of vesicular stomatitis virus is bound by a cellular protein reactive with anti-LA lupus antibodies. Cell *34*, 837–845 (1983).

Kurilla, M. G., Piwnica-Worms, H., Keene, J. D.: Rapid and transient localization of the leader RNA of VSV in the nuclei of infected cells. Proc. Natl. Acad. Sci. U.S.A. *79*, 5240–5244 (1982).

Lai, M. M. C., Stohlman, S. A.: The RNA of mouse hepatitis virus. J. Virol. *26*, 236–242 (1978).

Lai, M. M. C., Brayton, P. R., Armen, R. C., Patton, C. D., Pugh, C., Stohlman, S. A.: Mouse hepatitis virus A59 messenger RNA structure and genetic localization of the sequence divergence from the hepatropic strain MHV3. J. Virol. *39*, 823–834 (1981).

Lai, M. M. C., Patton, C. D., Stohlman, S. A.: Further characterization of mouse hepatitis virus: presence of common 5′-end nucleotides. J. Virol. *41*, 557–565 (1982 a).

Lai, M. M. C., Patton, C. D., Stohlman, S. A.: Replication of mouse hepatitis virus: negative-stranded RNA and replicative form RNA are of genome length, J. Virol. *44*, 487–492 (1982 b).

Lai, M. M. C., Patton, C. B., Barric, R. S., Stohlman, S. A.: Presence of leader sequences in the mRNA of mouse. J. Virol. *46*, 1027–1033 (1983).

Lamb, R. A.: The influenza virus RNA segments and their encoded proteins. In: Genetics of Influenza Viruses (Palese, P., Kingsbury, D. W., eds.), 21–98. Wien-New York: Springer 1983.

Lamb, R. A., Choppin, P. W.: Synthesis of influenza virus proteins in infected cells: translation of viral polypeptides, including three P polypeptides from RNA produced by primary transcription. Virology *74*, 504–519 (1976).

Lamb, R. A., Choppin, P. W.: The synthesis of Sendai virus polypeptides in infected cells. II. Intracellular distribution of polypeptides. Virology *81*, 371–381 (1977 a).

Lamb, R. A., Choppin, P. W.: The synthesis of Sendai virus polypeptides in infected cells. III. Phosphorylation of polypeptides. Virology *81*, 382–397 (1977 b).

Lamb, R. A., Choppin, P. W.: Determination by peptide mapping of the unique polypeptides in Sendai virions and infected cells. Virology 84, 469–478 (1978).

Lamb, R. A., Choppin, P. W.: Identification of a second protein (M₂) encoded by RNA segment 7 of influenza virus. Virology 112, 729–737 (1981).

Lamb, R. A., Choppin, P. W.: The gene structure and replication of influenza virus. Annual Review of Biochemistry 52, 467–506 (1983).

Lamb, R. A., Lai, C.-J.: Conservation of the influenza virus membrane protein (M₁) amino acid sequence and an open reading frame of RNA segment 7 encoding a second protein (M₂) in H₁N₁ and H₃N₂ strains. Virology 112, 746–751 (1981).

Lamb, R. A., Mahy, B. W. J., Choppin, P. W.: The synthesis of Sendai virus polypeptides in infected cells. Virology 69, 116–131 (1976).

Lamb, R. A., Lai, C.-J., Choppin, P. W.: Sequences of mRNAs derived from genome RNA segment 7 of influenza virus: colinear and interrupted mRNAs code for overlapping proteins. Proc. Natl. Acad. Sci. U.S.A. 78, 4170–4174 (1981).

Lange, J., Frank, H., Hunsmann, G., Moenerg, V., Wollmann, R., Schäfer, W.: Properties of mouse leukemia viruses: VI. The core of Friend virus, isolation, and constituents. Virology 53, 457–462 (1973).

Lascano, E. F., Berria, M. I.: Ultrastructure of Junin virus in mouse whole brain and mouse brain tissue cultures. J. Virol. 14, 965–974 (1974).

Lazarides, E.: Intermediate filaments as mechanical integrators of cellular space. Nature 283, 249–256 (1980).

Lazarowitz, S. G., Choppin, P. W.: Enhancement of the infectivity of influenza A and B viruses by proteolytic cleavage of the hemagglutinin polypeptide. Virology 68, 440–454 (1975).

Lazarowitz, S. G., Compans, R. W., Choppin, P. W.: Influenza structural and non-structural proteins in infected cells and their plasma membrane. Virology 46, 830–843 (1971).

Lazarowitz, S. G., Compans, R. W., Choppin, P. W.: Proteolytic cleavage of the hemagglutinin polypeptide of influenza virus: function of the uncleaved polypeptide HA. Virology 52, 199–212 (1973 a).

Lazarowitz, S. G., Goldberg, A. R., Choppin, P. W.: Proteolytic cleavage by plasmin of the HA polypeptide of influenza virus: host cell activation of serum plasminogen. Virology 56, 172–180 (1973 b).

Lazdins, I., Holmes, I. H.: Protein synthesis in Bunyamwera virus-infected cells. J. gen. Virol. 44, 123–133 (1979).

Lazzarini, R. A., Keene, J. D., Schubert, M.: The origins of defective interfering particles of the negative-strand RNA viruses. Cell 26, 145–154 (1981).

Leary, K., Blair, C. D.: Sequential events in the morphogenesis of Japanese encephalitis virus. J. Ultrastr. Res. 72, 123–129 (1980).

Leavitt, R., Schlesinger, S., Kornfeld, S.: Tunicamycin inhibits glycosylation and multiplication of Sindbis and vesicular stomatitis viruses. J. Virol. 21, 375–385 (1977 a).

Leavitt, R., Schlesinger, S., Kornfeld, S.: Impaired intracellular migration and altered solubility of nonglycosylated glycoproteins of vesicular stomatitis virus and Sindbis virus. J. Biological Chem. 252, 9018–9023 (1977 b).

Lehane, M. J., Leake, C.-J.: A kinetic and ultrastructural comparison of alphavirus infection of cultured mosquito and vertebrate cells. J. Trop. Med. Hyg. 85, 229–238 (1982).

Lehmann-Grübe, F.: Lymphocytic choriomeningitis virus. In: Virology Monographs, Vol. 10, 1–173. Wien-New York: Springer 1971.

Leibowitz, J. L., Weiss, S. R.: Murine coronavirus RNA. In: Biochemistry and Biology of Coronaviruses (ter Meulen, V., Siddell, S., Wege, H., eds.), 227–244. New York: Plenum Press 1981.

Leibowitz, H. L., Wihelmsen, K. C., Bond, C. W.: The virus-specific intracellular RNA species of two murine coronaviruses: MHV-A59 and MHV-JHM. Virology 114, 29–51 (1981).

Leibowitz, J. L., Weiss, S. R., Paavola, E., Bond, C. W.: Cell-free translation of murine coronavirus RNA. J. Virol. 43, 905–913 (1982).

Lenard, J., Wilson, T., Mancarella, D., Reidler, J., Keller, P., Elson, E.: Interaction of mutant and wild type M protein of vesicular stomatitis virus with nucleocapsids and membranes. In: The Replication of Negative Strand Viruses (Bishop, D. H. L., Compans, R. W., eds.), 865–870. New York-Amsterdam-Oxford: Elsevier/North-Holland 1981.

Lenard, J.: Virus envelopes and plasma membranes. Ann. Rev. Biophys. Bioengineering 7, 139–165 (1978).

Lennarz, W. J.: The Biochemistry of Glycoproteins and Proteoglycans. New York: Plenum 1980.

Leppert, M., Rittenhouse, L., Perrault, J., Summers, D. E., Kolakofsky, D.: Plus and minus strand leader RNAs in negative strand virus-infected cells. Cell 18, 735–747 (1979).

Leung, W.-C., Rawls, W. E.: Viron-associated ribosomes are not required for the replication of Pichinde virus. Virology 81, 174–176 (1977).

Li, E., Tabas, I., Kornfeld, S.: The synthesis of complex-type-oligosaccharides. I. Structure of the lipid-linked oligosaccharide precursor of the complex-type-oligosaccharides of the VSV G protein. J. Biol. Chem. 253, 7762–7770 (1978).

Li, K. K., Seto, J. T.: Electron microscope study of RNA of myxoviruses. J. Virol. 7, 524–530 (1971).

Lin, B. C., Lai, C. J.: The influenza virus nucleoprotein synthesized from cloned DNA in a simian virus 40 vector is detected in the nucleus. J. Virol. 45, 434–438 (1983).

Lin, D. C., Tobin, K. D., Grumet, M., Lin, S.: Cytochalasins inhibit nuclei-induced actin polymerization by blocking filament elongation. J. Cell Biol. 84, 455–460 (1980).

Lin, F. H., Thormar, H.: Absence of M protein in a cell-associated subacute sclerosing panencephalitis virus. Nature 285, 490–492 (1980).

Lindsey, J. D., Ellisman, M. H.: The varicose tubule: a direct connection between rough cytoplasmic reticulum and the Golgi apparatus. J. Cell Biol. In Press.

Liu, C.: Studies on influenza infection in ferrets by means of fluorescein-labelled antibody. 2. The role of soluble antigen in nuclear fluorescence and cross-reactions. J. Exp. Med. 101, 677–686 (1955).

Lodish, H. F., Porter, M.: Translational control of protein synthesis after infection by vesicular stomatitis virus. J. Virol. 36, 719–733 (1980 a).

Lodish, H. F., Porter, M.: Heterogeneity of vesicular stomatitis virus particles: implications for virion assembly. J. Virol. 33, 52–58 (1980 b).

Lodish, H. F., Porter, M.: Specific incorporation of host cell surface proteins into budding vesicular stomatitis virus particles. Cell 19, 161–169 (1980 c).

Lodish, H. F., Rothman, J. E.: The assembly of cell membranes. In: Molecules to Living Cells. Readings from Scientific American, 138–153. San Francisco: Freeman 1980.

Lodish, H. F., Kong, N.: Reversible block in intracellular transport and budding of mutant vesicular stomatitis virus glycoproteins. Virology 125, 335–348 (1983).

Lohmeyer, J., Talens, L. T., Klenk, H.-D.: Biosynthesis of the influenza virus envelope in abortive infection. J. gen. Virol. 42, 73–88 (1979).

Lomniczi, B., Morser, J.: Polypeptides of infectious bronchitis virus. I. Polypeptides of the virion. J. gen. Virol. 55, 155–164 (1981).

Lourenco, M. H., Nicholas, J. C., Cohen, J., Sherrer, R., Bricout, F.: Study of human rotavirus genome by electrophoresis: attempt of classification among strains isolated in France. Ann. Virol. 132, 161–173 (1981).

Lu, A. H., Soong, M. M., Wong, P. K. Y.: Maturation of Moloney murine leukemia virus. Virology 93, 269–274 (1979).

Lund, G. A., Salmi, A. A.: Purification and characterization of measles virus hemagglutinin protein G. J. gen. Virol. 56, 185–193 (1981).

Lyles, D. S.: Glycoproteins of Sendai virus are transmembrane proteins. Proc. Natl. Acad. Sci. U.S.A. 76, 5621–5625 (1979).

Lyons, M. J., Heyduk, J.: Aspects of the developmental morphology of California encephalitis virus in cultured vertebrate and arthropod cells and in mouse brain. Virology 54, 37–52 (1973).

Macnaughton, M. R., Davies, H. A., Nermut, M. V.: Ribonucleoprotein-like structures from corona-virus particles. J. gen. Virol. 39, 545–549 (1978).

Macnaughton, M. R., Davies, H. A.: Two particle types of avian infectious bronchitis virus. J. gen. Virol. 47, 365–372 (1980).

Maeno, K., Yoshii, S., Mita, K., Hamaguchi, M., Yoshida, T., Iinuma, M., Nagai, Y., Matsumoto, T.: Analysis of the inhibitory effect of canavanine on the replication of influenza RA/5+ virus. Virology 94, 128–137 (1979).

Maheshwari, R. K., Banerjee, D. K., Walchter, C. J., Olden, K., Friedman, R. M.: Interferon treatment inhibits glycosylation of a viral protein. Nature 287, 454–457 (1980).

Mahy, B. W. J., Hastie, N. D., Armstrong, S. J.: Inhibition of influenza virus replication by α-amanitin: mode of action. Proc. Natl. Acad. Sci. U.S.A. *69*, 1421–1424 (1972).

Mahy, B. W., Barrett, T., Briedis, D. J., Brownson, J. M., Wolstenholme, A. J.: Influence of the host cell on influenza virus replication. Philos. Trans. R. Soc. Lond. (Biol.) *288*, 349–357 (1980).

Mann, R., Mulligan, R. C., Baltimore, D.: Construction of a retrovirus packaging mutant and its use to produce helper-free defective retrovirus. Cell *33*, 153–159 (1983).

Mannweiler, K., Hohenberg, H., Bohn, W., Neumayer, U., Andresen, I., Schröder, S.: Development of plasma membrane alterations of Hela cells after measles virus infection studied with surface replicas from critical point dried and freeze-fractured cultures grown on coverslips. In: Electron Microscopy, Cell Biology, Vol. 2 (Bredero, P., de Priester, W., eds.), 392–393 (Proceedings of the Seventh European Congress on Electron Microscopy). The Hague, Netherlands: The Seventh European Congress on Electron Microscopy Foundation. 1980.

Markoff, L., Lai, C.-J.: Sequence of the influenza A/Udorn/72 (H3N2) virus neuraminidase gene as determined from cloned full-length DNA. Virology *119*, 288–297 (1982).

Markwell, M. A., Fox, C. F.: Protein-protein interactions within Paramyxoviruses identified by native disulfide bonding or reversible chemical cross-linking. J. Virol. *33*, 152–166 (1980).

Marsh, M., Wellsteed, J., Kern, H., Harms, E., Helenius, A.: Monensin inhibits Semliki Forest virus penetration into culture cells. Proc. Natl. Acad. Sci. U.S.A. *79*, 5297–5301 (1982).

Marsh, M., Bolzau, E., Helenius, A.: Penetration of Semliki Forest virus from acidic prelysosomal vacuoles. Cell *32*, 931–940 (1983).

Mason, B. B., Graham, D. Y., Estes, M. K.: *In vitro* transcription and translation of simian rotavirus SA11 gene products. J. Virol. *33*, 1111–1121 (1980).

Mason, B. B., Graham, D. Y., Estes, M. K.: Biochemical mapping of the simian rotavirus SA11 genome. J. Virol. *46*, 413–423 (1983).

Massalski, A., Coulter-Mackie, M., Dales, S.: Assembly of mouse hepatitis virus strain JHM. In: Biochemistry and Biology of Coronaviruses (ter Meulen, V., Siddell, S., Wege, H., eds.), 111–118. New York: Plenum Press 1981.

Massalski, A., Coulter-Mackie, M., Knobler, R. L., Buchmeier, M. J., Dales, S.: *In vivo* and *in vitro* models of demyelinating diseases. V. Comparison of the assembly of mouse hepatitis virus, strain JHM, in two murine cell lines. Intervirology *18*, 135–146 (1982).

Massey, R. J., Arthur, L. O., Nowinski, R. C., Schochetman, G.: Monoclonal antibodies identify individual determinants on mouse mammary tumor virus glycoprotein gp52 with group, class or type specificity. J. Virol. *34*, 635–643 (1980).

Matlin, K. S., Reggio, H., Helenius, A., Simons, K.: The pathway of vesicular stomatitis virus entry leading to infection. J. Molecular Biol. *156*, 609–631 (1982).

Matsumoto, T.: Assembly of Paramyxoviruses. Microbiol. Immunol. *26*, 285–320 (1982).

Matsumura, T., Stollar, V., Schlesinger, R. W.: Flavivirus RNA synthesis associated with vacuolar membranes. Virology *46*, 344 (1971).

Matsumura, T., Shiraki, K., Sashikata, T., Hotta, S.: Morphogenesis of Dengue-1 virus in cultures of a human leukemic leukocyte line (J-111). Microbiol. Immunol. *21*, 329–334 (1977).

Mayne, J. J., Rice, C. M., Strauss, E. G., Hunkapiller, M. W., Strauss, J. H.: Biochemical studies of the maturation of the small Sindbis virus glycoprotein E3. Virology *34*, 338–357 (1984).

McCarthy, M.: Nucleocapsid-associated RNA species from cells acutely or persistently infected by mumps virus. In: The Replication of Negative Strand Viruses (Bishop, D. H. L., Compans, R. W., eds.), 545–552. New York: Elsevier/North-Holland 1981.

McCrae, M. A., Faulkner-Valle, G. P.: Molecular biology of rotaviruses. I. Characterization of basic growth parameters and pattern of macromolecular synthesis. J. Virol. *39*, 490–496 (1981).

McCrae, M. A., McCorquodale, J. G.: The molecular biology of rotavirus. II. Identification of the protein-coding assignments of calf rotavirus genome RNA species. Virology *117*, 435–443 (1982).

McCrae, M. A., McCorquodale, J. G.: Molecular biology of rotaviruses. V. Terminal structure of viral RNA species. Virology *126*, 204–212 (1983).

McGavran, M. H., Easterday, B. C.: Rft Valley fever virus hepatitis. Light and electron microscopic studies in the mouse. Am. J. Path. *42*, 587–607 (1963).

McGeocq, D. J., Fellner, P., Newton, C.: Influenza virus genome consists of eight distinct RNA species. Proc. Natl. Acad. Sci. U.S.A. *73*, 3045–3049 (1976).

McGowan, J. J., Emerson, S. U., Wagner, R. R.: The plus-strand leader RNA of VSV inhibits DNA-dependent transcription of adenovirus and SV40 genes in a soluble whole-cell extract. Cell *28*, 325–333 (1982).

McNulty, M. S.: Rotaviruses. J. gen. Virol. *40*, 1–18 (1978).

McNulty, M. S., Allan, G. M., Pearson, G. R., McFerran, J. B., Curran, W. L., McCracken, R. M.: Reovirus-like agent (rotavirus) from lamb. Infect. Immun. *14*, 1332–1338 (1976).

McNulty, M. S., Curran, W. L., Allan, G. M., McFerran, J. B.: Synthesis of coreless, probably defective virus particle in cell cultures infected with rotavirus. Arch. Virol. *58*, 193–202 (1978).

McPhee, D. A., Della-Porta, A. J.: Biological and serological comparisons of Australian bunyaviruses belonging to the Simbu subgroup, In: The Replication of Negative Strand Viruses (Bishop, D. H. L., Compans, R. W., eds.), 93–101. New York: Elsevier/North-Holland 1981.

McSharry, J. J.: Viral membrane protein structure and function. In: Rhabdoviruses (Bishop, D. H. L., ed.), Vol. I, 161–168. Boca Raton, Fla.: CRC Press 1979.

McSharry, J. J., Compans, R. W., Choppin, P. W.: Proteins of vesicular stomatitis virus and of phenotypically mixed vesicular stomatitis virus–simian virus 5 virions. J. Virol. *8*, 722–729 (1971).

McSharry, J. J., Compans, R. W., Lacland, H., Choppin, P. W.: Isolation and characterization of the non-glycosylated membrane protein and a nucleocapsid complex from the paramyxovirus SV5. Virology *67*, 365–374 (1975).

Meier-Ewert, H., Compans, R. W.: Time course of synthesis and assembly of influenza virus proteins. J. Virol. *14*, 1083–1091 (1974).

Melnick, J. L.: Taxonomy of viruses. Prog. Med. Virol. *26*, 214–232 (1980).

Miller, C. A., Carrigan, D. R.: Reversible repression and activation of measles virus infection in neural cells. Proc. Natl. Acad. Sci. U.S.A. *79*, 1629–1633 (1982).

Minjou, W., Threlfall, G., Verhoeyen, M., Devos, R., Saman, E., Fang, R., Huylebroeck, D., Fiers, W., Barber, C., Carey, N., Emtage, S.: Complete structure of hemagglutinin gene from the human influenza A/Victoria/3/75 (H_3N_2) strain as determined from cloned DNA. Cell *19*, 683–696 (1980).

Montelaro, R. C., Sullivan, S. J., Bolognesi, D. P.: An analysis of type-C retrovirus polypeptides and their associations in the virion. Virology *84*, 19–31 (1978).

Mooren, H. W. D., Prins, F. A., Herbrink, P., Warnaar, S. O.: Electron microscopic studies on the role of the envelope antigens of R-MuLV-*ts*29 in budding. Virology *113*, 254–262 (1981).

Morein, B., Helenius, A., Simons, K., Pettersson, R., Kääriäinen, L., Schirrmacher, V.: Effective subunit vaccines against an enveloped animal virus. Nature *276*, 715–719 (1978).

Morrison, T. G.: Rhabdoviral assembly and intracellular processing of viral components. In: Rhabdoviruses (Bishop, D. H. L., ed.), Vol. II, 95–114. Boca Raton, Fla.: CRC Press 1980.

Morrison, T. G., Simpson, D.: Synthesis, stability, and cleavage of Newcastle disease virus glycoproteins in the absence of glycosylation. J. Virol. *36*, 171–180 (1980).

Morrison, T. G., Chatis, P. A., Simpson, D.: Conformation and activity of the Newcastle disease virus HN protein in the absence of glycosylation. In: The Replication of Negative Strand Viruses (Bishop, D. H. L., Compans, R. W., eds.), 471–477. New York: Elsevier/North-Holland 1981.

Morrongiello, M. P., Dales, S.: Characterization of cytoplasmic inclusions formed during influenza/WSN virus infection of chick embryo fibroblast cells. Intervirology *8*, 281–293 (1977).

Mountcastle, W. E., Choppin, P. W.: A comparison of the polypeptides of four measles virus strains. Virology *78*, 463–474 (1977).

Murphy, F. A.: Togavirus morphology and morphogenesis. In: The Togaviruses (Schlesinger, R. W., ed.), 241–316. New York: Academic Press 1980.

Murphy, F. A., Harrison, A. K., Tzianabos, T.: Electron microscopic observations of mouse brain infected with Bunyamwera group arboviruses. J. Virol. *2*, 1315–1325 (1968 a).

Murphy, F. A., Whitfield, S. G., Coleman, P. H., Calisher, C. H., Rabin, E. R., Jenson, A. B., Melnick, J. L., Edwards, M. R., Whitney, E.: California group Arboviruses: electron microscopic studies. Exp. Molec. Path. *9*, 44–56 (1968 b).

Murphy, F. A., Harrison, A. K., Whitfield, S. G.: Bunyaviridae: morphologic and morphogenetic similarities of Bunyamwera serologic supergroup viruses and several other arthropod-borne viruses. Intervirology *1*, 297–316 (1973).

Murphy, F. A., Webb, P. A., Johnson, K. M., Whitfield, S. G., Chappell, W. A.: Arenaviruses in Vero cells: ultrastructural studies. J. Virol. *6*, 507–518 (1970).

Murphy, F. A., Whitfield, S. G.: Morphology and morphogenesis of arenaviruses. Bull. World Health Organ. *52*, 409–419 (1975).

Murphy, F. A., Winn, W. C., Walker, D. H., Flemiter, M. R., Whitfield, S. G.: Early lymphoreticular viral tropism and antigenic persistence. Tamiami virus infection in the cotton rat. Lab. Invest. *34*, 125–140 (1976).

Murphy, F. A., Harrison, A. K.: Electron microscopy of the rhabdoviruses of animals. In: Rhabdoviruses (Bishop, D. H. L., ed.), Vol. I, 65–106. Boca Raton, Fla.: CRC Press 1979.

Murphy, M. F., Lazzarini, R. A.: Synthesis of viral mRNA and polyadenylate by a ribonucleoprotein complex from extracts of VSV-infected cell. Cell *3*, 77–84 (1974).

Nagai, Y., Klenk, H.-D.: Activation of precursor of both glycoproteins of Newcastle disease virus by proteolytic cleavage. Virology *77*, 125–134 (1977).

Nagai, Y., Yoshida, T., Yoshii, S., Maeno, K., Matsumoto, T.: Modification of normal cell surface by smooth membrane preparations from BHK-21 cells infected with Newcastle disease virus. Med. Microbiol. Immunol. *161*, 175–188 (1975).

Nagai, Y., Ogura, H., Klenk, H.-D.: Studies on the assembly of the envelope of Newcastle disease virus. Virology *69*, 523–538 (1976).

Nagai, Y., Yoshida, T., Hamaguchi, M., Iinuma, M., Maeno, K., Matsumoto, T.: The pathogeny of Newcastle disease virus isolated from migrating and domestic ducks and the susceptibility of the viral glycoproteins to proteolytic cleavage. Microbiol. Immunol. *24*, 173–177 (1980).

Nakamura, K., Compans, R. W.: The cellular site of sulfation of influenza viral glycoproteins. Virology *79*, 381–392 (1977).

Nakamura, K., Compans, R. W.: Effect of glucosamine, 2-deoxyglucose and tunicamycin on glycosylation, sulfation and assembly of influenza viral proteins. Virology *84*, 303–319 (1978 a).

Nakamura, K., Compans, R. W.: Glycopeptide components of influenza viral glycoproteins. Virology *86*, 432–442 (1978 b).

Narmanbetova, R. A., Bukrinskaia, A. G.: Localization and transport of influenza virus parental ribonucleoproteins in infected cells. Vopr. Virusol. *5*, 525–530 (1980).

Naso, R. B., Wu, Y.-H. C., Edbauer, C. A.: Antiretroviral effect of interferon: proposed mechanism. J. Interferon Res. *2*, 75–95 (1982).

Nayak, D. P.: The biology of Myxoviruses. In: The Molecular Biology of the Animal Viruses (Nayak, D. P., ed.), 281–348. New York: Marcel Dekker 1977.

Nermut, M. V., Frank, H., Schafer, W.: Properties of mouse leukemia viruses. III. Electron microscopic appearances as revealed after conventional preparation techniques as well as freeze-drying and freeze-etching. Virology *49*, 345–358 (1972).

Newcomb, W. W., Brown, J. C.: Role of vesicular stomatitis virus matrix protein in maintaining the viral nucleocapsid in the condense form found in native virions. J. Virol. *39*, 295–299 (1981).

Newcomb, W. W., Tobin, G. J., McGowan, J. J., Brown, J. C.: *In vitro* reassembly of vesicular stomatitis virus skeletons. J. Virol. *41*, 1055–1062 (1982).

Niemann, H., Klenk, H.-D.: Coronavirus glycoprotein E1, a new type of viral glycoprotein. J. Mol. Biol. *153*, 993–1010 (1981).

Niemann, H., Boschek, B., Evans, D., Rosing, M., Tamura, T., Klenk, H.-D.: Posttranslational glycosylation of coronavirus glycoprotein E1: inhibition by monensin. EMBO Journal *1*, 1499–1504 (1982).

Noonan, K. D., Burger, M. M.: An assay for labeled lectin binding to cell surfaces. In: Methods in Enzymology (Fleisder, S., Packer, L., eds.), Vol. 32. London-New York-Toronto-Sydney-San Francisco: Academic Press 1974.

Norrby, E., Chen, S.-N., Togashi, T., Shesberadaran, H., Johnson, K. P.: Five measles virus antigens demonstrated by use of mouse hybridoma antibodies in productively infected tissue culture cells. Arch. Virol. *71*, 1–11 (1982).

Novo, E., Esparza, J.: Composition and topography of structural polypeptides of bovine rotavirus. J. gen. Virol. *56*, 325–335 (1981).

Nusse, R., van der Ploeg, L., van Duijn, L., Michalides, R., Hilgers, J.: Impaired maturation of mouse mammary tumor virus precursor polypeptides in lymphoid leukemia cells, producing intracytoplasmic A particles and no extracellular B-type virions. J. Virol. *32*, 251–258 (1979).

Obijeski, J. F., Murphy, F. A.: Bunyaviridae: recent biochemical developments. J. gen. Virol. *37*, 1–14 (1977).

Obijeski, J. F., Bishop, D. H. L., Palmer, E. L., Murphy, F. A.: Segmented genome and nucleocapsid of La Crosse virus. J. Virol. *20*, 664–675 (1976).

Obijeski, J. F., McCauley, J., Skehel, J. J.: Nucleotide sequences at the termini of La Crosse virus RNAs. Nucleic Acids Res. *8*, 2431–2438 (1980).

Odenwald, W., Arnheiter, H., Dubois-Dalcq, M., Lazzarini, R.: The nucleocapsid coiling of vesicular stomatitis virus at the inner surface of plasmid membranes immunolocalization of the matrix protein. In: The Molecular Biology of Negative Strand Viruses (Bishop, D. H. L., Compans, R. W., eds.). New York: Academic Press. In Press.

Offit, P. A., Clark, H. F., Plotkin, S. A.: Response of mice to rotaviruses of bovine or primate origin assessed by radioimmunoassay, radioimmunoprecipitation and plaque reduction neutralization. Infect. Immun. *42*, 293–300 (1983).

Okada, Y.: Factors in fusion of cells by HVJ. Curr. Top. Microbiol. Immunol. *48*, 102–156 (1969).

Oker-Blom, C., Kalkkinen, N., Kääriäinen, L., Pettersson, R. F.: Rubella virus contains one capsid protein and three envelope glycoproteins, E_1, E_{2a} and E_{2b}. J. Virol. *46*, 964–973 (1983).

Olden, K., Bernard, B. A., Turner, W., White, S. L.: Effect of interferon on protein glycosylation and comparison with tunicamycin. Nature *300*, 290–292 (1982).

Oshiro, L. S.: Coronaviruses. In: Ultrastructure of Animal Viruses and Bacteriophages: An Atlas (Dalton, A. J., Haguenau, F., eds.), Chap. 18, 331–343. New York: Academic Press 1973.

Oshiro, L. S., Schieble, J. H., Lennette, E. H.: Electron microscopic studies of a coronavirus. J. gen. Virol. *12*, 161–168 (1971).

Oshiro, L. S., Dondero, D. V., Emmons, R. W., Lennette, E. H.: The development of Colorado tick fever virus within cells of the haemopoetic system. J. gen. Virol. *39*, 73–79 (1978).

Örvell, C., Grandien, M.: The effects of monoclonal antibodies on biologic activities of structural proteins of Sendai virus. J. Immunol. *129*, 2779–2787 (1982).

Ou, J.-H., Strauss, E. G., Strauss, J. H.: The 5'-terminal sequences of the genomic RNAs of several alphaviruses. J. Mol. Biol. *168*, 1–15 (1983).

Palese, P., Schulman, J. L.: Isolation and characterization of influenza virus recombinants with high and low neuraminidase activity. Virology *57*, 227–237 (1974).

Palese, P., Tobita, K., Ueda, M., Compans, R. W.: Characterization of temperature-sensitive influenza virus mutants defective in neuraminidase. Virology *61*, 394–410 (1974).

Palmer, E., Martin, M. L.: Further observations on the ultrastructure of human rotavirus. J. gen. Virol. *62*, 105–111 (1982).

Palmer, E. L., Obijeski, J. F., Webb, P. A., Johnson, K. M.: The circular, segmented nucleocapsid of an arenaviruses-Tacaribe virus. J. gen. Virol. *36*, 541–554 (1977).

Pastan, I. H., Willingham, M. C.: Journey to the center of the cell: role of the receptosome. Science *214*, 504–509 (1981).

Patton, J. T., Davis, N. L., Wertz, G. W.: Cell-free synthesis and assembly of vesicular stomatitis virus nucleocapsids. J. Virol. *45*, 155–164 (1983).

Payne, C. M., Ray, C. G., Yolken, R. H.: The 30 to 54 nm rotavirus-like particles in gastroenteritis: incidence and antigenic relationship to rotavirus. J. Med. Virol. *7*, 299–313 (1981).

Pearson, G. R., McNulty, M. S.: Ultrastructural changes in small intestinal epithelium of neonatal pigs infected with pig rotavirus. Arch. Virol. *59*, 127–136 (1979).

Pedersen, I. R.: Structural components and replication of arenaviruses. In: Advances in Virus Research (Lauffer, M. A., Bang, F. B., Maramorosch, K., Smith, K. M., eds.), Vol. 24, 277–330. New York-San Francisco-London: Academic Press 1979.

Pedersen, N. C., Ward, I., Mengeling, W. L.: Antigenic relationships of the feline infectious peritonitis virus to coronaviruses of other species. Arch. Virol. *58*, 45–53 (1978).

Peluso, R. W., Moyer, S. A.: Initiation and replication of vesicular stomatitis virus genome RNA in a cell-free system. Proc. Natl. Acad. Sci. U.S.A. *80*, 3198–3202 (1983).

Pennington, T. H., Pringle, C. R.: Negative strand viruses in enucleated cells. In: Negative Strand Viruses and the Host Cell (Mahy, B. W. J., Barry, R. D., eds.), 457–464. London: Academic Press 1978.

Pensaert, M. B., Debouck, P., Reynolds, D. J.: An immunoelectron microscopic and immunofluores-cent study on the antigenic relationship between the coronavirus-like agent CV777, and several coronaviruses. Arch. Virol. *68*, 45–52 (1981).

Pepinsky, R. B., Vogt, V. M.: Purification and properties of a fifth major viral *gag* protein from avian sarcoma and leukemia viruses. J. Virol. *45*, 648–658 (1983).

Pepinsky, R. B., Cappiello, D., Wilkowski, C., Vogt, V. M.: Chemical crosslinking of proteins in avian sarcoma and leukemia viruses. Virology *102*, 205–210 (1980).

Pesonen, M., Kääriäinen, L.: Incomplete complex oligosaccharides in Semliki Forest virus envelope proteins arrested within the cell in the presence of monensin. J. Mol. Biol. *158*, 213–230 (1982).

Pesonen, M., Saraste, J., Hashimoto, K., Kääriäinen, L.: Reversible defect in the glycosylation of the membrane proteins of Semliki Forest virus *ts*-1 mutant. Virology *109*, 165–173 (1981).

Petri, T., Meier-Ewert, H., Crumpton, W. M., Dimmock, N. J.: RNAs of influenza C virus strains. Arch. Virol. *61*, 239–243 (1979).

Petrie, B. L., Graham, D. Y., Estes, M. K.: Identification of rotavirus particle types. Intervirology *16*, 20–28 (1981).

Petrie, B. L., Graham, D. Y., Hanssen, H., Estes, M. K.: Localization of rotavirus antigens in infected cells by ultrastructural immunocytochemistry. J. gen. Virol. *63*, 457–467 (1982).

Petrie, B. L., Estes, M. K., Graham, D. Y.: Effects of tunicamycin on rotavirus morphogenesis and infec-tivity. J. Virol. *46*, 270–274 (1983).

Pettersson, R. F., von Bonsdorff, C. H.: Ribonucleoproteins of Uukuniemi virus are circular. J. Virol. *15*, 386–392 (1975).

Pettersson, R. F., Hewlett, M. J., Baltimore, D., Coffin, J. M.: The genome of Uukuniemi virus consists of three unique RNA segments. Cell *11*, 51–63 (1977).

Pfeffer, L. M., Wang, E., Tamm, I.: Interferon effects on microfilament organization, cellular fibronec-tin distribution, and cell motility in human fibroblasts. J. Cell Biol. *85*, 9–17 (1980).

Pinter, A., de Harven, E.: Protein composition of a defective murine sarcoma virus particle possessing the enveloped type A morphology. Virology *99*, 103–110 (1979).

Pinter, A., Honnen, W. J.: Topography of murine leukemia virus envelope proteins: characterization of transmembrane components. J. Virol. *46*, 1056–1060 (1983).

Pinter, A., Honnen, W. J., Tung, J. S., O'Donnell, P. V., Hammerling, U.: Structural domains of endo-genous murine leukemia virus gp60s containing specific antigenic determinants defined by mono-clonal antibodies. Virology *116*, 499–516 (1982).

Plotch, S. J., Krug, R. M.: Influenza virion transcriptase: synthesis *in vitro* of large, polyadenylic acid-containing complementary RNA. J. Virol. *21*, 24–34 (1977).

Plotch, S. J., Krug, R. M.: Segments of influenza virus complementary RNA synthesized *in vitro*. J. Virol. *25*, 579–586 (1978).

Plotch, S. J., Bouloy, M., Krug, R. M.: Transfer of 5′-terminal cap of globin mRNA to influenza viral complementary RNA during transcription *in vitro*. Proc. Natl. Acad. Sci. U.S.A. *76*, 1618–1622 (1979).

Pons, M. W., Schulze, I. T., Hirst, G. K.: Isolation and characterization of the ribonucleoprotein of influenza virus. Virology *39*, 250–259 (1969).

Portner, A.: Evidence for two different sites on the HN glycoprotein involved in neuraminidase and hemagglutinating activities. In: The Replication of Negative Strand Viruses (Bishop, D. H. L., Compans, R. W., eds.), 465–470. New York: Elsevier/North-Holland 1981.

Portner, A., Kingsbury, D. W.: Regulatory events in the synthesis of Sendai virus polypeptides and their assembly into virions. Virology *73*, 79–88 (1976).

Portner, A., Scroggs, R. A., Marx, P. A., Kingsbury, D. W.: A temperature-sensitive mutant of Sendai virus with an altered hemagglutinin-neuraminidase polypeptide: consequence for virus assembly and cytopathology. Virology *67*, 179–187 (1975).

Poste, G., Alexander, D. J., Reeve, P., Hewlett, G.: Modification of NDV release and cytopathogenicity in cells treated with plant lectins. J. gen. Virol. *23*, 255–270 (1974).

Pringle, C. R.: The genetics of vesiculoviruses. Arch. Virol. *72*, 1–34 (1982).

Quinn, P., Griffiths, G., Warren, G.: Dissection of the Golgi complex. II. Density separation of specific Golgi functions in virally infected cells treated with monensin. J. Cell Biol. *96*, 851–856 (1983).

Raine, C. S.: On the development of CNS lesions in natural canine distemper encephalomyelitis. J. Neurol. Sci. *30*, 13—28 (1976).

Raine, C. S., Byington, D. P., Johnson, K. P.: Subacute sclerosing panencephalitis in the hamster. Ultrastructure of the chronic disease. Lab. Invest. *31*, 355—368 (1974).

Ramig, R. F.: Isolation and genetic characterization of temperature sensitive mutants of simian rotavirus SA11. Virology *120*, 93—105 (1982).

Ramig, R. F.: Factors that affect genetic interaction during mixed infection with temperature-sensitive mutants of simian rotavirus SA11. Virology *127*, 91—99 (1983).

Ramos, B. A., Courtney, R. J., Rawls, W. E.: Structural proteins of Pichinde virus. J. Virol. *10*, 661—667 (1972).

Ranki, M., Pettersson, R. F.: Uukuniemi virus contains an RNA polymerase. J. Virol. *16*, 1420—1425 (1975).

Rawls, W. E., Leung, W.-C.: Arenaviruses. In: Comprehensive Virology (Fraenkel-Conrat, H., Wagner, R. R., eds.), Vol. 14, 157—192. New York-London: Plenum Press 1979.

Rees, P. J., Dimmock, N. J.: Electrophoretic analysis of influenza virus ribonucleoprotein from purified viruses and infected cells. In: The Replication of Negative Strand Viruses (Bishop, D. H. L., Compans, R. W., eds.), 341—344. New York: Elsevier/North-Holland 1981.

Reginster, M., Rentier, B., Dierickx, L.: Impairment of the M protein and unmasking of a superficial type-specific antigen by proteolytic treatment of influenza A virions with preservation of host-specific antigenicity. Intervirology *6*, 239—248 (1976).

Reginster, M., Joassin, L., Fontaine-Delcambe, P.: Ligands for antibody to M-protein are exposed at the surface of influenza virions: effect of proteolytic treatment on their activity. J. gen. Virol. *45*, 283—289 (1979).

Rentier, B., Hooghe-Peters, E., Dubois-Dalcq, M.: Electron microscopic study of measles virus infection. Cell fusion and hemadsorption. J. Virol. *28*, 567—577 (1978).

Rentier, B., Claysmith, A., Bellini, W. J., Dubois-Dalcq, M.: Chronic measles virus infection of mouse nerve cells *in vitro*. In: The Replication of Negative Strand Viruses (Bishop, D. H. L., Compans, R. W., eds.), 595—601. New York: Elsevier/North-Holland 1981.

Rice, C. M., Strauss, J. H.: Nucleotide sequence of the 26S mRNA of Sindbis virus and deduced sequence of the encoded virus structural proteins. Proc. Natl. Acad. Sci. U.S.A. *78*, 2062—2066 (1981).

Richardson, C. D., Scheid, A., Choppin, P. W.: Specific inhibition of paramyxovirus and myxovirus replication by N-termini of the F_1 or HA_2 viral polypeptides. Virology *105*, 205—222 (1980).

Robb, J. A., Bond, C. W., Leibowitz, J. L.: Pathogenic murine coronaviruses. III. Biological and biochemical characterization of temperature-sensitive mutants of JHMV. Virology *91*, 385—399 (1979).

Robertson, J. S.: 5′ and 3′ terminal nucleotide sequence of the RNA genome segments of influenza virus. Nucl. Acids Res. *6*, 3745—3757 (1979).

Robertson, J. S., Jennings, P. A., Finch, J. T., Winter, G.: Sequence rearrangements in influenza virus RNA and ribonucleoprotein structure. In: The Molecular Biology of Negative Strand Viruses (Bishop, D. H. L., Compans, R. W., eds.), New York: Academic Press. In Press.

Rodger, S. M., Holmes, I. H.: Comparison of the genomes of simian, bovine and human rotaviruses by gel electrophoresis and detection of genomic variation among bovine isolates. J. Virol. *30*, 839—846 (1979).

Rodger, S. M., Schnagl, R. D., Holmes, I. H.: Further biochemical characterization, including the detection of surface glycoproteins of human, calf and simian rotaviruses. J. Virol. *24*, 91—98 (1977).

Rodger, S. M., Bishop, R. F., Birch, C., McLean, B., Holmes, I. H.: Molecular epidemiology of human rotaviruses in Melbourne, Australia from 1973 to 1979 as determined by electrophoresis of genome ribonucleic acid. J. Clin. Microbiol. *13*, 272—278 (1981).

Rodriguez, M., Buchmeier, M. J., Oldstone, M. B. A., Lampert, P. W.: Ultrastructural localization of viral antigens in the CNS of mice persistently infected with lymphocytic choriomeningitis virus (LCMV). Amer. J. Path. *110*, 95—100 (1983).

Rodriguez-Boulan, E.: Membrane biogenesis, enveloped RNA viruses and epithelial polarity. Modern Cell Biol. *1*, 119—170 (1983).

Rodriguez-Boulan, E., Sabatini, D. D.: Asymmetric budding of viruses in epithelial monolayers: A model system for study of epithelial polarity. Proc. Natl. Acad. Sci. U.S.A. *75*, 5071—5075 (1978).

Rodriguez-Boulan, E., Pendergast, M.: Polarized distribution of viral envelope proteins in the plasma membrane of infected epithelial cells. Cell 20, 45—54 (1980).

Rodriguez-Boulan, E., Paskiet, K. T., Sabatini, D. D.: Assembly of enveloped viruses in Madin-Darby canine kidney cells: polarized budding from single attached cells and from clusters of cells in suspension. J. Cell Biol. 96, 866—874 (1983).

Rodriguez-Boulan, E., Paskiet, K. T., Salas, P. J. I., Bard, E.: Intracellular transport of influenza virus hemagglutinin to the apical surface of MDCK cells. J. Cell Biol. 98, 308—319 (1984).

Rodriguez-Toro, G.: Natural epizootic diarrhea of infant mice (EDIM). A light and electron microscopic study. Exp. Molec. Pathol. 32, 241—252 (1980).

Rogalski, A. A., Bergmann, J. E., Singer, S. J.: Intracellular transport and processing of an integral membrane protein destined for the cell surface is independent of the assembly status of cytoplasmic microtubules. J. Cell Biol. 95, 337a (1982).

Rohrschneider, J. M., Diggelmann, H., Ogura, H., Friis, R. R., Bauer, H.: Defective cleavage of a precursor polypeptide in a temperature-sensitive mutant of avian sarcoma virus. Virology 75, 177—187 (1976).

Roman, J. M., Simon, E. H.: Morphological heterogeneity in egg and monolayer-propagated Newcastle disease virus. Virology 69, 287—297 (1976 a).

Roman, J. M., Simon, E. H.: Defective interfering particles in monolayer-propagated Newcastle disease virus. Virology 69, 298—303 (1976 b).

Rose, J. K.: Ribosome recognition and translation of vesicular stomatitis virus messenger RNA. In: Rhabdoviruses (Bishop, D. H. L., ed.), Vol. II, 51—60. Boca Raton, Fla.: CRC Press 1980.

Rose, J. K., Gallione, C. J.: Nucleotide sequences of the mRNA's encoding the vesicular stomatitis virus G and M proteins determined from cDNA clones containing the complete coding regions. J. Virol. 39, 519—528 (1981).

Rose, J. K., Bergmann, J. E.: Expression from cloned cDNA of cell surface and secreted forms of the vesicular stomatitis virus glycoprotein in eucaryotic cells. Cell 30, 753—762 (1982).

Rose, J. K., Doolittle, R. F., Anilionis, A., Curtis, P. J., Wunner, W. H.: Homology between the glycoproteins of vesicular stomatitis virus and rabies virus. J. Virol. 43, 361—364 (1982).

Roseto, A., Escaig, J., Delain, E., Cohen, J., Scherrer, R.: Structure of rotaviruses as studied by the freeze-drying technique. Virology 98, 471—475 (1979).

Roth, M. G., Compans, R. W.: Delayed appearance of pseudotypes between vesicular stomatitis virus and influenza virus during mixed infection of MDCK cells. J. Virol. 40, 848—860 (1981).

Roth, M. G., Fitzpatrick, J. P., Compans, R. W.: Polarity of virus maturation in MDCK cells: lack for a requirement for glycosylation of viral glycoproteins. Proc. Natl. Acad. Sci. U.S.A. 76, 6430—6434 (1979).

Roth, M. G., Srinivas, R. V., Compans, R. W.: Basolateral maturation of retroviruses in polarized epithelial cells. J. Virol. 45, 1065—1073 (1983 a).

Roth, M. G., Compans, R. W., Giusit, L., Davis, A. R., Nayak, D. P., Gething, M.-J., Sambrook, J.: Influenza virus hemagglutinin expression is polarized in cells infected with recombinant SV40 viruses carrying cloned hemagglutinin DNA. Cell 33, 435—443 (1983 b).

Rothman, J. E.: The Golgi apparatus: two organelles in tandem. Science 213, 1212—1219 (1981).

Rothman, J. E., Lodish, H. F.: Synchronized transmembrane insertion and glycosylation of a nascent membrane protein. Nature (Lond.) 269, 778—780 (1977).

Rothman, J. E., Bursztyn-Pettegrew, H., Fine, R. E.: Transport of the membrane glycoprotein of vesicular stomatitis virus to the cell surface in two stages by clathrin-coated vesicles. J. Cell Biol. 86, 162—171 (1980).

Rott, R., Klenk, H.-D.: Structure and assembly of viral envelope. In: Virus Infection and the Cell Surface (Poste, G., Nicolson, D., eds.), 47—81. Amsterdam: North-Holland 1977.

Rottier, P. J. M., Spaan, W. J. M., Horzinek, M., van der Zeijst, B. A. M.: Translation of three mouse hepatitis virus (MHV-A59) subgenomic RNAs in Xenopus laevis oocytes. J. Virol. 38, 20—26 (1981a).

Rottier, P. J. M., Horzinek, M. C., van der Zeijst, B. A. M.: Viral protein synthesis in mouse hepatitis virus strain A59-infected cells: effect of tunicamycin. J. Virol. 40, 350—357 (1981b).

Rottier, P. J. M., van der Zeijst, B. A. M., Horzinek, M. (eds.): Coronaviruses: Molecular Biology and Pathogenesis. New York: Plenum Press. 1984.

Roux, L., Waldvogel, F. A.: Instability of the viral M-protein in BHK-21 cells persistently infected with Sendai virus. Cell 28, 293—302 (1982).

Roy, P., Bishop, D. L.: Initiation and direction of RNA transcription by vesicular stomatitis virus virion transcriptase. J. Virol. 11, 487—501 (1973).

Rozenblatt, S., Gesang, C., Lavie, V., Neumann, F. S.: Cloning and characterization of DNA complementary to the measles virus mRNA encoding hemagglutinin and matrix protein. J. Virol. 42, 790—797 (1982).

Russell, P. K., Brandt, W. E., Dalrymple, J. M.: Chemical and antigenic structure of flaviviruses. In: The Togaviruses (Schlesinger, R. W., ed.), 503—529. New York: Academic Press 1980.

Sabara, M., Babiuk, L. A., Gilchrist, J., Misra, V.: Effect of tunicamycin on rotavirus assembly and infectivity. J. Virol. 43, 1082—1090 (1982).

Sabatini, D. D., Kreibich, G., Morimoto, T., Adesnik, M.: Mechanisms for the incorporation of proteins in membranes and organelles. J. Cell Biol. 92, 1—22 (1982).

Saikku, P., von Bonsdorff, C. H., Brummer-Kornenkontio, M., Vaheri, A.: Isolation of a non-cuboidal ribonucleoprotein from Inkoo virus, a Bunyamwera supergroup arbovirus. J. gen. Virol. 13, 335—337 (1971).

Saleh, F., Gard, G.-P., Compans, R. W.: Synthesis of Tacaribe viral proteins. Virology 93, 369—376 (1979).

Samson, A. C. R., Fox, C. F.: A precursor protein for Newcastle disease virus. J. Virol. 12, 579—587 (1973).

Saraste, J., von Bonsdorff, C.-H., Hashimoto, K., Kääriäinen, L., Keränen, S.: Semliki Forest virus mutants with temperature sensitive transport of envelope proteins. Virology 100, 229—245 (1980 a).

Saraste, J., von Bonsdorff, C.-H., Hashimoto, K., Keränen, S., Kääriäinen, L.: Reversible transport defects of virus membrane glycoproteins in Sindbis mutant infected cells. Cell Biol. Int. Rep. 4, 279—286 (1980 b).

Sarkar, N. H., Nowinski, R. C., Moore, D. H.: Nucleocapsid structure of oncornaviruses. J. Virol. 8, 564—572 (1971).

Satake, M., Luftig, R. B.: Comparative immunofluorescence of murine leukemia virus-derived membrane-associated antigens. Virology 124, 259—273 (1983).

Sato, K., Inaba, Y., Shinozaki, T., Fuji, R., Matumoto, M.: Isolation of human rotavirus in cell cultures. Arch. Virol. 69, 155—160 (1981).

Sawicki, D., Sawicki, S.: Short-lived minus-strand polymerase for Semliki Forest virus. J. Virol. 34, 108—118 (1980).

Scheefers, H., Scheefers-Borchel, U., Edwards, J., Brown, D. T.: Distribution of virus structural proteins and protein-protein interactions in plasma membrane of baby hamster kidney cells infected with Sindbis or vesicular stomatitis virus. Proc. Natl. Acad. Sci., U.S.A. 77, 7277—7281 (1980).

Scheid, A., Choppin, P. W.: Identification of biological activities of Paramyxovirus glycoproteins. Activation of cell fusion, hemolysis and infectivity by proteolytic cleavage of an inactive precursor protein of Sendai virus. Virology 57, 475—490 (1974).

Scheid, A., Choppin, P. W.: Protease activation mutants of Sendai virus: activation of biological properties by specific proteases. Virology 69, 265—267 (1976).

Scheid, A., Choppin, P. W.: Two disulfide-linked polypeptide chains constitute the active F protein of paramyxoviruses. Virology 80, 54—66 (1977).

Scheid, A., Caliguiri, L. A., Compans, R. W., Choppin, P. W.: Isolation of paramyxovirus glycoproteins. Association of both hemagglutinating and neuraminidase activities with the layer SV5 glycoprotein. Virology 50, 640—652 (1972).

Scheid, A., Gravas, M. C., Silver, S. M., Choppin, P. W.: Studies on the structure and function of paramyxovirus glycoproteins. In: Negative Strand Viruses and the Host Cell (Mahy, B. W. J., Barry, R. D., eds.), 181—193. New York: Academic Press 1978.

Schidlovsky, G.: Structure of RNA tumor viruses. In: Recent Advances in Cancer Research: Cell Biology, Molecular Biology, and Tumor Virology (Gallo, R. C., ed.), Vol. I, 189—245. Boca Raton, Fla.: CRC Press 1977.

Schlesinger, M. J., Kääriäinen, L.: Translation and processing of alphavirus proteins. In: The Togaviruses (Schlesinger, R. W., ed.), 371—392. New York: Academic Press 1980.

Schlesinger, R. W.: Dengue viruses. In: Virology Monographs (Gard, S., Hallauer, C., eds.), Vol. 16. Wien-New York: Springer 1977.

Schlesinger, R. W. (ed.): The Togaviruses. Biology, Structure, Replication. New York: Academic Press 1980.

Schleuderberg, A., Chavanich, S., Lipman, M. B., Carter, C.: Comparative molecular weight estimates of measles and subacute sclerosing panencephalitis virus structural polypeptides by simultaneous electrophoresis in acrylamide gel slabs. Biochem. Biophys. Res. Comm. *58*, 647–651 (1974).

Schmidt, M. F. G.: Acylation of viral spike glycoproteins, a feature of enveloped RNA viruses. Virology *116*, 327–338 (1982 a).

Schmidt, M. F. G.: Acylation of proteins—a new type of modification of membrane proteins. Trends in Biochem. Sci. *7*, 322–324. *1982 b.*

Schmidt, M. F. G., Schlesinger, M. J.: Fatty acid binding to VSV glycoprotein, a new type of posttranslational modification of the viral glycoprotein. Cell *17*, 813–815 (1979).

Schmidt, M. F. G., Schlesinger, M. J.: Relation of fatty acid attachment to the translation and maturation of vesicular stomatitis and Sindbis virus membrane glycoproteins. J. Biol. Chem. *255*, 3334–3339 (1980).

Schmidt, O. W., Kenny, G. E.: Immunogenicity and antigenicity of human coronaviruses 229E and OC43. Infect. Immunity *32*, 1000–1006 (1981).

Schmidt, O. W., Kenny, G. E.: Polypeptides and functions of antigens from human coronaviruses 229E and OC43. Infect. Immunity *35*, 515–522 (1982).

Schnagl, R. D., Holmes, I. H.: Electron microscopy of Japanault and Tilligerry viruses: two proposed members of the orbivirus group. Aust. J. Biol. Sci. *28*, 425–432 (1975).

Schneider, J., Falk, H., Hunsmann, G.: Envelope polypeptides of Friend leukemia virus: purification and structural analysis. J. Virol. *33*, 597–605 (1980).

Schnitzer, T. J., Lodish, H. F.: Noninfectious vesicular stomatitis virus particles deficient in the viral nucleocapsid. J. Virol. *29*, 443–447 (1979).

Schnitzer, T. J., Hodes, D. S., Gerin, J., Camargo, E., Chanock, R. M.: Effect of 2-deoxy-D-glucose and glucosamine on the growth and function of respiratory syncytial and parainfluenza 3 viruses. Virology *67*, 306–309 (1975).

Scholtissek, C.: Influenza virus genetics. In: Advances in Genetics, Vol. 20, 1–36. New York: Academic Press 1979.

Scholtissek, C., Rott, R., Hau, G., Kaluza, G.: Inhibition of the multiplication of vesicular stomatitis and Newcastle disease viruses by 2-deoxy-D-glucose. J. Virol. *13*, 1186–1193 (1974).

Scholtissek, C., Rott, R., Klenk, H.-D.: Two different mechanisms of the inhibition of the multiplication of enveloped viruses by glucosamine. Virology *63*, 191–200 (1975).

Schubert, M., Keene, J. D., Herman, R. C., Lazzarini, R. A.: Site on the vesicular stomatitis virus genome specifiying polyadenylation and the end of the L gene mRNA. J. Virol. *34*, 550–559 (1980).

Schultz, A. M., Lockhart, S. M., Rabin, E. M., Oroszlan, S.: Structure of glycosylated and unglycosylated *gag* polyproteins of Rauscher murine leukemia virus: carbohydrate attachment sites. J. Virol. *38*, 581–592 (1981).

Schulze, I. T.: Structure of the influenza virion. Advances in Virus Research *18*, 1–55 (1973).

Schwalbe, J. C., Hightower, L. E.: Maturation of the envelope glycoproteins of Newcastle disease virus on cellular membranes. J. Virol. *41*, 947–957 (1982).

Schwartz, R. T., Klenk, H.-D.: Inhibition of glycosylation of influenza virus hemagglutinin. J. Virol. *14*, 1023–1034 (1974).

Schwendemann, G., Wolinsky, J. S., Hatzidimitriou, G., Merz, D. C., Waxham, M. N.: Postembedding immunocytochemical localization of paramyxovirus antigens by light and electron microscopy. J. Histochem. Cytochem. *30*, 1313–1319 (1982).

Sekikawa, K., Lai, C.-J.: Defects in functional expression of an influenza virus hemagglutinin lacking the signal peptide sequences. Proc. Natl. Acad. Sci. U.S.A., *80*, 3563–3567 (1983).

Sen, G. C., Pinter, A.: Interferon-mediated inhibition of production of Gazdar murine sarcoma virus, a retrovirus lacking *env* proteins and containing an uncleaved *gag* precursor. Virology *126*, 403–407 (1983).

Seto, J. T., Garten, W., Rott, R.: The site of cleavage in infected cells and polypeptides of representative paramyxoviruses grown in cultured cells of the chorioallantoic membrane. Arch. Virol. *67*, 19–30 (1981).

Shank, P. R., Linial, M.: Avian oncovirus mutant (SE21Q1b) deficient in genomic RNA: characterization of a deletion in the provirus. J. Virol. *36*, 450–456 (1980).

Shapiro, C., Brendt, W. E., Russell, P. K.: Change involving a viral membrane glycoprotein during morphogenesis of group B arboviruses. Virology *50*, 906–911 (1972).

Shapiro, D., Kos, K. A., Russell, P. K.: Japanese encephalitis virus glycoproteins. Virology *56*, 88–94 (1973).

Shapshak, P., Graves, M. C., Imagawa, D. T.: Polypeptides of canine distemper virus strains derived from dogs with chronic neurological diseases. Virology *122*, 158–170 (1982).

Sharon, N., Lis, H.: Glycoproteins: research looming on long-ignored, ubiquitous compounds. Chem. Eng. News *59*, 21–44 (1981).

Shaw, M. W., Compans, R. W.: Isolation and characterization of cytoplasmic inclusions form influenza A virus-infected cells. J. Virol. *25*, 605–615 (1978).

Shaw, M. W., Choppin, P. W., Lamb, R. A.: A previously unrecognized influenza B virus glycoprotein from a bicistronic mRNA that also encodes the viral neuraminidase. Proc. Natl. Acad. Sci. U.S.A. *80*, 4879–4883 (1983).

Shimizu, K., Ishida, N.: The smallest protein of Sendai virus: its candidate function of binding nucleocapsid to envelope. Virology *67*, 427–437 (1975).

Shimizu, K., Shimizu, Y. K., Kohama, T., Ishida, N.: Isolation and characterization of two distinct types of HVJ (Sendai virus) spikes. Virology *62*, 90–101 (1974).

Shope, R. E.: Medical significance of togaviruses: an overview of diseases caused by togaviruses in man and in domestic and wild vertebrate animals. In: The Togaviruses (Schlesinger, R. W., ed.), 47–82. New York: Academic Press 1980.

Siddell, S. G., Wege, H., Barthel, A., ter Meulen, V.: Coronavirus JHM. Cell-free synthesis of structural protein p60. J. Virol. *33*, 10–17 (1980).

Siddell, S. G., Barthel, A., ter Meulen, V.: Coronavirus JHM. A virion-associated protein kinase. J. gen. Virol. *52*, 235–243 (1981 a).

Siddell, S. G., Wege, H., Barthel, A., ter Meulen, V.: Coronavirus JHM. Intracellular protein synthesis. J. gen. Virol. *53*, 145–155 (1981 b).

Siddell, S., Wege, H., Barthel, A., ter Meulen, V.: Intracellular protein synthesis and the *in vitro* translation of coronavirus JHM mRNA. In: Biochemistry and Biology of Coronaviruses (ter Meulen, V., Siddell, S., Wege, H., eds.), 193–208. New York: Plenum Press 1981 c.

Siddell, S., Wege, H., ter Meulen, V.: The structure and replication of coronaviruses. Curr. Top. Microbiol. Immunol. *99*, 131–163 (1982).

Siddell, S., Wege, H., ter Meulen, V.: The biology of coronaviruses. J. gen. Virol. *64*, 761–776 (1983).

Simizu, B., Hashimoto, K., Ishida, I.: A variant of Western equine encephalitis virus with nonglycosylated E3 protein. Virology *125*, 99–106 (1983).

Simons, K., Garoff, H.: The budding mechanisms of enveloped animal viruses. J. gen. Virol. *50*, 1–21 (1980).

Simons, K., Helenius, A., Leonard, K., Sarvas, M., Gething, M. J.: Formation of protein micelles from amphiphilic membrane proteins. Proc. Natl. Acad. Sci. U.S.A. *75*, 5306–5310 (1978).

Simons, K., Garoff, H., Helenius, A.: Alphavirus proteins. In: The Togaviruses (Schlesinger, R. W., ed.), 317–333. New York: Academic Press 1980.

Simons, K., Garoff, H., Helenius, A.: How the virus comes in and out from the host cell. Sci. Amer. *246*, 57–66 (1982).

Sinarachatanant, P., Olson, L. C.: Replication of Dengue virus type 2 in Aedes albopictus cell culture. J. Virol. *12*, 275–283 (1973).

Skehel, J. J.: RNA-dependent RNA polymerase activity of the influenza virus. Virology *45*, 793–795 (1971).

Skehel, J. J.: Early polypeptide synthesis in influenza virus-infected cells. Virology *56*, 394–399 (1973).

Skehel, J. J., Waterfield, M.: Studies on the primary structure of the influenza virus hemagglutinin. Proc. Natl. Acad. Sci. U.S.A. *72*, 93–97 (1975).

Skehel, J. J., Hay, A. J.: Nucleotide sequences at the 5' termini of influenza virus RNAs and their transcripts. Nucl. Acids Res. *5*, 1207–1219 (1978).

Skehel, J. J., Bayley, P. M., Brown, E. B., Martin, S. R., Waterfield, M. D., White, J. M., Wilson, I. A., Wiley, D. C.: Changes in the conformation of influenza virus hemagglutinin at the pH optimum of virus-mediated membrane fusion. Proc. Natl. Acad. Sci. U.S.A. *79*, 968–972 (1982).

Smith, M. L., Lazdins, E., Holmes, I. H.: Coding assignments of double-stranded RNA segments of SA-11 rotavirus established by *in vitro* translation. J. Virol. *33*, 976—982 (1980).

Smith, G. L., Hay, A. J.: Synthesis of influenza virus RNAs. In: The Replication of Negative Strand Viruses (Bishop, D. H. L., Compans, R. W., eds.), 333—339. New York: Elsevier/North-Holland 1981.

Smith, G. W., Hightower, L. E.: Biochemical properties of the NDV P protein. In: The Replication of Negative Strand Viruses (Bishop, D. H. L., Compans, R. W., eds.), 479—484. New York: Elsevier/North-Holland 1981.

Smith, J. F., Pifat, D. Y.: Morphogenesis of Sandfly fever viruses (Bunyaviridae family). Virology *121*, 61—81 (1982).

Söderlund, H.: Kinetics of formation of the Semliki Forest virus nucleocapsid. Intervirology *1*, 354—362 (1973).

Söderlund, H., Kääriäinen, L., von Bonsdorff, C.-H., Weckström, P.: Properties of Semliki Forest virus nucleocapsid. II. An irreversible contraction by acid pH. Virology *47*, 753—760 (1971).

Soler, C., Musalem, C., Lorono, M., Espejo, R. T.: Association of viral particles and viral proteins with membranes in SA11-infected cells. J. Virol. *44*, 983—992 (1982).

Spaan, W. J., Rottier, P. J. M., Horzinek, M. C., van der Zeijst, B. A. M.: Isolation and identification of virus-specific mRNAs in cells infected with mouse hepatitis virus (MHV-A59). Virology *108*, 424—434 (1981).

Spaan, W. J. M., Rotter, P. J. M., Horzinek, M. C., van der Zeijst, B. A. M.: Sequence relationship between the genome and the intracellular RNA species 1, 3, 6, and 7 of mouse hepatitis virus strain A59. J. Virol. *42*, 432—439 (1982).

Spanier, B. B., Bratt, M. A.: The 50 S and 35 S RNAs from Newcastle disease virus infected cells. J. gen. Virol. *35*, 439—453 (1977).

Sprague, J., Condra, J. H., Arnheiter, H., Lazzarini, R. A.: Expression of a recombinant DNA gene coding for the vesicular stomatitis virus nucleocapsid protein. J. Virol. *45*, 773—781 (1983).

Srinivas, R. V., Melsen, L. R., Compans, R. W.: Effects of monensin on morphogenesis and infectivity of Friend murine leukemia virus. J. Virol. *42*, 1067—1075 (1982).

Stallcup, K. C., Fields, B. N.: The replication of measles virus in the presence of tunicamycin. Virology *108*, 391—404 (1981).

Stallcup, K. C., Raine, C. S., Fields, B. N.: Cytochalasin B inhibits the maturation of measles virus. Virology *124*, 59—74 (1983).

Stanley, P., Gandhi, S. S., White, D. O.: The polypeptides of influenza virus. VII. Synthesis of the hemagglutinin. Virology *53*, 92—106 (1973).

Stannard, L. M., Schoub, B. D.: Observations on the morphology of two rotaviruses. J. gen. Virol. *37*, 435—439 (1977).

Stern, D. F., Kennedy, S. I. T.: Coronavirus multiplication strategy. I. Identification and characterization of virus-specific RNA. J. Virol. *34*, 665—674 (1980 a).

Stern, D. F., Kennedy, S. I. T.: Coronavirus multiplication strategy. II. Mapping the avian infectious bronchitis virus intracellular RNA species to the genome. J. Virol. *36*, 440—449 (1980 b).

Stern, D. F., Sefton, B. M.: Coronavirus proteins: Structure and function of the oligosaccharides of the avian infectious bronchitis virus glycoproteins. J. Virol. *44*, 804—812 (1982).

Stern, D. F., Burgess, L., Sefton, B. M.: Structural analysis of virion proteins of the avian coronavirus infectious bronchitis virus. J. Virol. *42*, 208—219 (1982).

Stitz, M. R., Becht, H.: Studies on the inhibitory effect of lectins on myxovirus release. J. gen. Virol. *34*, 523—530 (1977).

Stohlman, S. A., Lai, M. M. C.: Phosphoproteins of murine hepatitis viruses. J. Virol. *32*, 672—675 (1979).

Stollar, V.: Defective interfering alphaviruses. In: The Togaviruses (Schlesinger, R. W., ed.), 427—457. New York: Academic Press 1980 a.

Stollar, V.: Togaviruses in cultured arthropod cells. In: The Togaviruses (Schlesinger, R. W., ed.), 584—621. New York: Academic Press 1980 b.

Stone, H. O., Kingsbury, D. W., Darlington, R. W.: Sendai virus-induced transcriptase from infected cells: polypeptides in the transcriptive complex. J. Virol. *10*, 1037—1043 (1972).

Strauss, E. G., Strauss, J. H.: Mutants of alphaviruses: genetics and physiology. In: The Togaviruses (Schlesinger, R. W., ed.), 393—426. New York: Academic Press 1980.

Strauss, E. G., Lenches, E. M., Stamreich-Martin, M. A.: Growth and release of several Alphaviruses in chick and BHK cells. J. gen. Virol. *49*, 297–307 (1980).

Strauss, E. G., Strauss, J. H.: Replication strategies of single stranded RNA viruses of eukaryotes. Curr. Topics Microbiol. Immunol. *105*, 1–98 (1983).

Strauss, E. G., Strauss, J. H.: Assembly of enveloped animal viruses. Ed. by Casjens-Sciencebook International. In Press.

Strauss, E. G., Tsukeda, H., Simizu, B.: Mutants of Sindbis virus IV. Heterotypic complementation and phenotypic mixing between temperature-sensitive mutants and wild type Sindbis and Western equine encephalitis virus. J. gen. Virol. *64*, 1581–1590 (1983).

Strauss, E. G., Rice, C. M., Strauss, J. H.: Complete nucleotide sequence of the genomic RNA of Sindbis virus. In Press.

Sturman, L. S.: The structure and behaviour of coronavirus A59 glycoproteins. Adv. Exp. Med. Biol. *142*, 1–18 (1981).

Sturman, L. S., Holmes, K. V.: Characterization of coronavirus. II. Glycoproteins of the viral envelope: tryptic peptide analysis. Virology *77*, 650–660 (1977).

Sturman, L. S., Holmes, K. V.: The molecular biology of coronaviruses. Adv. Virus Res. *28*, 35–112 (1983).

Sturman, L. S., Holmes, K. V.: Proteolytic cleavage of peplomeric glycoprotein E2 of MHV yields two 90 K subunits and activates cell fusion. In: Coronaviruses: Molecular Biology and Pathogenesis (Rottier, P. J. M., van der Zeijst, B. A. M., Horzinek, M., eds.), 25–36. New York: Plenum Press. 1984.

Sturman, L. S., Holmes, K. V., Behnke, J.: Isolation of coronavirus envelope glycoproteins and interaction with the viral nucleocapsid. J. Virol. *33*, 449–462 (1980).

Sugawara, K. E., Tashiro, M., Homma, M.: Intermolecular association of HANA glycoproteins of Sendai virus in relation to the expression of biological activities. Virology *117*, 444–455 (1982).

Sugiura, A., Ueda, M.: Neurovirulence of influenza virus in mice. I. Neurovirulence of recombinants between virulent and avirulent virus strains. Virology *101*, 440–449 (1980).

Sugiyama, K., Amano, Y.: Morphology and biological properties of a new coronavirus associated with diarrhea in infant mice. Arch. Virol. *66*, 241–251 (1981).

Suzuki, H., Kutsuzawa, T., Konno, T., Ebina, T., Ishida, N.: Morphogenesis of human rotavirus type 2 Wa strain in MA 104 cells. Arch. Virol. *70*, 33–41 (1981).

Sveda, M. M., Markoff, L. J., Lai, C.-J.: Cell surface expression of the influenza virus hemagglutinin requires the hydrophobic carboxy-terminal sequences. Cell *30*, 649–656 (1982).

Tabas, I., Kornfeld, S.: The synthesis of complex-type-oligosaccharides. III. Identification of an α-D-mannosidase activity involved in a late stage of processing of complex-type oligosaccharides. J. Biol. Chem. *253*, 7779–7786. 1978.

Takatsuki, A., Kohno, K., Tamura, G.: Inhibition of biosynthesis of polyisoprenol sugars in chick embryo microsomes by tunicamycin. Agric. Biol. Chem. *39*, 2089–2091 (1975).

Tanaka, H.: Precursor-product relationship between nonglycosylated polypeptides of A and B particles of mouse mammary tumor virus. Virology *76*, 835–850 (1977).

Tannenbaum, J.: Approaches to the molecular biology of cytochalasin action. In: Cytochalasins: Biochemical and Cell Biological Aspects (Tannenbaum, S. W., ed.), 521–559. New York: Elsevier/North-Holland 1978.

Tartakoff, A. M.: Perturbation of vesicular traffic with the carboxylic ionophore monensin. Cell *32*, 1026–1028 (1983).

Taylor, J. M., Hampson, A. W., Layton, J. E., White, D. O.: The polypeptides of influenza virus. V. An analysis of nuclear accumulation. Virology *42*, 744–752 (1970).

Tektoff, J., Dauvergne, M., Soulebot, J. P., Durafour, M.: Ultrastructural comparative study of rotavirus replication in cultured heterologous cells. In: 3rd International Symposium on Neonatal Diarrhea, Vido, Canada, 1980.

ter Meulen, V., Siddell, S., Wege, H. (eds.): Biochemistry and Biology of Coronaviruses. Adv. Exp. Med. Biol. *142* (1981).

Testa, D., Chanda, P. K., Barnejee, A. K.: Unique mode of transcription *in vitro* by VSV. Cell *21*, 267–275 (1980).

Tozawa, H., Watanabe, M., Ishida, N.: Structural components of Sendai virus. Serological and physicochemical characterization of hemagglutinin subunit associated with neuraminidase activity. Virology 55, 242–253 (1973).

Tycko, B., Maxfield, F. R.: Rapid acidification of endocytic vesicles containing α-2 macroglobulin. Cell 28, 643–657 (1982).

Tyrrell, J., Norrby, E.: Structural polypeptides of measles virus. J. gen. Virol. 29, 219–229 (1978).

Tyrrell, D. L. J., Ehrnst, A.: Transmembrane communication in cells chronically infected with measles virus. J. Cell Biol. 81, 396–402 (1979).

Tyrrell, D. L. J., Almeida, J. D., Berry, D. M., Cunningham, C. H., Hamre, D., Hofstad, M. S., Mallucci, L., McIntosh, K.: Coronaviruses. Nature 220, 650 (1968).

Tyrrell, D. L. J., Almeida, J. D., Cunningham, C. H., Dowdle, W. R., Hofstad, M. S., McIntosh, K., Tajima, M., Zakstelskaya, L. Y. A., Easterday, B. C., Kapikian, A., Bingham, R. W.: Coronaviridae. Intervirology 5, 76–82 (1975).

Tyrrell, D. L. J., Alexander, D. J., Almeida, J. D., Cunningham, C. H., Easterday, B. C., Garwes, D. J., Hierholzer, J. C., Kapikian, A., Macnaughton, M. R., MacIntosh, K.: Coronaviridae, 2nd report. Intervirology 10, 321–328 (1978).

Ulmanen, L.: Assembly of Semliki Forest virus nucleocapsid: detection of a precursor in infected cells. J. gen. Virol. 41, 353–365 (1978).

Ulmanen, I., Seppala, P., Pettersson, R. F.: In vitro translation of Uukuniemi virus specific RNAs: identification of a non-structural protein and a precursor to the membrane glycoproteins. J. Virol. 37, 72–79 (1981).

Ushijima, H., Clerx-van Haaster, C. M., Bishop, D. H. L.: Analyses of Patois group bunyaviruses: evidence for naturally occurring recombinant bunyaviruses and existence of immune precipitable and non-precipitable non-virion proteins induced in bunyavirus-infected cells. Virology 110, 318–332 (1981).

van Alstyne, D., Krystal, G., Kettyls, G. D., Bohn, E. M.: The purification of rubella virus (RV) and determination of its polypeptide composition. Virology 108, 491–498 (1981).

van Wyke, K. L., Hinshaw, V. S., Bean, W. J., Webster, R. G.: Antigenic variation of influenza A virus nucleoprotein detected with monoclonal antibodies. J. Virol. 35, 24–30 (1980).

van Wyke, K. L., Bean, W. J., jr., Webster, R. G.: Monoclonal antibodies to the influenza A virus nucleoprotein affecting RNA transcription. J. Virol. 39, 313–317 (1981).

Varghese, J. N., Laver, W. G., Colman, P. M.: Structure of the influenza virus glycoprotein antigen neuraminidase at 2.9 Å resolution. Nature 303, 35–40 (1983).

Varmus, H. E.: Form and function of retroviral proviruses. Science 216, 812–820 (1982).

Verwoerd, D. W., Huismans, H., Erasmus, B. J.: Orbiviruses. In: Comprehensive Virology, Vol. 14, Chapter 5, 285–346. New York: Plenum Press 1979.

Vezza, A. C., Gard, G. P., Compans, R. W., Bishop, D. H. L.: Structural components of the arenavirus Pichinde. J. Virol. 23, 776–786 (1977).

Vezza, A. C., Clewley, J. P., Gard, G. P., Abraham, N. Z., Compans, R. W., Bishop, D. H. L.: Virion RNA species of the arenaviruses Pichinde, Tacaribe, and Tamiami. J. Virol. 26, 485–497 (1978 a).

Vezza, A. C., Gard, G. P., Compans, R. W., Bishop, D. H. L.: Genetic and molecular studies of arenaviruses. In: Negative Strand Viruses and the Host Cell (Mahy, B. W., Barry, R. D., eds.), 73–90. New York-San Francisco-London: Academic Press 1978 b.

Vezza, A. C., Cash, P., Jahrling, P., Eddy, G., Bishop, D. H. L.: Arenaviruses recombination: the formation of recombinants between prototype Pichinde and Pichinde Munchique viruses and evidence that arenaviruses S RNA codes for N polypeptide. Virology 106, 250–260 (1980).

Vigne, R., Filippi, R., Querat, G., Sauze, N., Vitu, C., Russo, P., Deloir, P.: Precursor polypeptides to structural proteins of visna virus. J. Virol. 42, 1046–1056 (1982).

Vogt, V. M., Bruckenstein, D. A., Bell, A. P.: Avian sarcoma virus gag precursor polypeptide is not processed in mammalian cells. J. Virol. 44, 725–730 (1982).

Vogt, V. M., Wight, A., Eisenman, R.: In vitro cleavage of avian retrovirus gag proteins by viral proteins p15. Virology 98, 154–167 (1979).

Volk, W. A., Snyder, R. M., Benjamin, D. C., Wagner, R. R.: Monoclonal antibodies to the glycoprotein of vesicular stomatitis virus: comparative neutralizing activity. J. Virol. 42, 220–227 (1982).

von Bonsdorff, C.-H., Harrison, S. C.: Sindbis virus glycoproteins form a regular icosahedral surface lattice. J. Virol. *16*, 141—145 (1975).

von Bonsdorff, C.-H., Pettersson, R. F.: Surface structure of Uukuniemi virus. J. Virol. *16*, 1296—1307 (1975).

von Bonsdorff, C.-H., Saikku, P., Oker-Blom, N.: The inner structure of Uukuniemi and two Bunyamwera supergroup arboviruses. Virology *39*, 342—344 (1969).

Wadey, C. N., Westaway, E. G.: Structural proteins and glycoproteins of infectious bronchitis virus particles labelled during growth in chick embryo fibroblasts. Intervirology *15*, 19—27 (1981).

Wagner, R. R.: Reproduction of rhabdoviruses. In: Comprehensive Virology (Fraenkel-Conrat, H., Wagner, R. R., eds.), Vol. 4, 1—93. New York-London: Plenum Press 1975.

Wang, E., Wolf, B. A., Lamb, R. A., Choppin, P. W., Goldberg, A. R.: The presence of actin in enveloped viruses. In: Cell Motility (Pollard, T., Rosenbaum, J., eds.), 589—600. Cold Spring Harbor, N.Y.: Cold Spring Harbor Press 1976.

Watanabe, K.: Electron microscopic studies of experimental viral hepatitis in mice. II. J. Electron Micro. (Tokyo) *18*, 173—189 (1969).

Watanabe, S., Temin, H. M.: Encapsidation sequences for spleen necrosis virus, an avian retrovirus, are between the 5' long terminal repeat and the start of the *gag* gene. Proc. Natl. Acad. Sci. U.S.A. *79*, 5986—5990 (1982).

Watson, B. K., Coons, A. H.: Studies of influenza virus infection in the chick embryo using fluorescent antibody. J. Exp. Med. *99*, 419—428 (1954).

Wechsler, S., Fields, B. N.: Differences between the intracellular polypeptides of measles and subacute sclerosing panencephalitis virus. Nature *272*, 458—460 (1978).

Wechsler, S. L., Rustigian, R., Stallcup, K. C., Byers, K. B., Stuart, H. W., Fields, B. N.: Measles virus-specified polypeptide synthesis in two persistently infected HeLa cell lines. J. Virol. *31*, 677—684 (1979).

Wege, H., Siddell, S., Sturm, M., ter Meulen, V.: Coronavirus JHM: characterization of intracellular viral RNA. J. gen. Virol. *54*, 213—217 (1981).

Wege, H., Siddell, S., ter Meulen, V.: The biology and pathogenesis of coronaviruses. Curr. Top. Microbiol. Immunol. *99*, 165—200 (1982).

Wehland, J., Willingham, M. C., Dickson, R., Pastan, I.: Microinjection of anticlathrin antibodies into fibroblasts does not interfere with the receptor-mediated endocytosis of α_2-macroglobulin. Cell *25*, 105—119 (1981).

Wehland, J., Willingham, M. C., Gallo, M. G., Pastan, I.: The morphologic pathway of exocytosis of the vesicular stomatitis virus and G protein in cultured fibroblasts. Cell *28*, 831—841 (1982).

Weiss, R. A.: Rhabdovirus pseudotypes. In: Rhabdoviruses III (Bishop, D. H. L., ed.), 51—65. Boca Raton, Fla.: CRC Press 1980.

Weiss, R. A., Bennett, P. L. P.: Assembly of membrane glycoproteins studied by phenotypic mixing between mutants of vesicular stomatitis virus and retroviruses. Virology *100*, 252—274 (1980).

Weiss, S. R., Leibowitz, J. L.: Comparison of the RNAs of murine and human coronaviruses. Adv. Exp. Med. Biol. *142*, 245—260 (1981).

Welch, W. J., Sefton, B. M.: Two small virus-specific polypeptides are produced during infection with Sindbis virus. J. Virol. *29*, 1186—1195 (1979).

Welsh, R. M., Biron, C. A., Parker, D. C., Bukowski, J. F., Habu, S., Okumura, K., Haspel, M. V., Holmes, K. V.: Regulation and role of natural cell-mediated immunity during virus infection. In: Human Immunity to Viruses (Ennis, F. A., ed.). New York: Academic Press 1983.

Wengler, G., Wengler, G.: Terminal sequences of the genome and replicative-form RNA of the flavivirus West Nile virus: absence of poly (A) and possible role in RNA replication. Virology *113*, 544—555 (1981).

Wengler, G., Beato, M., Wengler, G.: *In vitro* translation of 42S virus-specific RNA from cells infected with the flavivirus West Nile virus. Virology *96*, 516—529 (1979).

Wengler, G., Boege, U., Wengler, G., Bischoff, H., Wahn, K.: The core protein of the alpha virus Sindbis virus assembles into core-like nucleoproteins with the viral genome RNA and with other single-stranded nucleic acids *in vitro*. Virology *118*, 401—410 (1982).

Westaway, E. G., Replication of flaviviruses. In: The Togaviruses (Schlesinger, R. W., ed.), 531—581. New York: Academic Press 1980.

Westaway, E. G., Ng, M. L.: Replication of flaviviruses: separation of membrane translation sites of Kunjin virus proteins and of cell proteins. Virology *106*, 107–122 (1980).

Whitfield, S. G., Frederick, A. M., Sudia, W. D.: St. Louis encephalitis virus: an ultrastructural study of infection in a mosquito vector. Virology *56*, 70–87 (1973).

Wiley, D. C., Skehel, J. J., Waterfield, M. D.: Evidence from studies with a cross-linking reagent that the hemagglutinin of influenza virus is a trimer. Virology *79*, 446–448 (1977).

Willingham, M. C., Pastan, I.: The visualization of fluorescent proteins in living cells by video intensification microscopy (VIM). Cell *13*, 501–507 (1978).

Willingham, M. C., Pastan, I.: The receptosome: an intermediate organelle of receptor-mediated endocytosis in cultured fibroblasts. Cell *21*, 67–77 (1980).

Wilson, I. A., Skehel, J. J., Wiley, D. C.: Structure of the haemagglutinin membrane glycoprotein of influenza virus at 3 Å resolution. Nature *289*, 366–373 (1981).

Wilusz, J., Kurilla, M. G., Keene, J. D.: A host protein (LA) binds to a unique species of minus sense leader RNA during the replication of vesicular stomatitis virus. Proc. Natl. Acad. Sci. U.S.A. In Press.

Witte, O. N., Baltimore, D.: Relationship of retrovirus polyprotein cleavages to virion maturation studied with *ts*-murine leukemia virus mutants. J. Virol. *26*, 750–761 (1978).

Witte, O. N., Tsukamoto-Adey, A., Weissman, I. L.: Cellular maturation of oncornavirus glycoproteins: topological arrangements of precursor and product forms in cellular membranes. Virology *76*, 539–553 (1977).

Wolinski, J. S., Baringer, J. R., Margolis, G., Kilham, L.: Ultrastructure of mumps virus replication in newborn hamster CNS. Lab. Invest. *31*, 403–412 (1974).

Wolinski, J. S., Hatzidimitriou, G., Waxham, M. N., Burke, S.: Immunocytochemical localization of mumps virus antigens *in vivo* by light and electron microscopy. In: The Replication of Negative Strand Viruses (Bishop, D. H. L., Compans, R. W., eds.), 609–614. New York-Amsterdam-Oxford: Elsevier/North-Holland. 1982.

Wyatt, R. G., Kalica, A. R., Mebus, C. A., Kim, H. W., London, W. T., Chanock, R. M., Kapikian, A. Z.: Reovirus-like agents (rotaviruses) associated with diarrheal illness in animals and man. In: Perspectives in Virology (Pollard, M., ed.), Vol. 10, 121–145. New York: Raven Press 1978.

Wyatt, R. G., James, W. D., Bohl, E. H., Theil, K. W., Saif, L. J., Kalica, A. R., Greenberg, H. B., Kapikian, A. Z., Chanock, R. M.: Human rotavirus type 2: cultivation *in vitro*. Science *207*, 189–191 (1980).

Wyatt, R. G., Greenberg, H. B., James, W. D., Pittman, A. L., Kalica, A. R., Flores, J., Chanock, R. M., Kapikian, A. Z.: Definition of human rotavirus serotypes by plaque reduction assay. Infect. Immun. *37*, 110–115 (1982).

Wyatt, R. G., James, H. D., jr., Pittman, A. L., Hushino, Y., Greenberg, H. B., Kalica, A. R., Flores, J., Kapikian, A. Z.: Direct isolation in cell culture of human rotaviruses and their characterization into four serotypes. J. Clin. Microbiol. *18*, 310–317 (1983).

Yeger, H., Kalnins, V. I., Stephenson, J. R.: Type-C retrovirus maturation and assembly: post-translational cleavage of the *gag*-gene coded precursor polypeptide occurs at the cell membrane. Virology *89*, 34–44 (1978).

Yoshida, T., Nagai, Y., Yoshii, S., Maeno, K., Matsumoto, T., Hoshino, M.: Membrane (M) protein of HVJ (Sendai virus): its role in virus assembly. Virology *71*, 143–161 (1976).

Yoshida, T., Nagai, Y., Maeno, K., Iinuma, M., Hamaguchi, M., Matsumoto, T., Nagayoshi, S., Hoshino, M.: Studies on the role of M protein in virus assembly using a *ts* mutant of HVJ (Sendai virus). Virology *92*, 139–154 (1979).

Yoshikura, H., Taguchi, F.: Mouse hepatitis virus strain MHVS: formation of pseudotypes with a murine leukemia virus envelope. Intervirol. *12*, 132–136 (1978).

Yoshinaka, Y., Luftig, R. B.: Murine leukemia virus morphogenesis: cleavage of p70 *in vitro* can be accompanied by a shift from a concentrically coiled internal strand ("immature") to a collapsed ("mature") form of the virus core. Proc. Natl. Acad. Sci. U.S.A. *76*, 3446–3450 (1977).

Yoshinaka, Y., Luftig, R. B.: A comparison of avian and murine retrovirus polyprotein cleavage activities. Virology *111*, 239–250 (1981).

Yoshinaka, Y., Luftig, R. B.: *In vitro* phosphorylation of murine leukemia virus proteins: specific phosphorylation of Pr6^{gag}, the precursor to the internal core antigens. Virology *116*, 181–195 (1982).

Young, P. R., Howard, C. R.: Fine structure analysis of Pichinde virus nucleocapsids. J. gen. Virol. *64*, 833–842 (1983).

Young, P. R., Chanas, A. C., Howard, C. R.: Analysis of the structure and function of Pichinde virus polypeptides. Dev. Cell Biol. *7*, 15–22 (1981).

Yuen, P. H., Wong, P. K. Y.: A morphological study on the ultrastructure and assembly of murine leukemia virus using a temperature-sensitive mutant restricted in assembly. Virology *80*, 260–274 (1977).

Zakowski, J. J., Wagner, R. R.: Localization of membrane-associated proteins in vesicular stomatitis virus by use of hydrophobic membrane probes and cross-linking reagents. J. Virol. *36*, 93–102 (1980).

Zakowski, J. J., Petri, W. A., jr., Wagner, R. R.: Role of matrix protein in assembling the membrane of vesicular stomatitis virus: reconstitution of matrix protein with negatively charged phospholipid vesicles. Biochemistry *23*, 3902–3907 (1981).

Zavada, J.: The pseudotypic paradox. J. gen. Virol. *63*, 15–24 (1982).

Zavada, J., Zavadova, Z., Russ, G., Polakova, K., Rajcani, J., Stencl, J., Loksa, J.: Human cell surface proteins selectively assembled into vesicular stomatitis virus-virions. Virology *127*, 345–360 (1983).

Ziemiecki, A., Garoff, H.: Subunit composition of the membrane glycoprotein complex of Semliki Forest virus. J. Mol. Biol. *122*, 259–269 (1978).

Ziemiecki, M., Garoff, H., Simons, K.: Formation of the Semliki Forest virus membrane glycoprotein complexes in the infected cell. J. gen. Virol. *50*, 111–123 (1980).

Zilberstein, A., Snider, M. D., Porter, M., Lodish, H. F.: Mutants of vesicular stomatitis virus blocked at different stages in maturation of viral glycoprotein. Cell *21*, 417–427 (1980).

Subject Index

The Molecular Biology of Poliovirus

By Dr. **Friedrich Koch** und Professor Dr. **Gebhard Koch,**
Abteilung Molekularbiologie, Universität Hamburg

1985. Approx. 130 figures. Approx. 500 pages. ISBN 3-211-81763-8

The book provides the first comprehensive review of an animal picorna-virus, and of the best studied of all animal viruses: the poliovirus. It summarizes research on poliovirus biology during the last three decades. Older data are reinterpreted with respect to more recent information. Poliovirus research has again attracted considerable interest within the last years. High points have been the elucidation of the complete nucleotide sequences of the RNAs of a wild type poliovirus and its vaccine strains; further characterization of the antigenic sites on the virus particle and of the antigenic drift; characterization of alternative conformational states of the virion capsid; observations on the role of the plasma membrane, cytoskeleton, and cytoplasmic membranes as mediators in the virus-induced redirection of the synthetic machinery of the host cell; and studies on enzymes involved in RNA replication.

The text is illustrated by more than 150 figures and tables, including construction plans for a poliovirus model and a list of the world laboratories presently engaged in research on the molecular biology of poliovirus. More than 2 000 references are summarized. Some unanswered questions concerning poliovirus structure and replication are listed at the end of the book.

Poliovirus research has provided impetus to our growing knowledge of the molecular mechanism in cell biology and of the cytological basis of viral diseases. The book should therefore be of interest not only to students and researchers in the field of virology, but also to cell biologists, physicians, and pharmacologists.

Springer-Verlag Wien NewYork